The Life and Death of Stars

Keivan G. Stassun, Ph.D.

THE
GREAT
COURSES®

PUBLISHED BY:

THE GREAT COURSES
Corporate Headquarters
4840 Westfields Boulevard, Suite 500
Chantilly, Virginia 20151-2299
Phone: 1-800-832-2412
Fax: 703-378-3819
www.thegreatcourses.com

Keivan G. Stassun, Ph.D.
Professor of Physics and Astronomy
Vanderbilt University

Professor Keivan G. Stassun is Professor of Physics and Astronomy at Vanderbilt University. He earned A.B. degrees in Physics and Astronomy as a Chancellor's Scholar at the University of California, Berkeley, and was selected valedictorian of his graduating class in 1994. He earned his Ph.D. in Astronomy in 2000 as a National Science Foundation (NSF) Graduate Research Fellow at the University of Wisconsin–Madison. His dissertation research focused on the birth of stars. Professor Stassun then served as assistant director of the NSF Graduate STEM Fellows in K–12 Education Program. This program connects graduate students in science, technology, engineering, and mathematics (STEM) with K–12 schools, both to enhance science teaching and to provide leadership development for future college and university faculty. Professor Stassun served for two years as a postdoctoral research fellow with the NASA Hubble Space Telescope Program, studying newborn eclipsing binary stars, before joining the faculty of Vanderbilt University in 2003. He also holds the title of Adjunct Professor of Physics at Fisk University.

Professor Stassun's research on the birth of stars, eclipsing binary stars, exoplanetary systems, and the Sun has appeared in the prestigious research journal *Nature*, has been featured on NPR's *Earth & Sky*, and has been published in more than 100 peer-reviewed journal articles. In 2007, the Vanderbilt Initiative in Data-intensive Astrophysics (VIDA) was launched as a $4 million pilot program in astro-informatics, with Professor Stassun as its founding director. He served as chair of the Sloan Digital Sky Survey exoplanet science team, is a member of the Large Synoptic Survey Telescope executive committee, and served on the National Research Council's Decadal Survey of Astronomy and Astrophysics.

Professor Stassun is a recipient of the prestigious CAREER Award from NSF and a Cottrell Scholar Award for excellence in research and university teaching from the Research Corporation for Science Advancement. In 2013, he was named a Fellow of the American Association for the Advancement of Science. Professor Stassun is also a national leader in initiatives to increase the number of underrepresented minorities earning doctoral degrees in science and engineering and a founding director of the Fisk-Vanderbilt Master's-to-PhD Bridge Program. In 2010, Professor Stassun served as an expert witness to Congress in its review of promising approaches for increasing American competitiveness in science and engineering. Today, he is a member of NSF's Committee on Equal Opportunities in Science and Engineering.

Professor Stassun's work has been published in several scholarly journals, including *The Astronomical Journal*, *The Astrophysical Journal*, the *American Journal of Physics*, and *Astronomy & Astrophysics*. He also serves as host for *Tennessee Explorers*, produced by Nashville Public Television, which highlights the lives and work of scientists and engineers in Tennessee to help inspire the next generation of scientific explorers. ■

Table of Contents

Table of Contents

Table of Contents

The Life and Death of Stars

Scope:

T he stars have always held a special place in the human experience and imagination. We look up at the night sky and behold their beauty— we even sing nursery songs about them and wonder what they are. While we know that our own Sun's light and warmth are essential to life on Earth, too often we regard the stars with a cold, distant remove. Yet we have much in common with the stars. Like us, stars are born, live out their lives, and then die. Like us, the lives and deaths of stars represent a circle of life, the ashes of dead stars becoming the raw material for new generations of stars and their systems of planets. And, most importantly, like us, the lives of stars can be seen as having a purpose, which is to wage a life-and-death struggle against the crush of gravity and, in the process, to transform the simplest elements in the universe into the full diversity of elements in the periodic table, upon which our lives and all of the material world around us depend. Indeed, not only are we like the stars, but we are also of them.

This course will tell the incredible story of the lives and deaths of stars. You will trace the arc of the stars' lives, from their births in gigantic nebulae of glowing gas and dust, stellar nurseries; to their "life's work" as stars in middle age, like the Sun; to their fiery deaths as planetary nebulae or supernova explosions. You will examine the bizarre corpses left behind by dead stars, including white dwarfs—diamonds in the sky—as well as neutron stars and black holes, representing the densest and strangest forms of matter in the universe. You will see how the ashes of dead stars are recycled into the next generation of stars, these newborn stars enriched thanks to the previous stellar generations in the types of elements that permit the formation of planets and of life on those planets. You will look at stars in isolation, like our Sun, as siblings influencing one another's future life course and as extended families called star clusters. To gain additional perspective on the stellar life cycle, you will also study the so-called brown dwarfs, stillborn stars representing what happens when the stellar birth process fails. And by examining current theories for the very first stars that populated the universe, you will see how the stellar life cycle got started in the first place. Bringing the story closer to home, the course will look at the magnetic nature of stars and how the Sun's magnetic storms can directly impact us here on Earth.

As a unifying theme for the course, you will take a physical perspective on the stellar life cycle, emphasizing how stars serve as agents of alchemy by which matter in the universe has been transformed from the simplest element, hydrogen, into the elements that matter most to us as carbon-based, oxygen-breathing, calcium-boned, iron-blooded life-forms. To understand these behemoths, stars, you will go down to the atomic level—both because what stars do in their furnaces is an atomic process and because our only means for studying them is to decode the information that's encoded in light, the information bearer of the stars. You will learn about the telescopes and instruments that astronomers use to collect and decode that light. You will also come to see the stars as living entities. By thinking of a star as more than merely an inanimate object, you are able to see our connection and relationship to the stars more clearly and fully. There is a beauty to thinking of stars not just in terms of physics but also in terms of their purpose, a purpose to which humans have always, and always will, relate. ■

Why the Stellar Life Cycle Matters
Lecture 1

I n this course, you will learn about the various stages of a star's life, from birth to death and back to birth—its life cycle—in detail. This lecture will begin by looking at the life cycle of a star in broad strokes, giving you the sweeping narrative of stellar life before the course zooms in on each fascinating feature. We now understand the stars from a physical perspective—not as mere lights in the sky, but as engines of matter, energy, and of the raw material of life itself.

The Life Cycle of a Star

- The birth of a star is a process that takes place within what we call stellar nurseries, which are enormous clouds of gas and dust that float through our galaxy, weighing tens of thousands of times what our Sun does. Within these massive nurseries, hundreds or thousands of stars can be birthed all at once.

- Gravity causes the onset of stellar birth within these gigantic clouds; it causes slightly denser portions of the cloud to fall in on themselves, generating an even denser region that then begins to collapse even faster. This runaway collapse process creates heat, so the embryonic star—called a protostar—begins to glow.

- The stellar birth process begins in these very dense clouds of gas and dust. After a few hundred thousand years, though, the cloud becomes increasingly incorporated into the fledgling stars themselves or else eroded away by the intense glare of the most massive stars that have ignited within the stellar nursery. At this point, the cocoons of individual stars become sufficiently exposed to be seen directly.

- The word "proplyds" is astronomers' shorthand for stars with protoplanetary disks, which are disks of gas and dust from which planets around stars are made. As seen through a telescope, proplyds are individual cocoons swimming in a sea of irradiated gas and dust, each one containing within it a new star encircled by the material from which its solar system of planets will be made.

- After about 10 million years, a star reaches adolescence. At this point, the star is fully exposed and is generating its own heat and light to warm and illuminate the entourage of planets that will encircle it.

- Next, the star enters a long, stable period of middle life. For a star like the Sun, this middle life stage lasts a very long time—about 10 billion years. And during all of that time, not much happens—at least nothing terribly dramatic.

- But in reality, it is during this time of adulthood that the star is slowly, patiently, and steadily doing its life's work. Deep in its core, the mature star sustains the process of nuclear fusion. The ongoing fusion reactions allow the star to create light and heat, which allows the star to wage a lifelong struggle against the inexorable crush of gravity.

- And in the process, it is performing the magic of alchemy—converting the simplest and most abundant element, hydrogen, into the other elements of the periodic table. An adult star fabricates new elements, such as carbon and oxygen and calcium and even iron, and it holds these elements in store until its death. At that point, it will give up these elements for the next generation of stars and planets and, potentially, their inhabitants.

- The stellar death is a quiet but dramatic one, as the star slowly disembodies itself, puffing off shells or rings of itself into the surrounding space as a so-called planetary nebula, which is the enduring legacy of a dead star.

- Stars can also end their lives more explosively as a supernova. These immense explosions, if several of them should go off in the same nearby region of the galaxy, can produce geysers of material that spew out above the disk of the galaxy and then rain back down into the galaxy.

- Thanks to the alchemy that those stars performed during their lives, this material, cascading down, has been enriched with heavier elements than what the exploding stars began with. Therefore, a new generation of stars in the galaxy inherits this material, and both the newborn stars themselves and the planets that form around them can now incorporate the types of elements that make life as we know it possible.

The open star cluster Pleiades is about 430 light-years from the solar system.

Types of Stellar Clusters

- Most stars are, in fact, not born single; rather, they most often are born with at least one sibling. Sibling stars grow up with one another and orbit one another through life and death. Indeed, in some cases there can be three or more sibling stars orbiting one another, and they are more present than you might think.

- Indeed, stellar siblings affect one another in important ways. If they get too close, a dramatic sibling rivalry can ensue such that, through the stars' disruptive gravitational influence, one or more of the stars can be prevented from forming solar systems. Even after death, the siblings can interact, as sometimes a living sibling will cause a dead sibling to reignite briefly but spectacularly in a special and important type of supernova explosion.

- Beyond having siblings, stars are also born into larger extended stellar families, which we call star clusters. There are two main types of stellar clusters. One type is called a globular cluster, which are massive, tight-knit stellar families with tens of thousands or even more members. These are also ancient families, left over from the formation of our galaxy. Because these families have so many members, their mutual gravitation attraction has been strong enough to keep these families together over the eons since our galaxy's beginnings.

- The other type of star cluster is what we call an open cluster. A good example of an open cluster is the Pleiades, or Seven Sisters. In fact, the Pleiades cluster has several hundred stellar members. These open clusters represent more modern stellar families, having formed in their stellar nurseries more recently than globular clusters. Unlike the globular clusters, open clusters do not have sufficient gravity to keep the stars together, so the stars in these families eventually drift apart.

The Elements of Life

- We can think of a star's purpose in life as being the creation of the elements that make all of life possible. The most abundant element in the universe—making up about 75% of matter in the universe and, therefore, 75% of the matter in stars—is the simplest element: hydrogen. Hydrogen is present in most of the compounds that we know, but on its own, hydrogen doesn't do anything, and you can't make anything out of just it. However, stars are made mostly of hydrogen and, over their life cycle, use it to make all of the other elements.

- The next element in the periodic table is helium. The matter in the universe started out as 75% hydrogen and 25% helium, and almost nothing else. But ever since the big bang, all of the new helium that has been created in the universe has been made by stars like the Sun, synthesized from hydrogen through the power of nuclear fusion.

- This relatively simple conversion of hydrogen to helium is the main process through which stars generate the light and heat that they will produce throughout their lifetimes. Stars are nuclear factories, using fusion to manufacture heavy elements from the lighter ones.

- Carbon is the most fundamental element in all of life. Every living cell, every organic compound, contains it as its building block. And every single carbon atom in the universe was created by stars like the Sun, synthesized from helium atoms. Without the stars, there would be no carbon, and without carbon, there would be no life.

- Oxygen is what we breathe, and it's also in every drop of water, so it's essential to life as we know it. Every atom of oxygen in the universe was created by stars, synthesized in their cores from carbon and helium atoms. The process of making oxygen atoms occurs as one of a dying star's final breaths. And in those dying breaths, stars make the stuff that we breathe to live.

- Silicon has become central to our technologically built world, and every silicon atom in the universe was created by stars.

- Sodium, phosphorus, potassium, and calcium are in things like salt and chalk, but they are also minor but critically important components in our blood, central to innumerable processes constantly at work in our bodies. Each of these elements, and indeed every single atom of them in the universe, was made by stars as they approached the ends of their lives.

- In our everyday experience, we regard iron as something so substantial and associated so strongly with the Earth that it may be difficult to imagine it in the context of a luminous being like a star. But, indeed, every atom of iron has been forged in the cores of stars. In fact, iron is special because it is the last element that a star can synthesize from lighter elements before conceding defeat to gravity.

- Copper, zinc, and iodine all play a role in our bodies, and all of these elements were also made by stars—or, rather, in the fiery explosive supernova deaths of massive stars.

- Altogether, the life cycle of stars is the universe's process for transforming the simple, basic matter with which the universe began—

hydrogen and helium—into the variety of elements that make life possible and almost every aspect of the world we've built around us.

The Importance of Stars

- Our connection to the stars is a profound one. We are made of the ashes of long-dead stars, ashes that represent the laborious effort of a behemoth struggling in a fight to the death with gravity. Interestingly, this connection to the stars is something humans have always seemed to sense—even before scientists understood the true physical nature of the stars and why they do what they do.

- Throughout human history, all civilizations have gazed on the stars and constructed mythologies to explain the apparent patterns—the constellations. There has always been an innate understanding that the stars, whatever they might be, are important.

- As fantastic as these mythological stories may have been, constellations are not real. In 3-D space, most stars that look to us like they're in the same constellation are actually thousands of light-years apart, and if we could view the night sky from a different angle, we'd see a whole different set of "constellations." So there are no patterns in the stars beyond those that people imagine are there.

Suggested Reading

O'Dell, *The Orion Nebula.*

Smith, Stassun, and Bally, "Opening the Treasure Chest."

Questions to Consider

1. What are the parallels—literal and metaphorical—between the life cycle of the stars and that of human life?

2. Were the mythological conceptions of the stars by early human societies "wrong," even if not physically accurate?

Why the Stellar Life Cycle Matters
Lecture 1—Transcript

I'd like you to think back to the birth of someone dear to you, perhaps your child or grandchild, or a niece or nephew. Think about that experience and hold it in your mind. Recall the moments around that child's birth; remember how powerfully you felt the sense of immense investment in that life and the enormous hope for a purpose to the life ahead, the feeling that this was the start of something wonderful. That's exactly how I felt at the birth of my first son James. Holding him in my arms when he was just minutes old remains one of the most significant, transcendent moments of my life. And the same was true of my second son, Emilio, although I did dress more appropriately the second time around.

I distinctly remember feeling with the birth of both of my sons, that everything my wife and I had worked to achieve up to that moment, and everything we would do for the rest of our lives, was an investment in the future of these precious children. Yet as we'll see, when we consider the cosmic investment required to make the raw substance of human existence possible, life as we know it, is in a real sense not the beginning, but rather the culmination, the ultimate product of a grand machinery that, over the course of billions of years, has slowly but steadily transformed the bland, raw material of the Universe into the living, breathing, sentient, and self-aware stuff of life. And at the heart of that grand machinery—the engines that make it all go—are the stars.

That new life that you first cradled in your hands is all the more miraculous when you consider that it is literally made up of stars. The stars, in their lives, have manufactured the elements that make up that baby's precious little body, that make up the lifeblood coursing through him, and that now fill his lungs with breath. And the stars, in their deaths, have sprinkled the sky with the life-giving material, so that it might be gathered up and fashioned into our Sun, our little world, our salty oceans, and our air, and the host of plants and creatures and people that walk upon it, gazing up, wondering whence they came.

Seen as givers of life, or at least as the makers of the stuff of life, the stars are all the more interesting when we consider the parallels between their life cycle and our own. Put simply, stars are born, they live, and eventually, they die. But this is not to suggest mere existence or a mere passing transience. Yes, like us, the stars are impermanent. But like us, they can also be seen as self propagating. From the stars themselves, new generations of stars are born, and these stellar offspring inherit what their parents worked so hard to produce over the course of their lives—material enriched with all the elements of the periodic table. Each successive generation of stars picks up where the previous generation left off, further enriching the cosmos with life-giving material, further extending the legacy of the stars.

But it's not easy being a star. In fact, you could say that from the very start the "pressure" to perform is overwhelming. I mean it literally. From the moment a star is born, it finds itself locked in a mighty struggle with gravity, bearing down, crushing, smothering, exerting enormous pressure against which the star must somehow push back or else be squished, and not like a bug, but squished into non existence. So for a star it's the hard-knocks life—something that perhaps many of us can relate to. But that struggle to survive is also, in a sense, the star's will to live. The star's purpose in life, if you want to think of it that way, is to ensure that it makes it long enough to provide for its offspring. To build from nothing a wealth and an inheritance that will leave the next generation enriched and better prepared to extend the legacy.

In this course we'll be learning about the various stages of a star's life, from birth to death and back to birth—its life cycle—in detail.

But today I'd like to start by looking at the life cycle of a star in broad strokes. I'd like to give you the sweeping narrative of stellar life before we zoom in on each fascinating feature. It's an amazingly beautiful story. Because the life cycle of a star is, well, cyclical, we could jump in to the narrative at just about any point. But let's start with the birth process. That process takes place within what we call stellar nurseries. These are enormous clouds of gas and dust that float through our Galaxy, weighing tens of thousands of times the mass of our Sun. Within these massive nurseries, hundreds or thousands of stars can be birthed all at once.

One such stellar nursery is the Eagle Nebula, which includes the famous region known as the pillars of creation. Each pillar in this image is part of a much larger, gigantic cloud of gas and dust. In fact, the entire Eagle Nebula stretches some 50 light-years across. The pillars themselves are about five light-years in extent, and they are shaped by the intense radiation that erodes the giant gas cloud. There are many individual stars forming deep within these pillars. All around the pillars, we see a bright nebular glow, which is the result of heat and light from the newborn stars illuminating the pillars and the surrounding nebula from within.

What you don't see directly in this image is the clutch of very massive, hot stars that are off the image to the upper right. The pillars point the way to those hot, massive stars. And so, what we are seeing with these pillars is the overall cloud of gas and dust being literally eroded by the intense radiation of those stars, a process we call photoevaporation. These stellar nurseries can be turbulent places. Finally, all around the peripheries of the pillars, we observe small protrusions. These are, in fact, not small at all; they are the cocoons of individual stars and entire solar systems forming in the pillars.

What causes the onset of stellar birth within these gigantic clouds? The answer, one which we will see time and time again in this course, is gravity. Gravity causes slightly denser portions of the cloud to fall in on themselves, generating an even denser region that then begins to collapse even faster. This runaway collapse process creates heat, and so the embryonic star—called a protostar—begins to glow. We can't see the protostar directly because it's enshrouded in a cocoon of the surrounding cloud of gas and dust. But we can see the glow from its heat in the infrared, an unmistakable signature of a beating heart within. Indeed, as we'll see in upcoming lectures, using different types of telescopes to see different wavelengths of light actually shows us different aspects of what we're studying, in this case, different aspects of the nebula.

Here we are seeing a comparison of an infrared image taken with the Spitzer Space Telescope and a visible light image. The infrared image allows us to peer into the pillar, revealing a litter of stars forming within. In contrast, in the visible light image, those stars are entirely obscured by the dust in the pillar, so visible light would not allow us to see the newborn stars that are actually there. The visible light image is very useful, however, because it emphasizes the erosion process that goes on in a stellar nursery. The evidence of that process can be seen in the bright glow that the visible light image shows all around the pillar.

So the stellar birth process begins in these very dense clouds of gas and dust. After a few hundred thousand years, though, the cloud becomes increasingly incorporated into the fledgling stars themselves, or else eroded away by the intense glare of the most massive stars that have ignited within the nursery. At this point, the cocoons of individual stars become sufficiently exposed to be seen directly.

The word proplyds is astronomers' shorthand for stars with protoplanetary disks, which are disks of gas and dust from which planets around stars are made. As seen through a telescope, proplyds are individual cocoons swimming in a sea of irradiated gas and dust, each one containing within it a new star encircled by the material from which its solar system of planets will be made. In this amazing gallery of images from the Hubble Space Telescope, we are seeing four different proplyds, cocoons in the Orion Nebula stellar nursery. Within each, we see a baby star at the center of the cocoon, birthing within. We also see the dark silhouette of a protoplanetary disk of gas and dust encircling each baby star. These protoplanetary disks are the material from which planets will be formed, eventually comprising the stars' solar systems. It's always a challenge in astronomy to get the right sense of spatial scale, so let me emphasize the point: Each one of these seemingly tiny cocoons is, in fact, the size of our entire solar system.

After about 10 million years, a star reaches adolescence. At this point, the star is fully exposed, and is generating its own light and heat to warm and illuminate the entourage of planets that will encircle it. Piecing together information that astronomers have obtained from studying these systems at various light wavelengths, we can visualize what one of these adolescent systems looks like. What we have is the young star shining at the center of its protoplanetary disk. That protoplanetary disk represents the residual gas and dust from which the star itself was born. At this point, planets may begin to form from the material in this disk, orbiting the star as they take shape. These forming planets sweep up the disk material, incorporating it into themselves, and so, over time, the disk dissipates as the forming solar system uses up the material.

So let's review the stellar life cycle so far. Within a large stellar nursery, a fledgling star collapses under gravity, heats and lights up, and forms a disk of gas and dust from which planets may be formed around it. Upon completing this process, the star has reached an adolescent stage. It is fully formed, but its life is really just beginning. Next, the star enters a long, stable period of middle life. For a star like the Sun, this middle life stage lasts a very long time, about 10 billion years. And during all that time, not much happens, at least, nothing terribly dramatic.

But in reality, it is, during this time of adulthood, that the star is slowly, patiently, and steadily doing its life's work. Deep in its core, the mature star sustains the process of nuclear fusion. The ongoing fusion reactions allow the star to create light and heat, which allows the star to wage a lifelong struggle against the inexorable crush of gravity. And in the process, it is performing the magic of alchemy—converting the simplest and most abundant element, hydrogen, into the other elements of the periodic table. To understand this, we will spend plenty of time in later lectures delving deep into the beating heart of our closest star, the Sun, and examining the physics at the nuclear level.

For now, it's important only to note that an adult star fabricates new elements, such as carbon and oxygen and calcium and even iron, and that it holds these elements in store until its death. At that point, it will give up these elements for the next generation of stars and planets, and potentially, their inhabitants.

That stellar death is a quiet but dramatic one, as the star slowly disembodies itself, puffing off shells or rings of itself into the surrounding space as a so-called planetary nebula. Dramatic pictures of the Helix Nebula help us understand this process. At center of the planetary nebula we see the glowing remnant of the now-dead star—a fantastic object we call a white dwarf. The surrounding ring with its glowing colors is the disembodied material from the dead star. That material, enriched in heavier elements produced by the star during its life, expands into the surrounding space. As a result, the elements created during the star's life can be incorporated into the next generation of stars. So in a very real sense, a planetary nebula is the enduring legacy of a dead star.

Within the expanding material of the dead star, we can zoom in and see new cocoons that have formed all around the periphery of the planetary nebula. The presence of these cocoons shows in particularly dramatic fashion that the graveyards of stars can also be the breeding grounds of new stars.

Stars can also end their lives even more explosively as a supernova. These immense explosions, if several of them should go off in the same nearby region of the galaxy, can literally produce geysers of material that spew out above the disk of the galaxy and then rains back down into the galaxy. Thanks to the alchemy that those stars performed during their lives, this material, cascading down, has been enriched with heavier elements than what the exploding stars began with. And so a new generation of stars in the galaxy inherits this material, and both the newborn stars themselves and the planets that form around them can now incorporate the types of elements that make life as we know it possible.

So far in our sweeping portrait of the stellar life cycle, we've been discussing stars as solitary entities with their individual lives and deaths. But as we'll see in this course, most stars are, in fact, not born single. Rather, they most often are born with at least one sibling. Sibling stars grow up with one another and orbit one another through life and death. Indeed, in some cases there can be three or more sibling stars orbiting one another, and they are more present than you might think. For example, one of the brightest stars in the sky, Castor, one of the two brothers in the constellation Gemini, is in fact six stars—an amazing sextuple system.

Looking in detail at this amazing sextuple system, we see just how intricately the siblings have arranged themselves. Two siblings are orbited by two other siblings, and those four siblings are, in turn, orbited by yet another two siblings. As we'll see, this arrangement is not an accident, but in fact, reveals much about how stellar siblings form and how they influence one another. Indeed, stellar siblings affect one another in important ways. If they get too close, a dramatic sibling rivalry can ensue, such that, through the stars' disruptive gravitational influence, one or more of the stars can be prevented from forming solar systems. Even after death, the siblings can interact, as sometimes a living sibling will cause a dead sibling to reignite briefly but spectacularly in a special and important type of supernova explosion.

Beyond having siblings, stars are also born into larger, extended stellar families, what we call star clusters. There are two main types of stellar clusters that we'll learn about. One type is called a globular cluster. These are massive, tight-knit stellar families with tens of thousands, or even more, members. These are also ancient families, left over from the formation of our galaxy itself. Because these families have so many members, their mutual gravitation attraction has been strong enough to keep these families together over the eons since our galaxy's beginnings.

The other type of star cluster is what we call an open cluster. A good example of an open cluster is the Pleiades, or Seven Sisters. In fact, the Pleiades cluster has several hundred stellar members. These open clusters represent more modern stellar families, having formed in their stellar nurseries more recently than globular clusters. In fact, in this image of the Pleiades cluster, we can still see some of the remnant gas and dust from the cluster's nursery as faint blue wisps of material among the stars. But unlike the globular clusters, these open clusters do not have sufficient gravity to keep the stars together. And so eventually, the stars in these families will drift apart.

Now, whether we look at a star in the context of its family or in isolation, we have to account for one unavoidable force of nature; I mean gravity. Throughout the star's life, gravity is an ever-present force. Initially, it is gravity that gives the fledgling star the embrace that brings it to life. Then, it is gravity that impels the star to perform its alchemy as the star struggles to prevent gravity from crushing it into non-existence. In the end, though, it is gravity that wins the battle, causing the explosion that returns the products of the star's life back to the cosmos. And then it will be gravity that will reconstitute those stellar ashes into new generations of stars and planets, the stellar life cycle beginning anew.

As I mentioned earlier when I introduced you to James and Emilio, we can think of a star's purpose in life as being the creation of the elements that make all of life possible. Let's meet some of these elements. No doubt you're familiar with some of them already, but let's introduce them in the context of the stellar life cycle.

The most abundant element in the universe, making up some 75% of matter in the universe, and therefore, 75% of the matter in stars, is the simplest element of all, hydrogen. Hydrogen is present in most of the compounds that we know. For example, in water, two of the atoms are hydrogen, and one is oxygen, H_2O. But on its own, hydrogen is not very interesting. It doesn't do anything, and you can't make anything out of just it. But a star can. Stars are made mostly of hydrogen, and over their life cycle use it to make all of the other elements.

The next element in the periodic table is helium. It is slightly more interesting. For example, we use it in balloons and in many industrial applications, including as a cryogen for keeping things extremely cold. The matter in the universe started out as 75% hydrogen and 25% helium, and almost nothing else. But ever since the big bang, all of the new helium that has been created in the universe has been made by stars like the Sun, synthesized from hydrogen through the power of nuclear fusion. This relatively simple conversion of hydrogen to helium is the main process through which stars generate the light and heat that they will produce through their lifetimes. Stars are literally nuclear factories, using fusion to manufacture heavy elements from the lighter ones.

Our next stop on the periodic table is carbon. This is the most fundamental element in all of life. Every living cell, every organic compound, contains it as its building block. And every single carbon atom in the universe was created by stars like the Sun, synthesized from helium atoms. Without the stars, no carbon, and without carbon, no life.

And what about oxygen? You know that's important; you're breathing it right now. And it's also in every drop of water, so essential to life as we know it. Every atom of oxygen in the universe was created by stars, synthesized in their cores from carbon and helium atoms. The process of making oxygen atoms occurs as one of a dying star's final breaths. And in those dying breaths, stars make the stuff that we breathe to live.

Silicon, it's become central to our technologically built world. And you guessed it; every silicon atom in the universe was created by stars. The next time you tap out a message on your phone, remember that the substance of that phone in your hands owes its existence to a long-dead star. Sodium, phosphorus, potassium, calcium, you recognize many of these elements in things like salt and chalk. You may also know these elements as minor but critically important components in our blood, central to innumerable processes constantly at work in our bodies. Each of these elements, indeed, every single atom of them in the universe, was made by stars as they approach the ends of their lives.

And what about iron? In our everyday experience, we regard iron as something so substantial and associated so strongly with the earth, that it may be difficult to imagine it in the context of a luminous being, like a star. But, indeed, iron, every atom of it, has been forged in the cores of stars. In fact, iron is special, because it is the last element that a star can synthesize from lighter elements before conceding defeat to gravity.

Of course there are still more elements, some of them obscure to us, but some of them familiar, like copper and zinc and iodine, all of which play a role in our bodies. If nothing else, you probably recognize these from the label of your daily multi-vitamin. And all of these elements were also made by stars, or rather, in the fiery explosive supernova deaths of massive stars. Altogether, the life cycle of stars is the universe's process for transforming the simple, basic matter with which the universe began—hydrogen and helium—into the variety of elements that make life possible and almost every aspect of the world we've built around us.

This is the work that stars do. Their life's work. And from the standpoint of life on Earth, their purpose, if you will, our connection to the stars is a profound one. And it is literal. We are made of the ashes of long-dead stars, ashes that represent the laborious effort of a behemoth struggling in a fight to the death with gravity.

Interestingly, this connection to the stars is something humans have always seemed to sense, even before scientists understood the true physical nature of the stars and why they do what they do. Throughout human history, all civilizations have gazed on the stars and constructed mythologies to explain the apparent patterns—the constellations. There has always been an innate understanding that the stars, whatever they might be, are important. And in most mythologies, this sense of importance was conveyed through a kind of connect-the-dots storytelling involving characters, creatures, beings imagined to be depicted in the heavens.

Not surprisingly, these depictions often involved representations of struggle and the essence of human existence. Take the constellation Orion, for example. It's one of my personal favorites because of the amazing stellar nursery we discussed, which is in Orion's sword. But in mythology, Orion was seen as a mighty hunter, decked out with belt and sword, raised in battle against the raging bull, Taurus, its angry red eye focused with fury on the hunter. That fiery red eye of the bull, a star called Aldebaran, is, in fact, a red giant star, a star like our Sun, only nearing the end stages of its life cycle. We might imagine that, like Taurus, the bull of mythology, that star will soon give up its life, defeated by a relentless hunter called gravity. I don't know what supposedly became of Taurus in the mythological telling. But Aldebaran, when it dies, will provide for a new generation of mighty beings to begin the toil anew.

Suffice it to say that, as fantastic as these mythological stories may have been, constellations are not real. In three-dimensional space, most stars that look to us like they're in the same constellation are actually thousands of light-years apart, and if we could view the night sky from a different angle, we'd see a whole different set of "constellations." So there are no patterns in the stars beyond those that people imagine are there. The stars are far more interesting and eminently more important than mere dots defining pictures in the sky. Now we understand the stars from a physical perspective, not as mere lights in the sky, but as engines of matter, energy, and of the raw material of life itself.

I think back to that moment when my James and Emilio emerged, holding their small, living, breathing bodies, made of carbon and calcium and iron, inhaling the oxygen—gifts literally from the stars. When we think about the life and the death of stars this way, then their relationship to the human story matters even more than our ancestors imagined.

The Stars' Information Messenger
Lecture 2

In this lecture, you will learn that light is the ultimate information bearer of the stars. Imprinted in the light we receive is crucial information—quantifiable information—about the stars: their brightnesses, their temperatures, their motions, and their elemental compositions. The ability to sense the stars and everything about them with nothing more than light is foundationally important to everything that you will learn about throughout the rest of this course.

How Light Interacts with Matter

- The Sun appears yellowish to our eyes, but in fact it emits light of all colors. Scientists can create a light spectrum of sunlight by spreading out the Sun's intrinsic rainbow of light into its constituent colors. When we look at this spectrum, we see all of the colors that we know—from red at one end to purple, or violet, at the other. The yellows and greens are the brightest colors (which is why the Sun mainly looks to be that color), but overall, we see a smooth, continuous rainbow of many colors.

- However, we also see a curious pattern of very specific colors that are "missing" from the Sun's spectrum. These show up as dark patches in the spectrum. In other words, we see specific colors that have been absorbed from the Sun's otherwise continuous spectrum.

- There are essentially four ways that light and matter interact: emission, absorption, transmission, and reflection. Emission refers to the process by which light is produced by an object—whether it is a solid, liquid, or gas. Essentially, emission is the radiance, or "glow," of an object. All things glow, just not necessarily in ways that we can see with our eyes. The Sun is an example of an object that does emit light we can see. And that light, when it reaches the Earth, interacts with the things it encounters.

- Some things absorb the sunlight and heat up as a result. Other objects are transparent to the light—such as a windowpane—so we say that the light is transmitted through it. Still other objects reflect the Sun's light, which means that the light "bounces" off the object.

- Most things do a combination of these things. For example, they might absorb some of the light and reflect the rest. Most of the things around us that have colors that we see are not actually glowing that color; rather, they absorb most of the colors that are present in the light they receive and only reflect certain specific colors.

- Whatever color an object reflects is the color our eyes perceive. For example, the green leaves on a tree appear green not because they are glowing that color; rather, they appear green because the chlorophyll in the leaves absorbs the reds and the yellows from the sunlight and reflects only the greens and the blues.

Properties of Light and Wave-Particle Duality
- Light has a very peculiar nature, which is referred to as its wave-particle duality. This means that we can equally think of light as being like a wave or as a particle. In some cases, it will be more helpful to talk about it as a wavelike phenomenon—like sound—and, in other cases, it will be more helpful to describe it as if it were a discrete particle—like an electron.

- The wavelike version of light can be thought of as being similar to waves in water. If you drop a pebble into a pond, the water's surface is disturbed, causing it to undulate up and down where the pebble plopped in. From that point, those undulations travel outward in all directions—they propagate—and they do so with a characteristic speed and with a characteristic distance between successive undulations.

- We refer to the distance between undulations as the wavelength of the wave. With the wave traveling outward at a certain speed, and with a characteristic wavelength between successive waves, we can also define a frequency—the rate at which the waves pass by a fixed point. A good analogy for this is waves in the ocean washing up to shore. The distance between successive swells is the wavelength. Those swells move toward the shore with a certain speed. When the waves reach the shore and crash down onto the beach, they do so with a steady cadence—the frequency of the waves.

- For any wave phenomenon, including waves at the beach, there is a specific relationship between the wave's speed, wavelength, and frequency. For light, the speed is the speed of light, which is a fundamental constant of nature.

- Because the speed of light is constant, that means that a light wave's frequency and wavelength are always inversely related. In other words, a light wave of a higher frequency has a shorter wavelength, and a light wave of a lower frequency has a longer wavelength. Think again about the waves at the beach. If the swells are close together (short wavelength), successive waves will arrive at the shore with a higher cadence (higher frequency).

- Just as our ears perceive different frequencies of sound waves as different pitches of sound, our eyes perceive different frequencies of light as different colors. Higher-frequency light (or shorter-wavelength light) is blue, while lower-frequency light (or longer-wavelength light) is red.

- Just as there are other fundamental particles in nature—like protons and electrons—light can be thought of as a discrete particle, which we call a photon. And as a discrete particle, light carries a discrete amount of energy. This energy, it turns out, is directly related to the light's wavelike properties. Higher-frequency, shorter-wavelength photons carry more energy, and lower-frequency, longer-wavelength photons carry less energy. Blue light is more energetic, and red light is less energetic.

The Electromagnetic Spectrum
- Light comes in different varieties, what we ordinarily think of as color. But the colors that we're familiar with from everyday life

are just a tiny sliver of a much larger range of colors that exist in nature. The full range of colors, including those invisible to our eyes, is what we refer to more generally as the electromagnetic spectrum of light.

- The part of the electromagnetic spectrum that we can see is referred to as the visible part of the spectrum, and it contains the red-to-violet rainbow. The wavelengths of light that are responsible for that band of visible colors are typically about 1 micron in length, or about the size of a bacterium.

- But light comes in an infinite range of other wavelengths, both shorter and longer than those we can see. Ultraviolet light has wavelengths that are just a little bit shorter than the deepest violet we can see. In other words, this light is "beyond violet," which is the literal meaning of ultraviolet.

- If you continue beyond ultraviolet, you come to the X-ray part of the electromagnetic spectrum. X-ray light has even shorter wavelengths and even higher energies. Beyond that, the shortest wavelengths and highest energies are the gamma rays. These photons are so energetic that on Earth they are generally only encountered in nuclear explosions.

- Coming back to the visible part of the electromagnetic spectrum, there are photons with longer wavelengths as well. At wavelengths just longer than the reddest red that we can see is the infrared, meaning "below red." Infrared light is sometimes referred to as "thermal radiation," because most things that have temperatures of the kinds we experience every day glow most strongly in infrared light.

- At wavelengths longer than infrared, and at the lowest energies, are radio waves, a portion of which are called microwaves. Radio waves are photons with extremely long wavelengths, as long as a football field or even longer. In addition to its applications in radios and microwave ovens, our understanding of light's properties enables us to build the scientific tools with which we explore the cosmos.

Properties of Matter and Atomic Fingerprints

- All atoms consist of the same basic parts: a nucleus composed of some number of protons and neutrons, orbited by a number of electrons. All of the mass of the atom is essentially contained in the nucleus, because the electrons weigh next to nothing. The number of protons in the nucleus is what determines which element it is.

An atom consists of a nucleus of protons and neutrons surrounded by a cloud of electrons.

- The simplest element is hydrogen. It is simply a proton that, in its neutral and most common state, is orbited by a single electron. We say that the orbital energy of the electron is "quantized," because it is only permitted to have energies of certain discrete quantities.

- An electron possessing the minimum permitted amount of orbital energy is in the ground energy state—its energy is at the lowest possible level. When an electron gains energy in an amount sufficient for it to have one of the higher permitted energy levels—suppose it got bumped by another atom—then the electron is in an excited energy state or level. When that happens, the electron will, after a short time, spontaneously drop back down to the ground energy level.

- Because the electron loses energy in the course of dropping from an excited energy level to a lower energy level, that same amount of energy must be released by the atom, and it is released in the form of a photon of exactly that amount of energy. In other words, the atom emits a photon of specific wavelength, because that photon carries away a specific amount of energy exactly equal to the amount of energy lost by the electron when it dropped down.

- Each element has a unique set of permitted energy levels for its electrons, and therefore, each element's electrons can only perform certain discrete energy jumps. Each element will emit only certain specific wavelengths of light corresponding to those discrete energy jumps. In other words, every element in the periodic table emits a distinct pattern of wavelengths of light—a unique light fingerprint.

- If you examine the light emission patterns for a few different elements, you will see how distinct each one is from the others. And because emission patterns are like fingerprints, you can tell which elements a star contains by looking at its emission pattern.

- The emission process also happens in reverse, and it provides a complementary way of identifying the presence of specific elements in a star.

- We can discern the presence of an element either from the appearance of light at specific wavelengths or from the absence of light at those specific wavelengths. In both cases—an atom emitting its light fingerprint in one case or an atom absorbing its light fingerprint in the other case—we know that atom is there because we know that only that type of atom can interact with those specific wavelengths of light.

Suggested Reading

NASA Goddard Space Flight Center, "Electromagnetic Spectrum."

Ptable, "Periodic Table."

1. How does the astronomical study of the stars differ from most other sciences, in which experiments can be performed under controlled conditions in a laboratory and in which the objects of study can be directly manipulated?

2. What is the relationship between studies of elements in chemistry laboratories and our interpretation of the light spectra of the elements from astronomical objects?

The Stars' Information Messenger
Lecture 2—Transcript

The astronomer's most fundamental physical tool for understanding the stars is light. Virtually everything we know about the Sun and the stars is based on the light that we detect from them. In our everyday lives we are accustomed to the luxury of sensing our world in multiple ways—sight, smell, touch, taste, sound—giving us an amazingly rich amount of information about the world around us. But to understand the stars, we are limited to one sense alone—sight. And what we see is the light that the stars emit, which travels across space to our eyes and telescopes. But light is no meager informant. In this lecture, we'll see that, in fact, light is the ultimate information bearer of the stars. Imprinted in the light we receive is crucial information, quantifiable information about the stars, their brightnesses, their temperatures, their motions, and their elemental compositions.

The ability to sense the stars and everything about them with nothing more than light is foundationally important to everything that we'll be talking about throughout the rest of this course. That's because to understand how stars live out their lives, we'll need to understand their physical characteristics and how those characteristics change throughout their lives. And the way we'll infer their characteristics is through the power of light. We'll also rely on the fact that stars emit, primarily, different types of light at different stages of their lives. For example, we'll see that when stars are first being born, they are enveloped in cocoons of gas and dust that obscure them from sight, but we can see them, nonetheless, by peering into their birth cocoons using different varieties of light that have the power to penetrate through the obscuring gas and dust.

In *The Mighty Sky*, an album of astronomically themed songs by award-winning songwriter Beth Nielsen Chapman, is a song called "You Can See the Blues." Let me share some of that song's lyrics with you, and no, I'm not going to try to sing it. The song says,

> The stars are various colors, shining up in the sky
> Studying their wavelengths is how they're classified

And you can see their blues, and other energies too
The secret's in the starlight, and their photons give us clues

By the time we're done today, I promise you'll understand what that song means. But where do we begin? Well, as a first step, we'll need to learn how to read the light spectrum of a star the way that an astronomer does, to infer from the patterns imprinted in a star's light spectrum the physical properties of the matter that produced that light.

Let's start by looking at the Sun. Think of it as a light bulb in the sky. It appears yellowish to our eyes, but in fact, it emits light of all colors. Scientists can create a light spectrum of Sunlight by spreading out the Sun's intrinsic rainbow of light into its constituent colors. When we look at this spectrum, we see all of the colors that we know—from red at one end to purple or violet at the other. The yellows and greens are the brightest colors, which is why the sun mainly looks to be that color to eyes, but overall, we see a smooth, continuous rainbow of many colors.

However we also see a curious pattern of very specific colors that are missing from the Sun's spectrum. These show up as dark patches in the spectrum. In other words, we see specific colors that have been absorbed from the Sun's otherwise continuous spectrum. By the time we're done, all of these features will make sense, and in our next lecture, we'll also understand how astronomers use these properties of sunlight to infer the Sun's temperature and its chemical makeup.

Now let's take a step back and review some of the basics of light and the ways that it interacts with the world around us. There are essentially four ways that light and matter interact: emission, absorption, transmission, and reflection. Emission refers to the process by which light is produced by an object, be it a solid, or liquid, or gas. Essentially, emission is the radiance or glow of an object. All things glow, just not necessarily in ways that we can see with our eyes. The Sun is an example of an object that does emit the kind of light we can see, and that light, when it reaches the Earth, interacts with the things that it encounters.

Some things absorb the sunlight and heat up as a result. Other objects are transparent to the light, like a window pane, and so we say that the light is transmitted through it. Still other objects reflect the Sun's light, which is to say that the light bounces off the object.

Now, and this is important, most things do a combination of these, For example, they might absorb some of the light and reflect the rest. Most of the things around us which have colors that we see are not actually glowing that color. Rather, they absorb most of the colors that are present in the light they receive and only reflect certain specific colors.

Whatever color an object reflects, that's the color our eyes perceive. For example, a chair with red upholstery is not actually glowing red. Instead, it appears red because it absorbs most of the colors that are present in the light it receives from the Sun through the window or from a light bulb in the room. The dyes in the upholstery absorb the blues and the greens and the yellows and reflect only the reds, hence, it appears red.

Similarly, the green leaves on a tree appear green not because they are glowing that color. Ask yourself whether the leaves of a tree appear to have any color at all when it's dark out. Rather, they appear green because the chlorophyll in the leaves absorbs the reds and the yellows from the sunlight and reflects only the greens and the blues. In the fall when the leaves die, the leaves no longer are able to absorb those reds and yellows, and instead, they reflect those colors, giving us a spectacular display of fall foliage.

I am emphasizing this point because later on in our next lecture we will see that we can use the colors of stars to learn something fundamentally important about them, namely, their temperatures. But this works for stars because their colors are real in the sense that they actually are glowing those colors. This is not the case for most of the things around us, whose colors are, instead, the result of absorption and reflection of light from other sources. The things around us glow with true colors as well, just not colors we can see. But we'll come back to those "invisible" colors in a moment.

Let's turn now to a description of what light itself actually is. Light has a very peculiar nature, which is referred to as its wave-particle duality.

This means that we can equally think of light as being like a wave or as a particle. In some cases it will be more helpful to talk about it as a wave-like phenomenon, like sound, and in other cases it will be more helpful to describe it as if it were a discrete particle, like an electron.

Let's start with the wave-like description of light. Most of us have experienced waves in water, and the wave-like version of light can be thought of in a similar way. Imagine dropping a pebble into a pond. What happens? The water's surface is disturbed, causing it to undulate up and down where the pebble plopped in. From that point, those undulations travel outward in all directions—they propagate—and they do so with a characteristic speed and with a characteristic distance between successive undulations.

We refer to this distance between undulations as the wavelength of the wave. With the wave traveling outward at a certain speed, and with a characteristic wavelength between successive waves, we can also define a frequency, that's rate at which the waves pass by a fixed point. A good analogy for this is waves in the ocean washing up to shore. The distance between successive swells is the wavelength. Those swells move toward the shore with a certain speed. When the waves reach the shore and crash down onto the beach, they do so with a steady cadence, that's the frequency of the waves.

It turns out that for any wave phenomenon, including waves at the beach, there is a specific relationship between the wave's speed, wavelength, and frequency. For light, the speed is, well, the speed of light. And as you probably know, the speed of light is a fundamental constant of nature. Since the speed of light is constant, that means that a light wave's frequency and wavelength are always inversely related. In other words, a light wave of a higher frequency has a shorter wavelength, and a light wave of a lower frequency has a longer wavelength. Think again about the waves at the beach. If the swells are close together—short wavelength—successive waves will, of course, arrive at the shore with a higher cadence—higher frequency.

And it turns out that just as our ears perceive different frequencies of sound waves as different pitches of sound, so do our eyes perceive different frequencies of light as different colors. Higher frequency light—or shorter wavelength light—is blue; while lower frequency light—or longer wavelength light—is red. Remember that; short wavelength is blue, long wavelength is red. We'll be making use of that basic relationship throughout the course.

Now, how about light as a particle? Well, just as there are other fundamental particles in nature, like protons and electrons, so can light be thought of as a discrete particle, which we call a photon. And as a discrete particle, light carries a discrete amount of energy. And this energy, it turns out, is directly related to the light's wave-like properties. Higher frequency, shorter wavelength photons carry more energy, and lower frequency, longer wavelength photons carry less energy. Or to bring it back to color, blue light is more energetic, and red light is less energetic.

So now we have all of the ingredients for describing light, and what you should remember are basically two things; first, light is simultaneously like a wave and a particle. Like a wave, light possesses a frequency and a wavelength and propagates with a certain speed. And like a particle, which we call a photon, light carries a discrete amount of energy. The second thing you should remember is that shorter wavelength light is more energetic than longer wavelength light. Blue equals shorter wavelength, equals more energy; and red equals longer wavelength, equals less energy.

From what we've talked about so far, you're probably already getting the idea that light comes in different varieties, what we ordinarily think of as color. But the colors that we're familiar with from everyday life are just a tiny sliver of a much larger range of colors that exist in nature. The full range of colors, including those invisible to our eyes, is what we refer to more generally as the electromagnetic spectrum of light.

Let's review the electromagnetic spectrum together in order to familiarize ourselves with all of the varieties of color of the real rainbow. The part of the electromagnetic spectrum that we can see is referred to as the visible part of the spectrum, and it contains the red-to-violet rainbow that we all know so well. The wavelengths of light that are responsible for that band of visible colors are typically about one micron in length, or about the size of a bacterium. But light comes in an infinite range of other wavelengths, both shorter and longer than those which we can see. Ultraviolet light has wavelengths that are just a little bit shorter than the deepest violet that we can see. In other words, this light is beyond violet, which is the literal meaning of ultraviolet.

Now, if you continue beyond ultraviolet, you come to the X-ray part of the electromagnetic spectrum. X-ray light has even shorter wavelengths and even higher energies, which is why your dentist can use X-rays to examine your teeth. Those wavelengths are very small, as small as a single atom. Beyond that, the shortest wavelengths and highest energies are the gamma rays. These photons are so energetic that on Earth they are generally only encountered in nuclear explosions.

Coming back to the visible part of the electromagnetic spectrum, there are photons with longer wavelengths as well. At wavelengths just longer than the reddest red that we can see, is the infrared, meaning literally below red. Infrared light is sometimes referred to as thermal radiation, because most things that have temperatures of the kinds we experience every day, such as our bodies at room temperature, glow most strongly in infrared light. We'll talk more about that later. At wavelengths longer than infrared, and at the lowest energies, are radio waves, a portion of which are called microwaves. Radio waves are photons with extremely long wavelengths, as long as a football field or even longer.

We often associate the word radio with sound, but that's just because the radios that we use to listen to music operate by receiving—or seeing—light signals at radio wavelengths from a radio light bulb (the radio tower) that emits those radio light waves. These waves carry the information that our radios then turn into sound.

And what about those microwaves? We, of course, associate them with our microwave ovens at home. That's because inside the oven is a microwave light bulb that shines light of microwave wavelengths onto the food inside, causing the water molecules in the food to vibrate, and thereby heat the food. The regular light bulb inside the oven is there for convenience, to enable us to see what's cooking inside the oven. The microwave light that is doing the cooking does not escape the oven, because there is a metal screen in the door of the oven that acts as a reflector for the microwave photons, ensuring that they stay safely within. In addition to its applications in radios and in microwave ovens, our understanding of light's properties enables us to build the scientific tools with which we explore the cosmos. We'll talk more about how light waves are reflected and focused when we discuss telescopes and how they work in a later lecture.

So now we understand what light is, and we've learned about its different varieties—the electromagnetic spectrum. The final piece is to discuss how light interacts with individual atoms, the elements, because this will help us understand how astronomers can decode the patterns of light from the stars to determine what the stars are actually made of. And understanding what the stars are made of—their elemental compositions—is one of the most important things that we'll need to understand the life cycles of stars.

Let's recall what the basic structure of an atom is. All atoms consist of the same basic parts, a nucleus, composed of some number of protons and neutrons, orbited by a number of electrons. All of the mass of the atom is essentially contained in the nucleus, because the electrons weigh next to nothing. The number of protons in the nucleus is what determines which element it is. For example, a hydrogen atom has a single proton in its nucleus, that's what makes it hydrogen. And helium always has two protons. An atom with two protons in its nucleus, regardless of how many neutrons it has, is always called helium. So looking at the periodic table of the elements, we are really just looking at the different types of atoms ordered by the number of protons in their nuclei; carbon has 6 protons, oxygen has 8, iron has 26.

Now let's come back to consider the simplest element of all, hydrogen, once again. It is simply a proton that, in its neutral and most common state, is orbited by a single electron. A crucially important thing to understand is that this electron in its orbit about the proton is not permitted to have just any amount of energy. Rather, it is permitted to have only certain specific orbital energies. We say that the orbital energy of the electron is quantized, because it is only permitted to have energies of certain discrete quantities, which can be represented in a diagram as discrete lines, or levels, arranged vertically.

An electron possessing the minimum permitted amount of orbital energy is said to be in the ground energy state, because its energy is at the lowest possible level. When an electron gains energy in an amount sufficient for it to have one of the higher permitted energy levels, suppose it got bumped by another atom, then the electron is said to be in an excited energy state, or in an excited energy level. And when that happens, the electron will, after a short time, spontaneously drop back down to the ground energy level.

And now here's the important part. Because the electron loses energy in the course of dropping from an excited energy level to a lower energy level, that same amount of energy must be released by the atom, and it is released in the form of a photon of exactly that amount of energy. In other words, the atom emits a photon of specific wavelength, because that photon carries away a specific amount of energy exactly equal to the amount of energy lost by the electron when it dropped down. Every time an electron in a hydrogen atom drops down from that excited energy level to that lower energy level, it will emit a photon of exactly that same energy, exactly that same wavelength. That's important to remember.

Now, the electron in a hydrogen atom has many different energy levels that it is permitted to be in, so there are many different jumps that an electron can make from an excited energy level to a lower energy one. For example, if the electron finds itself in the fourth energy level, it could drop straight down to the ground energy level, or it could first jump down to the second energy level, followed by a drop from there to the ground energy level. In any event, there are only certain permitted energy jumps because there are only certain permitted energy levels that the electron can ever be in.

The energy levels that are permitted for the electron in a hydrogen atom are unique to hydrogen. The electrons in a helium atom also have certain permitted energy levels, but they are different energies than the ones for hydrogen, and they are unique to helium. This is true of every element. Each element has a unique set of permitted energy levels for its electrons, and therefore, each element's electrons can only perform certain discrete energy jumps.

And why is that important? Because it means that each element will emit only certain specific wavelengths of light corresponding to those discrete energy jumps. In other words, and this is the essential point, every element in the periodic table emits a distinct pattern of wavelengths of light, a unique light fingerprint, if you will. And once we know what that distinct light pattern is for a given element, we can recognize it when we see it, no matter where or how far away it may be, as long as we can detect the light.

If you examine the light emission patterns for a few different elements, you will immediately see how distinct each one is from the others. And since emission patterns are like fingerprints, you can tell what elements a star contains by looking at its emission pattern. If you see hydrogen's pattern of light emission from a distant nebula, you can be certain that that nebula contains hydrogen. If you see iron's pattern of light emission from a distant supernova explosion, you can be certain the supernova contains iron.

The process we just described for the emitted pattern of light from an atom also happens in reverse, and it provides a complementary way of identifying the presence of specific elements in a star. Let's consider that hydrogen atom once again. We've just described how we can detect the presence of hydrogen atoms in a star based on their distinct emission patterns. But now, instead of visualizing the hydrogen atom in isolation, emitting its unique light fingerprint, let's imagine that it is bathed in light from a nearby source, such as a star. Now, the star is emitting a continuous rainbow spectrum of light at all wavelengths. Essentially the atom is being bombarded by photons of every conceivable wavelength. What will happen to those incoming photons?

Well, the vast majority of them will pass right on through, undisturbed, because those photons do not carry the right amount of energy to kick the electron up from its current energy level to one of the higher permitted energy levels. However, some photons of very specific energies, or wavelengths, will be able to kick the electron up by just the right amount, and so those photons will be absorbed as they impart their energy to the atom. Again, the essential point here is that only a very specific set of photons, only those with precisely the right energies corresponding to the permitted energies of the electrons in the hydrogen atoms, can be absorbed.

As seen from the other side, almost all of the originally incident light passes right on through to our eye or telescope, but those specific wavelengths of light that were absorbed by the hydrogen will be missing. So if we look at the absorption spectrum for hydrogen, what do we see? We see light at all wavelengths, but darkness at just those specific wavelengths that were absorbed. And that absorption pattern is just like the emission pattern that we talked about before. It is a distinct pattern unique to hydrogen. So if you see a star that is emitting light at all wavelengths, except at those wavelengths unique to hydrogen are missing, you can be certain there is hydrogen in that star.

It might seem counterintuitive that we can discern the presence of an element like hydrogen either from the appearance of light at specific wavelengths or from the absence of light at those specific wavelengths. But in both cases—an atom emitting its light fingerprint in the one case or an atom absorbing its light fingerprint in the other case—we know that atom is there because we know that only that type of atom can interact with those specific wavelengths of light.

The emission case applies generally to atoms seen in isolation, such as in the diffuse nebulae of stellar nurseries that we'll talk about in later lectures. The absorption case applies generally to atoms seen against a bright backdrop of continuous rainbow, such as the atoms in the outer layers of a star bathed in the white light emerging from the deeper layers of the star.

So let's close by thinking about that beautiful light spectrum of the Sun one last time. Can we make sense of what we see? We see a continuous rainbow spectrum of light at all wavelengths that the Sun has emitted. But we also see a large number of wavelengths missing. Those wavelengths have been absorbed by elements in the outer layers of the Sun, and by matching those missing wavelengths with the known light patterns of various elements, we can account for each of those missing wavelengths and infer the elemental makeup of the Sun. Here is calcium; here is sodium; here is magnesium, and many, many others.

In this lecture, we've learned how astronomers read the spectrum of light from a star to determine the elements that are within it, emitting and absorbing the light that we ultimately see with our eyes and telescopes. In our next lecture we'll extend what we've discussed here to understand how astronomers use the colors of starlight and the patterns of light emission and absorption to quantify the basic physical properties of the stars. Properties such as temperature, which are critical to understanding the stellar life cycle. In the meantime, I hope you'll never look at the colors of stars the same way again. They're still as beautiful, but now those colors mean so much more.

Measuring the Stars with Light
Lecture 3

T he information that we glean from starlight is the key with which we can unlock the mysteries of the stellar life cycle. In this lecture, you will learn a few more essential details about the way astronomers use light to understand the physical properties of stars. Specifically, this lecture will focus on those physical attributes of stars that you will need throughout this course to truly understand the stellar life cycle.

Kirchhoff's Laws and Wien's Law

- Kirchhoff's laws are a set of physical laws that together describe the ways in which different types of objects emit and absorb light. Kirchhoff's first law describes the radiation emitted by any opaque body. It says that an opaque object emits a continuous spectrum of light at all wavelengths.

- Kirchhoff's first law states that the rainbow of light emitted by an opaque object is continuous. That means that it emits all wavelengths of light—from gamma rays to radio waves, and everything in between. That doesn't mean it emits the same amount at all wavelengths, just that it does emit some light at all wavelengths.

- One of the simplest but most important pieces of information we can glean from the stars is color. The color of light is associated with its energy. Bluer light is more energetic light; redder light is less energetic. More energetic objects emit more energetic light. Therefore, more energetic objects emit bluer, more energetic light, whereas less energetic objects emit redder, less energetic light.

- The basic measure of an object's energy is its temperature. A hotter object contains within it more energy in the form of heat—what physicists call thermal energy. A colder object contains less heat, or less thermal energy, and its temperature is lower.

- It turns out that we can quantify this relationship between color and temperature precisely. The relationship is called Wien's law, which is a simple formula that can be dissected as follows: The wavelength of the light at which an object emits most strongly— in other words, the brightest wavelength of light in the continuous rainbow emitted by the object—is a constant divided by that object's temperature. That is, the wavelength of peak light emission is inversely proportional to the temperature of the emitting body.

- The light spectrum emitted by an opaque body such as a star is continuous—some amount of light is emitted at all wavelengths, but not the same amount of light at all wavelengths. Rather, any given opaque body will emit most of its light in one part of the spectrum and less (but some) light at all other wavelengths.

- The point of Wein's law together with Kirchhoff's first law is that we have a straightforward way of determining the temperature of a star. All we have to do is measure the star's continuous light spectrum using a spectrograph attached to a telescope and record the wavelength at which most of the light is being emitted, which is nothing more than looking at which wavelength of the spectrum is brightest. Then, we can calculate the temperature of that star's emitting surface.

- Kirchhoff's second and third laws relate to nonopaque, or transparent, gases and how they interact with light. Kirchhoff's second law states that a transparent gas will emit not a continuous rainbow of light, but it will emit light at only certain discrete wavelengths. These specific, discrete wavelengths of emission correspond to the energy levels in the atom that its electrons are permitted to jump between.

- This means that each type of atom in a gas—each element—emits a distinctive fingerprint. It emits only certain wavelengths of light, and that pattern of light emission is unique to that element.

- Kirchhoff's third law states that a transparent gas, when exposed to a continuous spectrum of light, will absorb a specific, discrete set of those photons. It will be the same wavelengths of light that this gas would emit in the Kirchhoff's-second-law scenario. The only difference is that, whereas in Kirchhoff's second law the elements in the gas are seen in isolation, emitting their permitted wavelengths of light, in the third law we are seeing the elements against the backdrop of a bright, continuous source of white light, which the elements can absorb.

Gustav Robert Kirchhoff (1824–1887) was a German physicist who first announced Kirchhoff's laws in 1845.

- We can directly identify which elements are absorbing the light by recognizing the specific pattern of wavelengths being absorbed out from the otherwise continuous spectrum impinging on the absorbing gas.

- When we look at the light spectrum emitted by actual stars, we mainly see a continuous rainbow of light. But we also see imprinted on that continuous spectrum a set of dark absorption bands, from which we can determine the elemental makeup of the star.

The Doppler Effect
- Another essential piece of information we can learn about a star from its light spectrum is its motion. All stars in our galaxy move relative to one another. The stars orbit about the center of our galaxy. In addition, pairs of stars are often orbiting one another—what we call binary star systems—and we can learn still more by studying those motions.

- The astronomer's basic tool for measuring the speed of a star's motion is the Doppler effect, which was first studied in the context of sound waves, which in their physical behavior bear some similarities to light waves. The effect states that when an object is moving toward you, the emitted waves are bunched up and register as a higher frequency. When an object is receding from you, the emitted waves are stretched out and register as a lower frequency.

- The light waves emitted by an approaching object are bunched together, which we register as a higher frequency of light, which means a higher energy, or bluer color of light. Conversely, the light waves emitted by a receding object are stretched, which we register as a lower frequency of light, which means a lower energy, or redder color of light. The color of the light is shifted, to the blue or to the red, depending on whether the object is moving toward or away from us.

- The Doppler effect has a formula associated with it that allows us to precisely calculate the speed of a star from the exact amount by which its light spectrum has been shifted. The formula says that the percentage by which the star's wavelengths are shifted is the same as the percentage of the speed of light that the star is moving.

Measuring Luminosity and Parallax

- The total intrinsic light output of a star is its luminosity. A star's luminosity together with its temperature allow us to piece together its entire life cycle and to pinpoint exactly where in its life cycle a given star is at any moment.

- In principle, measuring the intrinsic luminosity of a star is simple: Just measure how bright it appears, measure how far away it is, and adjust its apparent brightness for its distance. How bright something appears depends both on how bright it actually is and how far away it is.

- In fact, light obeys what is referred to as an inverse-square law: The farther away a light is, the dimmer it appears, and it gets dimmer by the square of the distance.

- The problem is that, while measuring the apparent brightness of the star is trivial, measuring its distance is generally very difficult. That's because most stars are so far away, so incredibly distant, that we have no way of getting a vantage point to sense whether they are relatively close or far.

- Fortunately, there are some stars that are nearby enough that we can use a method of triangulation, also known as parallax, to measure their distances directly. We can discern relative distances of objects by virtue of how much of a jump our brain notices between the apparent position of an object as seen by each eye. A relatively large jump means that the object is relatively close; a small jump means that it must be farther away.

- Our ability to do this is somewhat limited, however, because our two eyes are not very far apart. If they were twice as far apart, we'd be able to perceive these positional jumps half as big, allowing us to have twice the depth perception.

- We can do the same thing with the stars, now with "eyes" as far apart as the diameter of the Earth's orbit about the Sun. By measuring the amount by which the position of a star appears to jump relative to the more distant stars in a 6-month period, for example, we can directly determine its distance. Given the size of the Earth's orbit, our depth perception on the stars is pretty good, but it's not great.

- We can use this technique to directly measure the distances to stars as far away as about 1,500 light-years. That's not very far in the grand scheme of things—our galaxy is 100,000 light-years across—but it is far enough that we can measure the distances to hundreds of stars and use their properties to figure out the properties of all other stars.

- By measuring the parallax of a star, and thereby determining its distance, we can now use the brightness of the star that we see on Earth to determine its intrinsic luminosity. That luminosity and the temperature that we can measure with Wien's law allow us to determine almost everything about a star's life and death.

Suggested Reading

University of Nebraska at Lincoln, "Doppler Shift Demonstrator."

Wikipedia, "Stellar Classification."

Questions to Consider

1. What are the most important physical properties of the stars, with respect to the life cycles of stars, that we can measure using the light spectra they emit?

2. How does the actual relationship between color and temperature for physical objects, as described by Wien's law, compare to our everyday usage of colors to describe temperatures? For example, what color do you usually associate with cold, and what color do you associate with hot?

Measuring the Stars with Light
Lecture 3—Transcript

In our last lecture, we learned about the basic properties of light. We learned that light comes in a variety of forms, from the most energetic gamma rays to the least energetic radio waves, and we learned that the ways in which a body emits or absorbs light tells us about the elemental makeup of that body. In this lecture, we'll extend our understanding of light one step further. We'll look at how astronomers actually use light from the stars to measure specific physical properties of the stars—their temperatures, their motions, their distances, and their elemental compositions.

But you may be wondering, how does this relate to the life cycle of stars? Well, the answer is that, as we'll see in subsequent lectures, the life cycles of stars depend simply and elegantly on a few basic physical properties of stars, properties such as the stars' temperatures and luminosities, and their masses and diameters. But the stars don't reveal these crucial physical properties to us directly. Rather, we must learn how to translate the information that they do send us directly in the form of light of many different wavelengths into those crucially important physical properties.

To understand the stars, we have to understand the light that they emit. After all, one of the most basic things that stars do is to generate light, lots of it, and lots of different kinds. Most importantly, because the stars are so far away, we can't touch them or probe them directly with our instruments. Instead, nature has provided us with a rich, powerful tool in the form of the light that the stars send streaming to us across the vastness of space and time.

Light is the stars' information messenger. Using basic physical laws about light and the ways in which it interacts with matter, laws learned in the laboratory, we can use the starlight we measure to determine almost everything about a distant star. And as we'll see time and again throughout this course, the information that we glean from starlight is the key through which we can unlock the mysteries of the stellar life cycle.

So today, let's learn a few more essential details about the way astronomers use light to understand the physical properties of stars, focusing specifically on those physical attributes of stars that we'll need throughout this course to truly understand the stellar life cycle. Let's start by discussing a set of physical laws that together describe the ways in which different types of objects emit and absorb light. These laws are called Kirchhoff's Laws, and there are three of them. Why do physical laws always come in sets of three? Never mind that. But let's talk about Kirchhoff's First Law.

This first law describes the radiation emitted by any opaque body. It says that an opaque object emits a continuous spectrum of light at all wavelengths. That's important to remember, so let's dissect it. First of all, Kirchhoff's First Law states that the rainbow of light emitted by an opaque object is continuous. That means that it emits all wavelengths of light, all wavelengths of light, from gamma rays to radio waves, and everything in between. Secondly, Kirchhoff's First Law applies to any opaque object. Obviously, a solid is opaque. Take a light bulb for example. The solid filament inside the bulb is an opaque piece of solid metal. Heated by the electrical current within, it emits a continuous rainbow of light, much of which is light that we can see with our eyes. But the light bulb doesn't emit only visible light. Kirchhoff's First Law states that all wavelengths of light are emitted, and indeed, a typical incandescent light bulb actually emits most of its light in the infrared. It emits some amount of light at radio wavelengths too, and even some at X-ray and gamma-ray wavelengths. Again, the rainbow spectrum of an opaque body is continuous. That doesn't mean it emits the same amount at all wavelengths, just that it does emit some light at all wavelengths.

Liquids can be opaque too, and this is the important part as far as the stars are concerned. Even a gas can be opaque, if it is sufficiently dense. Indeed, stars are opaque, gaseous bodies, and so Kirchhoff's First Law applies to them as well. Stars emit a continuous rainbow of light, emitting most strongly in some parts of the electromagnetic spectrum, but emitting some light at all parts of the spectrum, from radio waves to gamma rays.

Now that we've discussed Kirchhoff's First Law, let's step away from Kirchhoff for a moment to talk about one of the simplest but most important pieces of information we can glean from the stars—color. From our last lecture you already know that the color of light is associated with its energy. Bluer light is more energetic light, redder light is less energetic. It probably already makes sense to you that more energetic objects emit more energetic light. So putting two and two together, more energetic objects emit bluer, more energetic light; whereas less energetic objects emit redder, less energetic light. Now, what does it mean for an object to be more or less energetic? The basic measure of an object's energy is its temperature. You know this intuitively; a hotter object contains within it more energy in the form of heat, what physicists call thermal energy. A colder object contains less heat, less thermal energy, and its temperature is lower.

It turns out that we can quantify this relationship between color and temperature precisely. The relationship is called Wien's Law. Let's look at Wien's Law together. Wien's Law is a simple formula that can be read as follows: The wavelength of the light at which an object emits most strongly, in other words, the brightest wavelength of light in the continuous rainbow emitted by the object, is a constant divided by that object's temperature. That is, the wavelength of peak light emission is inversely proportional to the temperature of the emitting body.

As we saw in Kirchhoff's First Law, the light spectrum emitted by an opaque body, such as a star, is continuous—some amount of light is emitted at all wavelengths, but not the same amount of light at all wavelengths. Rather, any given opaque body will emit most of its light in one part of the spectrum, and less, but some light, at all other wavelengths.

Let's think about our bodies as emitting objects. At room temperature, our bodies are at about 300 degrees Kelvin (that is 300 degrees above absolute zero). Putting that temperature into Wien's Law, we calculate that the wavelength of maximum emission of our bodies is at about 10 micrometers, a wavelength corresponding to the middle of the infrared part of the electromagnetic spectrum. When we see one another with our eyes, we are not seeing visible light emitted by our bodies, but rather reflected visible light from the hot light bulbs in the room. Turn off the lights, and we don't see one another. Even though our bodies do technically emit some amount of visible light—because Kirchhoff's First Law says we emit a continuous spectrum—the amount of visible light that we emit is miniscule. However, put on some night-vision goggles, which have special detectors to see infrared light, and our bodies glow brightly! Each of us, in fact, radiates about 100 watts of light in the infrared.

And what about that incandescent light bulb I mentioned a moment ago? The temperature of the hot metal filament is about 1500 degrees. Put that temperature into Wien's law, and we calculate a wavelength of maximum emission of about one micrometer. That wavelength is just on the infrared side of the visible wavelength range that our eyes can see. So the light bulb, emitting a continuous spectrum, emits enough visible light to illuminate a room for the benefit of our eyes. Interestingly, though, the actual majority of the bulb's light is essentially wasted in the form infrared light we don't see, but which we can feel as heat.

The point for us of Wien's Law, together with Kirchhoff's First Law, is that we have a straightforward way of determining the temperature of a star. All we have to do is measure the star's continuous light spectrum through a spectrograph attached to a telescope and record the wavelength at which most of the light is being emitted, which is nothing more than looking at which wavelength of the spectrum is brightest. And voila, we can calculate the temperature of that star's emitting surface.

Let's come back now to Kirchhoff's Second and Third Laws. These two laws relate to non-opaque, or transparent, gases, and how they interact with light. Kirchhoff's Second Law states that a transparent gas will emit not a continuous rainbow of light; rather, it will emit light at only certain discrete wavelengths. These specific, discrete wavelengths of emission correspond to the energy levels in the atom that its electrons are permitted to jump between.

As we discussed in the last lecture, this means that each type of atom in a gas—each element—emits a distinctive fingerprint, if you will. It emits only certain wavelengths of light, and that pattern of light emission is unique to that element. The light spectrum emitted by hydrogen atoms is recognizable as being hydrogen and hydrogen alone. And the same goes for every other element in the periodic table. Each has its own unique pattern of wavelengths of light that it can emit.

Kirchhoff's Third Law states that a transparent gas, when exposed to a continuous spectrum of light, will absorb a specific, discrete set of those photons. It will be the same wavelengths of light that this gas would emit in the Kirchhoff's Second Law scenario. The only difference is that, whereas in Kirchhoff's Second Law the elements in the gas are seen in isolation, emitting their permitted wavelengths of light, here in the Third Law, we are seeing the elements against the backdrop of a bright, continuous source of white light, which the elements can absorb. And again, we can directly identify which elements are absorbing the light by recognizing the specific pattern of wavelengths being absorbed out from the otherwise continuous spectrum impinging on the absorbing gas.

And so here is the essential point of Kirchhoff's Laws all together. When we look at the light spectrum emitted by actual stars, we mainly see a continuous rainbow of light. But we also see imprinted on that continuous spectrum a set of dark absorption bands from which we can determine the elemental makeup of the star. Kirchhoff's Third Law in action.

Let's put all this together by considering once more the light spectrum of the Sun. Actually, it's just the part of the Sun's spectrum in the visible portion of the electromagnetic spectrum. In fact, the Sun emits light at all wavelengths, from radio waves to gamma rays, as Kirchhoff's First Law says it should. In any event, when we examine the light spectrum of the Sun, here's what we see. We see a continuous rainbow from red to blue, and it is, mostly, continuous, from the deepest red that our eyes can see, through to the oranges and the yellows and the greens, all the way to the deepest violet that our eyes can see.

But notice that there's one part in the middle of this continuous band of color that is somewhat brighter than the other colors. The brightest part of the Sun spectrum is yellow. You know this from seeing the yellow Sun in the sky. Can we understand this now using the tools that we've discussed? Let's put the Sun's temperature, which at its surface is about 6000 degrees, into Wien's Law. So, let me walk you through that calculation. The constant in Wien's Law, that 2.9, let's call it 3, divide that by the 6000 degrees temperature of the Sun, that's about half of one one-thousandth. More precisely, it's half of one one-thousandth of a millimeter. That translates to half of a micrometer. Half of a micrometer is 500 nanometers, and that wavelength, 500 nanometers, corresponds to the yellow part of the visible part of the electromagnetic spectrum.

We also see in the spectrum of the Sun, those dark bands that I mentioned before, specific wavelengths that have been absorbed out of the Sun's otherwise continuous spectrum of light. These patterns are due to the different elements that are present in Sun's atmosphere. Toward the red part of the spectrum we see a dark band that is due to the absorption by hydrogen atoms. In the yellow, we see an absorption band that is due to sodium atoms, also present in the Sun's atmosphere.

And so here, in the Sun, we see Kirchhoff's Third Law in action. The opaque interior of the Sun shines a continuous white-light spectrum at all colors, but that continuous spectrum bathes and passes through the outer transparent atmosphere, which mostly allows those wavelengths of light to pass through, but absorbs out specific wavelengths of light caused by the specific elements present in the Sun's atmosphere.

So you see, already with just these basic laws of physics, we can infer the surface temperature of a star and its elemental composition. This is how it's done; it's just light. But imprinted in the light is a tremendous amount of valuable information.

But wait, there's more! Another essential piece of information we can learn about a star from its light spectrum is its motion. All stars in our galaxy move relative to one another. The stars orbit about the center of our galaxy. In addition, as we'll discuss in later lectures, pairs of stars are often orbiting one another, what we call binary star systems, and we can learn still more by studying those motions. The astronomer's basic tool for measuring the speed of a star's motion is the Doppler Effect. The Doppler Effect was actually first studied in the context of sound waves, which in their physical behavior bear some similarities to light waves.

So let's start with the sound analogy first. Think about the sound of an ambulance driving toward you on the road. As the ambulance approaches, you hear its high-pitched siren. Then, as it passes you, the sound suddenly drops a bit in pitch. It's still loud, because the ambulance is close, but there is a perceptible change to a lower tone as the ambulance goes from approaching to leaving. That's the essence of the Doppler Effect. The motion of the object emitting the sound waves causes the waves to be bunched closer together in the direction of motion and stretched apart in the direction away from the motion. That bunching up of the waves in one direction means that the waves arrive with a higher frequency, which in the case of sound, we hear as a higher pitch. The stretching out of the waves in the other direction means a lower frequency, which in the case of sound, we register as a lower pitch. So the essential thing to remember with the Doppler Effect is, when an object is moving toward you, the emitted waves are bunched up and register as a higher frequency. When an object is receding from you, the emitted waves are stretched out and register as a lower frequency.

So let's translate that into the language of light. The light waves emitted by an approaching object are bunched together, which we register as the higher frequency of light, which means a higher energy, or bluer color of light. Conversely, the light waves emitted by a receding object are stretched, which we register as a lower frequency of light, which means a lower energy, or redder color of light. So the color of the light is shifted to the blue or to the red depending on whether the object is moving toward, or away from us. Approaching equals blue shift; receding equals red shift.

This is getting a little bit ahead of ourselves, but just to give you an idea of how the Doppler Effect is actually applied to the motions of stars, let's briefly look at an example of a binary star system—two stars orbiting one another. When we look at the light spectrum of a star that is actually two stars orbiting one another, we will see in that light spectrum essentially two different spectra superimposed on one another, one from each star. As the two stars orbit, at any given moment one star will be mainly moving toward us as the other is mainly moving away. At that moment, the spectrum of the one approaching will be shifted to bluer, or shorter, wavelengths; and the one moving away will be shifted to redder, or longer, wavelengths. As the stars proceed in their mutual orbit, they will end up swapping places, now the other approaching and the other receding. And so their spectra will swap also, now the one being bluer, and the other being redder. This is, in fact, how we measure the orbits of distant star systems, even if we can't directly see the two stars separately.

The Doppler Effect has a formula associated with it, and this formula allows us to precisely calculate the speed of a star from the exact amount by which its light spectrum has been shifted. The formula says that the percentage by which the star's wavelengths are shifted—that's Delta lambda divided by lambda—is just the same as the percentage of the speed of light that the star is moving—that's the velocity, v, divided by the speed of light, c. So for example, if a star is moving toward us at 1% of the speed of light, its light spectrum will be shifted to bluer (shorter) wavelengths by 1%.

You might be wondering, how do we know, or measure, the amount by which a star's light has been shifted? We see the light as we see it; how do we know that's not just the way the star actually emitted it, as opposed to the light being compressed or stretched by its motion? This is another way in which the presence of those absorption bands in the star's spectrum that we talked about can come in handy. When we see the pattern of absorbed colors due to hydrogen, for instance, what we are recognizing is the unique pattern, or spacing, between the different absorption bands. Once that pattern is recognized, we can compare the wavelengths at which the individual bands occur to the wavelengths that would occur if the star weren't moving. It's the percentage shift of those absorption bands that allows us to precisely determine the amount by which the entire spectrum has been shifted, and through the Doppler formula, to calculate the star's motion toward or away from us.

So let's check in on how far we've come. From a star's light spectrum we can use its color to measure its temperature. From the pattern of light absorption we can measure the elemental composition. And from the shift of the spectrum, we can measure the star's motion. That's a lot of valuable physical information, and from something we can only see from a great distance.

There is one more aspect of how we use light to measure the stars that we need to discuss. And that is, the total intrinsic light output of a star, what we call its luminosity. As we'll see in our upcoming lectures, a star's luminosity, together with its temperature, allows us to piece together its entire life cycle and to pinpoint exactly where in its life cycle a given star is at any moment. In principle, measuring the intrinsic luminosity of a star is simple: Just measure how bright it appears, measure how far away it is, and adjust its apparent brightness for its distance. As you know, how bright something appears depends both on how bright it actually is together with how far away it is. For example, a light bulb right in front of your face can appear quite a lot brighter than a distant light house, but you know that light house is actually putting out a lot more light.

In fact, light obeys what we refer to as an inverse-square law. The farther away a light is, the dimmer it appears, and it gets dimmer by the square of the distance. So if you move a light two times further away, it will appear four times dimmer; three times further away, nine times dimmer, and so on. So, just measure how bright the star appears, measure its distance, and you have what you need to figure out its true luminosity, right? The problem is that, while measuring the apparent brightness of the star is trivial, measuring its distance is generally very hard. That's because most stars are so far away, so incredibly distant, that we have no way of getting a vantage point to sense whether they are relatively close or far. They're just... far!

Fortunately, there are some stars that are nearby enough that we can use a method of triangulation, also known as parallax, to measure their distances directly. Let's look at how this works. Let's do a little exercise together, and you don't even have to get up out of your seat, but please do this; it'll be fun. First, make a thumbs up, like this. Hold out thumb at arm's length. Now, close one eye and position your thumb so that it covers my face on the screen. Now, without moving your hand, switch eyes. Notice how your thumb appears to jump. Now, you can blink your eyes back and forth, left-right, left-right, a few times, and watch your thumb appear to jump back and forth. Now, one more thing, move your thumb closer to face, and blink your eyes back and forth one more time. Notice this time that your thumb appears to jump by a much larger amount. What you've just experienced is the miracle of stereoscopic vision. This is why we have two eyes. We can discern relative distances of objects by virtue of how much of a jump our brain notices between the apparent position of an object as seen by each eye. A relatively large jump means the object is relatively close; a small jump means it must be farther away. Our ability to do this is somewhat limited, however, because our two eyes are not very far apart. If they were twice as far apart, we'd be able to perceive these positional jumps half as big, allowing us to have twice the depth perception.

Well, we can do the same thing with the stars, now, with eyes as far apart as the diameter of the Earth's orbit about the Sun. Imagine the Earth's orbit being like our face. And imagine our eyes on two opposite sides of the orbit. Now, as the Earth goes around its year-long orbit about the Sun, imagine that we look at the position of a star at one time, noting its apparent position relative to the backdrop of much more distant stars. Then, six months later, as the Earth has orbited half way around, we're on the other side of the Sun, we look toward the star again. By waiting six months for the Earth to traverse in its orbit, it's like we're blinking, first one eye and then the other. By measuring the amount by which the position of the star appears to jump relative to the more distant stars in that six-month period, we can directly determine that star's distance.

This is the method of triangulation, or what astronomers call parallax. Given the size of the Earth's orbit, our depth perception on the stars is pretty good, but not great. We can use this technique to directly measure the distances to stars as far away as about 1500 light-years; that's not very far in the grand scheme of things—our galaxy is 100,000 light-years across! But it is far enough that we can measure the distances to hundreds of stars and use their properties to figure out the properties of all other stars.

By measuring the parallax of a star, and thereby determining its distance, we can now use the brightness of the star that we see here on Earth to determine that star's intrinsic luminosity. As we'll see in an upcoming lecture, that luminosity, together with the temperature that we know how to measure with Wien's Law, allows us to determine almost everything about a star's life and death. Together, the temperature and the luminosity of a star allow us to pinpoint precisely where it is along its entire arc of life.

Now that we understand the importance of light as a tool for quantitatively measuring the fundamental physical properties of the stars, we're ready to jump in and use this tool to tell the story of the life cycle of the stars. In our next lecture, we'll use light in its various manifestations to peer into the first stage in a star's life cycle—its birth—as we examine the majestic and beautiful stellar nurseries in which stars are born.

Stellar Nurseries
Lecture 4

In this lecture, you will begin a detailed investigation of the places where stars begin their lives—stellar nurseries. You will learn that the properties of light can be used to infer the nature of these spectacular places. You will also learn that stellar birth is intimately connected to stellar death, as massive stars, firstborns in these nurseries, exert both a destructive and a creative influence on the surrounding clouds of gas and dust, eroding them, shaping and sculpting them, and helping to trigger the onset of stellar birth both in their immediate vicinity and across the galaxy.

The Carina Nebula

- Stellar nurseries involve enormous nebulae, clouds of gas and dust illuminated from within by a thousand forming stars, with the massive stars at the center sculpting the surrounding nursery. The process of stellar birth that is presented in this lecture is not unique to our corner of the universe; it is a general description of how stars everywhere come to be.

- In images of the Carina Nebula, one of the most striking patterns is the presence of pillar-like structures across the image. These structures do not point in random directions. Instead, they seem to be pointing up toward the top. In fact, there is something very bright at the top: Eta Carinae, which is an intensely luminous star within the Carina stellar nursery.

- Eta Carinae is a very massive star that evidently formed very early on within the larger stellar nursery. In fact, it may very well have been the first star to form here. And this firstborn of the Carina Nebula is no runt. It is one of the most massive stars in this neighborhood of the galaxy, probably weighing in at 30 times the mass of our Sun or more. It is having a tumultuous birth, its immense luminosity blowing it apart so that it is already entering its death throes.

The Carina Nebula, containing striking pillar-like structures, is a stellar nursery.

- The bubble-like features immediately surrounding Eta Carinae, called the Homunculus, are parts of that star that are being expelled back into the surrounding larger nebula. The Carina Nebula is about 8,000 light-years away from us, so what we are witnessing in images of the Carina Nebula actually occurred 8,000 years ago. But that is a snapshot in what is a million-year-long process, so we really are watching this happen effectively in real time.

- Eta Carinae is not just blowing itself apart. The intensity and harshness of its radiation and of its powerful wind is shredding the surrounding cloud of gas and dust in the nebula as a whole, sculpting it, compressing it, driving waves of pressure through it that in turn may be serving as the trigger for the thousands of smaller stars forming throughout the nursery.

- This is a crucially important aspect of how the stellar birth process appears to work in these stellar nurseries. The early formation of extremely massive, luminous stars within the nursery, and the near immediate self-destruction of those stars, leads simultaneously to the destruction of the gigantic cloud of gas and dust and the triggering of stellar birth throughout the nursery. Stellar life and death are intimately related from the start.

- In images of the Carina Nebula, you can see glowing wisps of material surrounding the pillar structure. This is hot, evaporated gas from the destruction of the surrounding cloud of gas and dust as it was boiled away by the intense radiation from the massive star Eta Carinae. That boiled-off gas now fills the space within the larger cloud of gas and dust that has now been evacuated by Eta Carinae. It is like a blister within the cloud, carved into the larger body of the cloud by the massive star and now filled with boiling, hot gas.

The Trifid Nebula
- Another example of a stellar nursery where you can see an expanding blister filled with hot gas is in images of the Trifid Nebula. The beautiful colors of these magnificent stellar nurseries reveal some of the major elemental constituents of the gas out of which the entire stellar nursery is made and which are, therefore, becoming incorporated into the stars as they form within.

- For example, in the Trifid Nebula, you can see the red glow of hydrogen. That is the result of hydrogen atoms in the cloud being heated by the intense radiation of the central hot star, which excites their electrons to higher energy levels. Then, as the electrons within the hydrogen atoms make a permitted jump back down to a lower level, the atoms emit a photon of a specific red wavelength.

The Orion Nebula
- When you look at images of the Orion Nebula stellar nursery, you can see a striking blue color. This is the result of oxygen atoms, ionized not just once but twice, and their electrons making a permitted jump down in energy that emits this specific color of light.

- Interestingly, it took a long time for astronomers to recognize this emission as being the result of oxygen in these stellar nurseries, because oxygen had never been seen to emit that particular color of light in the laboratory. So, early on, astronomers assumed that this must be some new element not before seen on Earth, and it was dubbed "nebulium."

- Later, experiments showed that oxygen twice ionized can emit this particular color of light as long as the gas is extremely rarefied, which is the case in outer space. In the laboratory, it requires vacuum conditions that are very difficult to achieve, explaining why it had not been seen on Earth.

- Incidentally, a similar thing occurred with helium. The light fingerprint of that element had not been seen in the laboratory, because helium is such a rare element in Earth's atmosphere. So, when the light fingerprint of this mystery element was first detected in a solar eclipse in 1868, it was dubbed "helium," from the Greek word *helios*, meaning "Sun." Of course, helium is not unique to the Sun; rather, it is in abundance throughout the universe, second only to hydrogen. Indeed, the new stars forming in these stellar nurseries are being made of fully 25% helium.

- But where are these amazing stellar nurseries found, and what do they look like before they light up as nebulae? These nurseries, in fact, begin as immense clouds of gas and dust that move through our galaxy, material gathered up by gravity after the deaths of previous generations of stars, and they appear as cold, dark clouds drifting through space.

- When we look at galaxies similar to our own, such as the Andromeda Galaxy, we see these gigantic clouds as dark swaths obscuring the light from the billions of stars shining within the disk of the galaxy. The clouds can be seen as long streams extending along the disk of the galaxy, demarcating the spiral arms that we have come to associate with galaxies such as our own.

The Milky Way Galaxy, consisting of several billion stars, is home to our Sun.

- In fact, the spiral arms of galaxies like ours represent places within the galaxy where gravity is relatively stronger, due to a buildup of mass at those locations in the form of stars, gas, and dust. The additional strength of gravity here compresses the gas and dust, forming the relatively tight streams of material that we see and within which gravity can further do its work to initiate the birth of stars.

- We can see the same types of giant clouds of gas and dust within our own Milky Way Galaxy. In fact, if you've been fortunate enough to see the sky on a dark, clear night from a remote location far from any city lights, you probably noticed the band of light stretching across the sky. And you probably noticed the dark swaths cutting through that bright band, obscuring it deeply in some places.

- Those dark swaths are the result of many giant clouds of gas and dust within the disk of our galaxy, blocking our view of distant stars. If you study these clouds as seen in the Milky Way on a clear night sky, you'll notice a great deal of complexity and variety. Some of the clouds are long, like tendrils; some are straight and jutting; others are curved; and some appear as thin, wispy veils.

- One of the clouds, known as Barnard 72, looks like an S-shaped snake against the background of stars. It's named after the famed 19th-century astronomer and astrophotographer E. E. Barnard. In *visible* light, these clouds appear in dark silhouette against the backdrop of bright distant stars. But in *infrared* light, they appear as mammoth luminous beings all their own.

- The cloud complex in the constellation Orion is a wonderful example. In the head of Orion, we once again see the intimate connection between stellar death and new stellar birth. In this case, a massive star first born within the giant cloud has already lived out its brief, dramatic life, in the span of just a million years or so, and in its death has sculpted the surrounding cloud, weathered it and compressed it, triggering the onset of stellar birth all around its fiery grave.

- Indeed, the distribution of cloud material and the instigation of new stellar nurseries across the entire galaxy are influenced by the life cycles of giant stars. The sculpting action of the lives and deaths of massive stars on the surrounding clouds is more than a local phenomenon. The entire galaxy is influenced by these life-giving stellar deaths.

Suggested Reading

Astronomy Picture of the Day, "Stellar Nurseries."

HubbleSource, "Orion Nebula Fly-Through."

1. The most massive stars in a stellar nursery are clearly very important for influencing the birth process of other stars in the nursery and for shaping the nursery itself. Why do you suppose that the most massive stars tend to be found at the centers of nurseries?

2. What other processes can you think of in nature in which death is essential to new life?

Stellar Nurseries
Lecture 4—Transcript

In the stellar life cycle, stars are born, they live, and they die. But in their deaths, old stars provide the material, the conditions, and in some cases, the very trigger for new stars to be born. Nowhere is this intimate relationship between death and new life more evident than in magnificent stellar nurseries.

In this lecture, we'll start our detailed exploration of the stellar birth process by surveying these breeding grounds of the stars. And we'll start using what we've already learned about light in order to interpret the incredible colors and appearances that we see, and thereby, truly understand what's going on in these stellar nurseries.

Of course, when thinking about the stellar life cycle as a repeating cycle from birth to death and back to new life, we immediately encounter a kind of chicken-and-egg problem. Did the stellar life cycle begin with stars forming in stellar nurseries—the egg? Or were the very first stars somehow different, without a birth process, such as the one we're about to describe? As we'll see later in the course, the answer appears that with stars, the chicken probably came first, the very first generation of stars being totally unlike what we see now. But let's hold off on that part of the story, and instead, jump right in to the stellar birth process in stellar nurseries as it occurs today.

It's been said that a picture is worth a thousand words, so let's begin with a purely visual tour of some examples of the amazing stellar nurseries closest to us in the neighboring regions of our galaxy. Each of the examples we're about to look at involves an enormous glowing cloud of gas and dust, illuminated by the many stars forming within. Such a glowing cloud of gas and dust is what we refer to, generally, as a nebula.

Let's start by looking at the Orion Nebula. What we see in this image immediately are these striking red and blue colors together that tell us about the kinds of elements present in this stellar nursery that are emitting these beautiful colors. Surrounding the nebula and within it are obscuring shrouds of dust, and we also see that these dust shrouds have cusps and edges to them, as if they're being, literally, sculpted. In fact, they are. They are being sculpted by the massive stars that we see toward the center of this image. Overall, to give you a sense of scale, this nebula is about three light-years across and about 1500 light-years away from us.

Here is the Rosette Nebula. Again, we see these striking colors that we saw in the Orion Nebula. And again we see these shrouds of dust. In this case, we also see protruding pillars pointing toward the center, aiming right at the massive stars that, again, we find toward the center of the nebula.

Here's the Lagoon Nebula. Again, massive stars brilliantly shine at the center of the nebula, sculpting the surrounding gas and dust, producing these ribbons of dust that we see within and around the nebula. And surrounding the nebula as a whole [are] enormous walls of dust that seem to contain the nebula. And once again, the striking red and blue colors that we see here in the Sharpless 2-106 Nebula. As before, at the center of the nebula are massive, luminous stars. In this case, however, the massive, luminous stars at the center of the nebula are obscured from our view by a belt of dust that stands between us and those massive stars at the center. But we see here a beautiful, wonderful scalloping of the gas and dust in the surrounding nebula

The Carina Nebula is one of the most striking examples of the pillar-like structures that we've seen in some of the other examples, and we see them all over. And we also see, in very dramatic fashion, that these pillar-like structures are all pointing in a common direction. And that's because there's a very massive star up toward the top of the image toward which all of these pillar-like structures are pointing.

Here, the Eagle Nebula, that we've looked at in a previous lecture, we see these enormous pillars of gas and dust that are pointing in the direction of massive, brilliant stars up off the top of the image. And those massive stars are sculpting these pillars and bathing those pillars in the glow of the nebula that we see all around.

The Horsehead Nebula is a wonderful example of an individual massive pillar of gas and dust that is pointing in the direction of massive stars that are sculpting and scalloping it into the wonderful patterns that we see.

The Tarantula Nebula, here we see those red and blue hues spread out throughout the nebula, and we see enormous groupings of bright, massive stars all around that are performing that sculpting and scalloping. That last one, the Tarantula Nebula, is not actually in our own galaxy. It was spied by the Hubble Space Telescope in one of our neighboring galaxies. So already we've learned something important. Stellar nurseries are the same everywhere; they involve enormous nebulae—clouds of gas and dust illuminated from within by a thousand forming stars—and with the massive stars at the center sculpting the surrounding nursery. The process of stellar birth that we'll be discussing today is not unique to our corner of the universe; it is a general description of how stars everywhere come to be.

Having taken a quick stroll through the galactic zoo of stellar nurseries, let's return to one in particular, so that we can examine what is happening in some detail. Let's start with the Carina Nebula. Look again at this incredible image taken with the Spitzer Space Telescope. Take just a moment to run your eyes around this image and see what patterns you notice. One of the most striking patterns in this image is the presence of pillar-like structures across the entire image. Once you spot one, you'll quickly start spotting them all over. And you're probably noticing that these structures do not point in random directions. Instead, they seem to be pointing up toward the top of the image, as if pointing accusingly to something up there.

Well, there *is* something up there, something very important. Look again toward the top of the image, and you'll notice a hint of something very bright, just off the top edge of the image. Let me show you what's up there. It's called Eta Carinae, an intensely luminous star within the Carina stellar nursery. It was deliberately excluded from the main image because its brightness is so high that it could have damaged the sensitive detectors on the Spitzer Space Telescope.

Eta Carinae is a very massive star that evidently formed very early on within the larger stellar nursery. In fact, it may very well have been the first star to form here. And this firstborn of the Carina Nebula is no runt. It is one of the most massive stars in this neighborhood of the galaxy. It probably weighs in at 30 times the mass of our Sun or more. And as you can see in this image, it is having a tumultuous birth, its immense luminosity literally blowing it apart, so that it is already entering its death throes.

The bubble-like features immediately surrounding Eta Carinae itself, called the Homunculus, are parts of that star that are being expelled back into the surrounding larger nebula. The Carina Nebula is about 8000 light-years away from us, so what we are witnessing actually occurred 8000 years ago. But that is a snapshot in what is a million-year-long process, so we really are watching this happen, effectively, in real time. This tumultuousness is very important. You see, Eta Carinae is not just blowing itself apart, the intensity and harshness of its radiation and of its powerful wind is shredding the surrounding cloud of gas and dust in the nebula as a whole, sculpting it, compressing it, driving waves of pressure through it that, in turn, may be serving as the trigger for the thousands of smaller stars forming throughout the nursery.

This is a crucially important aspect of how the stellar birth process appears to work in these stellar nurseries. The early formation of extremely massive, luminous stars within the nursery and the near immediate self-destruction of those stars leads simultaneously to the destruction of the gigantic cloud of gas and dust and the triggering of stellar birth throughout the nursery. Stellar life and death are intimately related from the start.

Let's zoom in on one of the pillar-like structures in the Carina stellar nursery to see in detail the simultaneously destructive and creative influence of Eta Carinae on its surroundings. By looking at this image, can you guess in which is Eta Carinae itself? If you're guessing it's up toward the top, you're probably right. Multiple finger-like protrusions within these pillars are evident in this amazing image from the Hubble space telescope. And at the tips of those finger-like protrusions are individual stars, individual solar systems forming at those tips, being now exposed through the erosive influence of Eta Carinae at the top.

The original cloud that made up this stellar nursery was large and continuous, like a wall. But now, that wall has been mostly shredded away by Eta Carinae's death throws, expelling material out, and plowing into the surrounding nebulae. If these pillar-like structures of the Carina nebula remind you of similar-looking structures here on Earth, well, it's not a coincidence. The strange rock structures rising up beautifully but strangely in a place like Monument Valley in Utah also are the result of erosive forces. These rock pillars didn't rise up out of the ground, though they might look like that way on first impression. Rather, they are what is left over of an originally much more massive expanse of rock that was steadily eroded by water and wind. It's the same with the pillars in a stellar nursery like Carina. Only, it wasn't water and wind doing the eroding, but rather, the intense radiation and the intense stellar wind of Eta Carinae.

Coming back to our examination of these structures in the Carina Nebula, we see surrounding the pillar structure all around it, glowing wisps of material. This is hot, evaporated gas from the destruction of the surrounding cloud of gas and dust as it was boiled away by the intense radiation from the massive star Eta Carinae. That boiled-off gas now fills the space within the larger cloud of gas and dust that has now been evacuated by Eta Carinae. It is like a blister within the cloud, carved into the larger body of the cloud by the massive star, and now filled with boiling hot gas.

Zooming back out to the larger view of the Carina Nebula, try to imagine the three dimensional nature of the image. Some of the pillar structures are closer to us, pointing back toward Eta Carinae, while others are on the back side, pointing in our direction toward Eta Carinae. The glow in this image filling the evacuated space between the pillars is that blister cavity that I mentioned, surrounded on all sides by the pillars that remain as Eta Carinae continues to shred the cloud, the cavity filling further with boiling gas, expanding outward, further shredding the surrounding cloud. That shredding and sculpting further compresses the remaining gas and dust within the pillars, encouraging gravity's contraction to form still more stars within.

Let's look at another example of a stellar nursery where we can see, perhaps even more clearly, an example of an expanding blister filled with hot gas. This is the Trifid Nebula. What we notice immediately is a round, red, bubble-like structure. This is a blister of expanding hot gas that I mentioned before. All around the sides of the blister bubble are pillars, and we see these pillars of gas and dust out in front, and so the obscuration of the dust in those pillars blocks some of the light of the hot bubble within. At the center of that hot bubble is a very hot, massive star that is illuminating the bubble, heating the gas, and boiling it.

Why does it glow red, in particular? Well, hydrogen emits this particular hue of red light when it is heated and excited and is permitted to emit these specific wave lengths of light. Why do we see blue off to the side? The answer is that what we're seeing in those blue colors is reflective light from the hot star at the center of the nebula being scattered and reflected by dust particles in the surrounding gas and dust. This is the same reason that our own sky on Earth is blue. The dust particles in the Earth's atmosphere preferentially reflect and scatter blue wavelengths of light from the Sun.

More generally, the beautiful colors of these magnificent stellar nurseries reveal some of the major elemental constituents of the gas out of which the entire stellar nursery is made, and which are therefore becoming incorporated into the stars as they form within. For example, in the Trifid Nebula, we just saw the red glow of hydrogen. That is the result of hydrogen atoms in the cloud being heated by the intense radiation of the central hot star, which excites their electrons to higher energy levels. Then, as the electrons within the hydrogen atoms make a permitted jump back down to a lower energy level, the atoms emit a photon of a specific red wavelength of light.

When you look at images of the Orion Nebula stellar nursery, you can see a striking blue color. This is the result of oxygen atoms, ionized not just once, but twice, and their electrons making a permitted jump down in energy that emits this specific color of light. Interestingly, it took a long time for astronomers to recognize this emission as being the result of oxygen in these stellar nurseries, because oxygen had never been seen to emit that particular color of light in the laboratory. And so, early on astronomers assumed this must be some new element not before seen on Earth, and it was dubbed nebulium. Later, experiments showed that oxygen twice ionized can emit this particular color of light as long as the gas is extremely rarefied, which is the case in outer space. In the laboratory it requires vacuum conditions that are very difficult to achieve, explaining why it had not been seen on Earth.

Incidentally, a similar thing occurred with helium. The light fingerprint of that element had not been seen in the laboratory, because helium is such a rare element in Earth's atmosphere. So when the light fingerprint of this mystery element was first detected in a solar eclipse in 1868, it was dubbed helium, from the Greek word *helios*, meaning Sun. Now of course, we know that helium is not unique to the Sun, but is, rather, in abundance throughout the universe, second only to hydrogen. Indeed, the new stars forming in these stellar nurseries are being made of fully 25% helium. But where are these amazing stellar nurseries found, and what do they look like before they light up as nebulae? These nurseries, in fact, begin as immense clouds of gas and dust that move through our galaxy, material gathered up by gravity after the deaths of previous generations of stars, and they appear as cold, dark clouds drifting through space.

When we look at galaxies similar to our own, such as the Andromeda galaxy, we see these gigantic clouds as dark swaths obscuring the light from the billions of stars shining within the disk of the galaxy. The clouds can be seen as long streams extending along the disk of the galaxy, demarcating the spiral arms that we have come to associate with galaxies such as our own. In fact, the spiral arms of galaxies, like ours, represent places within the galaxy where gravity is relatively stronger due to a buildup of mass at those locations in the form of stars, and gas, and dust. The additional strength of gravity here compresses the gas and dust, forming the relatively tight streams of material that we see, and within which, gravity can further do its work to initiate the birth of stars. We'll talk in more detail about this gravitational process in our next lecture.

Bringing us back close to home, we can see the same types of giant clouds of gas and dust within our own Milky Way galaxy. In fact, if you've been fortunate enough to see the sky on a dark, clear night from a remote location far from any city lights, you probably noticed the band of light stretching across the sky, as you've probably noticed, the dark swaths cutting through that bright band, obscuring it deeply in some places. Those dark swaths are the result of many giant clouds of gas and dust within the disk of our galaxy, blocking our view of distant stars. If you study these clouds, as seen in the Milky Way on a clear night sky, you'll notice a great deal of complexity and variety. Some of the clouds are long, like tendrils, some straight and jutting, others curved, some appearing as thin wispy veils. One of the clouds, known as Barnard 72, looks like an S-shaped snake against the background of stars.

It's named after the famed 19th century astronomer and astrophotographer E. E. Barnard, who was one of my great predecessors at Vanderbilt University.

Now, in visible light, these clouds appear in dark silhouette against the backdrop of bright, distant stars. But in infrared light, they appear as mammoth, luminous beings all their own. The cloud complex in the constellation Orion is a wonderful example, so let's look at it together. In this image, let's first look at what the constellation of Orion looks like in familiar and visible light. We see that familiar pattern of bright stars, that shape of Orion, the hunter. And beyond that, we notice, perhaps, a faint, red glow of the Orion Nebula down in the sword of Orion, but there's not much else to see here. However, when we look at the same constellation in infrared light, we see something completely different; what we see is the warmth of the dust filling this region of space ablaze from the heat provided by the stars forming throughout the stellar nursery complex here.

The Orion Nebula itself looks totally different. Now we see the heat from the heated gas and dust in the Orion Nebula glowing brightly as a blister heated from within. And we also notice, in the head of Orion, a large ring, or shell. This is the result of a massive star that already lived out its life, has died and exploded, and pushed out the material that was forming in the surrounding stellar nursery. And we see pillars pointing inward in that ring, much as we saw in the Carinae Nebula. And finally, within that expanding material, we find new stars forming all around the rim.

So, in the head of Orion, we once again see the intimate connection between stellar death and new stellar birth. In this case, a massive star first born within the giant cloud has already lived out its brief, dramatic life in the span of just a million years or so, and in its death has sculpted the surrounding cloud, weathered it, and compressed it, triggering the onset of stellar birth all around its fiery grave.

Indeed, the distribution of cloud material and the instigation of new stellar nurseries across the entire galaxy is influenced by lives and deaths of massive stars throughout the galaxy. In this image of a galaxy similar to our own, called NGC 891, we can see in exquisite detail how the clouds of gas and dust are distributed within the disk of the galaxy.

Now, this galaxy is a particularly good one to look at, because we happen to be seeing, by chance, the galaxy edge on. And because we're seeing it edge on, most of the visible light from the billions of stars within, are obscured by the dust within the disk of the galaxy. Within that disk of the galaxy, we see a large number of small, or apparently small, blue knots spread out all throughout. Each one of those little blue knots is, in fact, an entire stellar nursery, just like the many examples that we've seen in this lecture. We're seeing light from the hot stars forming within, heating the surrounding nurseries, and we also see the reflected light, preferentially blue light, from those stellar nurseries; those are the blue knots that we see. We also see the clouds being pushed vertically above and below the disk of the galaxy. This is the action of many massive stars dying within these various stellar nurseries, sculpting those surrounding nurseries, and spreading the material in those clouds up and out of the plane of the galaxy.

So you see, the sculpting action of the lives and deaths of massive stars on the surrounding clouds is more than a local phenomenon. The entire galaxy is influenced by these life-giving stellar deaths.

For some stellar nurseries, the contents and structure have been studied and mapped in sufficient detail to permit a full three-dimensional picture of the nursery. The Orion Nebula, being just 1500 light-years away, and therefore the closest of the massive stellar nurseries to us, is perhaps the best studied. Let's look at a full computer simulation mock-up of the Orion Nebula nursery and summarize what we see. Everything in the mock-up is true to life, from the positions of the individual stars to the detailed structure of the surrounding gas and dust.

We enter the nebula through a cavity in a larger cloud of gas and dust, its dark rim slightly illuminated at the bottom and the upper left. This cavity is the womb, which the massive luminous stars seen at the center have excavated with their intense radiation. Surrounding those massive stars is an entourage of smaller stars still enveloped in their cocoons, the proplyds. Now we approach one of those cocoons; we see it being ablated by the intense radiation impinging upon it from those massive stars. And finally, zooming in, we see the baby star kicking within, ringed by the disk of gas and dust from which it will make its own solar system, and spewing out jets of gas with which it is kicking its way out of its enveloping cocoon.

So you see, these stellar nurseries are remarkable places. Tucked within massive clouds of gas and dust, entire litters of stars are being born. The massive stars, the firstborns of the litter, exert an enormous influence on the rest of nursery surrounding them, sculpting and eroding it, and ringing these massive stars, entire entourages of smaller stars, each kicking within its cocoon, each one a solar system in the making.

In this lecture, we have started our detailed investigation of the places where stars begin their lives—stellar nurseries. We've seen that we can use the properties of light to infer the nature of these spectacular places. We've also seen that stellar birth is intimately connected to stellar death as massive stars first born in these nurseries exert both a destructive and a creative influence on the surrounding clouds of gas and dust, eroding them, shaping and sculpting them, and helping to trigger the onset of stellar birth, both in their immediate vicinity and across the galaxy.

Gravitational Collapse and Protostars
Lecture 5

S tars are born from immense clouds of gas and dust several light-years in extent, what we call stellar nurseries. These nurseries are like the neonatal ward at the hospital, with many babies being tended to all together. But, just like the neonatal ward, the individual birth of each one of those stellar babies is intimate and special, and it involves a birth process that plays out very quickly—just a few million years. While the last lecture focused on the nursery as a whole, this lecture will focus on the individual babies within the nursery.

The Formation of Protostars

- Like an obstetrician using sonograms to peer into a mother's womb, the astronomer can use infrared light to peer into the stellar womb, otherwise hidden from our view by the obscuring gas and dust from which the baby star is forming. We refer to the embryonic star within its cocoon of gas and dust as a protostar.

- At the earliest stages of the process, before there even is a protostar, there is just a knot of enhanced density within the cloud. These dense knots, or cloud cores, are the regions where we see the most intense radio light, indicating that gravity has coalesced the gas densely enough there for a new star to form.

- For reference, each of these dense knots is about the size of our solar system, including not only the Sun and planets, but also the larger Oort cloud of comets that swarm around the outskirts of the solar system.

- These very dense knots of gas and dust, buried deep within the dark cloud, are the places where the next stars will form. Indeed, it is this very material that will be gathered up by gravity to fashion these next stars. You could say that there is not yet an embryo, not yet a beating heart, but conception has occurred, and the stellar birth process is definitively underway here, each one an entire solar system to be.

- Whereas many, if not most, stars are formed in these dense cores of the most massive clouds, some stars are formed within relatively isolated cloudlets, called Bok globules for the astronomer Bart Bok, who spent much of his career studying them. And because they are isolated, these cloudlets offer a unique opportunity to understand the stages of the process of stellar gestation in detail.

- Clouds like the Bok globule B68 are entirely devoid of stars. These clouds, therefore, represent the very beginning stages of the stellar birth process. This is the stellar womb prior to conception. But gravity will continue acting upon the cloud, compressing and shaping it, making it denser still.

- Eventually, this will lead to the central part of the cloud becoming dense and warm enough to become a stellar embryo, glowing with its own heat. When this happens, the protostar makes itself known. Even if not directly visible, you might say that the baby's kicks are definitely felt by the parent cloud. That kicking baby star within the parent cloud has clear consequences that we can see even from the outside.

- Often, a protostar can barely be perceived in visible light, being deep within its parent cloud. However, a protruding jet reveals the kicking baby within. And in infrared light, like the obstetrician's sonogram, we can see the kicking baby directly. The jets of gas that it spews in opposite directions bore through the surrounding cloud, as the baby pushes to free itself and emerge from the womb.

- The jets that newly forming stars shoot help the newborn star to blow away the remnant cloud material from which it formed. Like a baby chick hatching from its egg, the fledgling star pokes holes in its surrounding cocoon. After a million years or so, this is how the baby star will come to be fully revealed.

- It appears likely that all stars go through a tumultuous start. These babies don't coo so much as kick and scream. Call it a bad case of colic. But these colicky babies have fits and bouts, as evidenced by the knots of ejected gas. What causes these colicky fits is indigestion.

- From detailed studies, we now also know that these "burp" episodes correspond directly to the pace at which the baby star ingests material from its parent cloud. In other words, when it eats too much too fast, it spits up.

Accretion and Magnetic Fields

- At this stage of the infant star's development, it is actively feeding from the swirling disk of gas and dust that encircles it. This process, which is called accretion, is how the star continues to feed and bulk itself up.

- Importantly, the accretion process is mediated by the star's magnetic field, which acts to funnel the material in the disk onto the star. Like the Earth's magnetic field that causes compasses to point north, stars are born with magnetic fields that can influence the motions of electrically charged material near the stars. As gravity and friction cause the material in the disk to swirl in toward the star, like water in a tub spiraling in toward the drain, the star's magnetic field directs the incoming material onto the star itself.

- However, that same magnetic field also acts to launch and funnel a portion of that material away from the star, confining and focusing the material into a tight, fast stream. The star's magnetic field is stretched outward by the outflowing material and twisted around by the star's spin so that it takes on a shape like a nozzle on a fire hose. As a result, episodes of enhanced accretion onto the star lead to episodes of enhanced ejection, as bursts of inflowing material are converted into bursts of ejected material.

- The material in the disk of gas and dust swirls in toward and onto the baby star at the center. That material is clumpy, so the baby star's feeding from that material occurs in little bursts that are then manifested in bursty clumps in the outflowing jets. Fortunately for the star, less gets ejected than is accreted so that the star can and does grow. In any case, it's clear that at those times when the baby feeds most heartily it also burps most powerfully.

- Occasionally, these episodic ejections can be extraordinarily dramatic, giving rise to what is called an FU Orionis event. In this type of event, the protostar suddenly and dramatically brightens by a factor of about 100 and can sustain this new elevated brightness for months or years. The event is named for a star in the Orion constellation that in 1937 suddenly became about 100 times brighter than it was before.

- The current thinking is that these dramatic eruptions are also related to the star's magnetic field. The idea is that, most of the time, the star's magnetic field acts like a dam with a flood gate, holding back the gas and dust in the disk swirling around the star, only letting a small amount through at a time. But every now and then, for reasons that are not clear, the dam breaks, and the gas and dust from the encircling disk come crashing down onto the star, heating it and brightening it temporarily. These dam breaks may also be what drive the knots we see in the jets.

- In fact, those knots of ejected material in the jets very often appear to have a regular spacing, as if the star were ejecting them with a certain cadence. Measurements of the speeds of those knots expelled in the jets, and of the amount of space between them, imply that for most stars, the knots are ejected every 50 years or so. Estimates for the often-dramatic FU Orionis eruptions are similarly every 50 years or so. So it really does appear that those dam breaks are responsible for the knot ejections.

Protoplanetary Disks
- There remains much about this general picture that we don't fully understand; the role of magnetism in the stellar birth process remains one of the frontiers of stellar astrophysics research. In any event, it

is clear that the interrelated processes of accretion and ejection are self-regulated in such a way that the baby star feeds, grows, and—

through the feedback of its jets on the cloud from which it feeds—shuts off its own food supply after a few million years.

Protoplanetary disks can be studied by carefully measuring the amount of light emitted by the disk at different wavelengths.

- So, the baby star has a few million years' worth of time before it is weaned from the bottle. At that point, the star has fully emerged from its cocoon, now an adolescent star ringed with just enough remnant gas and dust from its parent cloud to allow the formation of planets that will orbit it as a solar system.

- These so-called protoplanetary disks can be seen in silhouette against the backdrop of a bright stellar nursery. Their structures can also be studied in detail by carefully measuring the amount of light emitted by the disk at different wavelengths. That's because the gas and dust in the disk that is closer to the star will be warmer than the material farther from the star, and consequently, that closer material will radiate most strongly at bluer wavelengths than the farther-out material.

- Increasingly, astronomers are discovering numerous examples of protoplanetary disks that have certain light wavelengths missing from their emissions. Because the temperature of the material in the protoplanetary disk declines with distance from the central illuminating star, we can map a specific temperature of the material to a specific distance from the star. And because there is a direct relationship between the temperature of a material and its peak wavelength of emission, we can in turn map a specific wavelength of emission to a specific distance from the star.

• So, the fact that specific wavelengths of emission from the protoplanetary disk are missing indicates that the disk is devoid of gas and dust at specific distances from the star. In other words, something is carving out gaps in the protoplanetary disks. Almost certainly, these are newly formed planets orbiting within the disk and sweeping out large gaps.

Suggested Reading

Astrobiology Magazine, "The Stuff Stars Are Made Of."

European Space Agency, "Born in Beauty."

Questions to Consider

1. What are some of the similarities and differences between the wombs in which baby stars and human babies are born?

2. In what ways are the stellar birth process and the planet-formation process intimately linked?

Gravitational Collapse and Protostars
Lecture 5—Transcript

Stars are born from immense clouds of gas and dust several light-years in extent, what we call stellar nurseries. These nurseries are like the neonatal ward at the hospital, with many babies being tended to all together. But just like the neonatal ward, the individual birth of each one of those stellar babies is intimate and special, and it involves a birth process that plays out very quickly—just a few million years.

In our last lecture we focused on the nursery as a whole, and we examined the macroscopic processes at work to shape and sculpt and trigger the nurseries as a whole. In this lecture, we'll look inward and turn our attention to the individual babies within the nursery. This is stellar birth, one baby at a time, well, sometimes two, or even three.

Like an obstetrician using sonograms to peer into a mother's womb, so can the astronomer use infrared light to peer into the stellar womb, otherwise hidden from our view by the obscuring gas and dust from which the baby star is forming. We refer to the embryonic star within its cocoon of gas and dust as a protostar.

Let's start by looking at what the structure of a giant star-forming cloud looks like at the earliest stages of the process, before there is even a protostar, instead, just a knot of enhanced density within the cloud. We're going to need to utilize the power of light at several different wavelengths to truly see what is happening.

Stretching across this image, we're seeing a large cloud of gas and dust in the constellation of Aquila. The image is of a region that is 10 light-years across. This is what we call a false-color image, because the image is taken at infrared wavelengths, so the colors correspond to what the image would look like to our eyes if we could see infrared wavelengths. Blue corresponds to shorter wavelength infrared light; red corresponds to longer wavelength infrared light. Blue corresponds to hotter objects, red to cooler ones. The white lines in the image have been added to draw your attention to particular regions. The dotted line around the perimeter corresponds to a rough outline of the overall cloud as a whole. The dashed line toward the lower left is the part that we mainly will focus on here, representing the densest part of the overall cloud where we also see a dark silhouette feature that we'll come back to in a moment.

There are different things happening in different parts of the cloud. The blue-colored spots are baby stars within the cloud that have already started to turn on, glowing hot and bright. Near the top we see a very bright blister of hot gas, a cavity within the cloud caused by the excavation of that part of the cloud by a massive star that has turned on and is beginning to sculpt and shred the cloud there. Similar to the Orion Nebula or Trifid Nebula blister regions that we saw in our previous lecture, we see the intense glow of the heated hydrogen and oxygen gas filling that cavity. Within the cavity are large numbers of baby stars are turning on.

But toward the lower left we see an extended dark silhouette with few to no blue dots. In this part of the cloud, there are not yet many, if any, baby stars that have ignited. This region has a greater density, and so generally, we cannot see into or through that region. Whereas that bright blister cavity at the top has relatively low density, having been largely evacuated, the dark silhouette region at the lower left has a relatively high density. The orange spots indicate the regions of the very highest density, as measured in radio light, and indeed, these reside entirely along the dense, dark silhouette. The densest knots are massive cores within the cloud.

These dense knots, or cloud cores, are the regions where we see the most intense radio light, indicating that gravity has coalesced the gas densely enough there for a new star to form. For reference, each of these dense knots is about the size of our solar system, including not only the Sun and planets, but also the larger Oort cloud of comets that swarm around the outskirts of the solar system. These very dense knots of gas and dust, buried deep within the dark cloud, are the places where the next stars will form. Indeed, it is this very material that will be gathered up by gravity to fashion these next stars. So you could say that there is not yet an embryo, not yet a beating heart, but conception has occurred, and the stellar birth process is definitively underway here, each one an entire solar system to be.

Whereas many, if not most, stars are formed in these dense cores of the most massive clouds, some stars are formed within relatively isolated cloudlets, called Bok globules for the astronomer Bart Bok, who spent much of his career studying them. And because they are isolated, these cloudlets offer a unique opportunity to understand the stages of the process of stellar gestation in detail.

The images you are seeing now are of the Bok globule B68, one in visible light, the other in infrared light. Here, what you're seeing is a true-color image in wavelengths of light that our own eyes can see. And what we see all around the periphery are background stars in distant parts of our galaxy, and these stars have different colors representing the different temperatures of the stars. Remember what we learned before; colors of stars indicate to us the temperatures of stars. Redder stars are cooler, and bluer stars are hotter.

The globule that you see at the center is so dense that it completely blocks out the visible light from those background stars, and so we don't see any of those stars' light penetrating through to us. Note also that the globule has a roughly spherical shape, but not perfectly so; it's more kidney shaped. And that's because gravity is just now beginning to collapse it down to the eventual spherical shape of the star that it will eventually form. Right now this cloud is about 10,000 times the size of Earth's orbit around our Sun, comparable to the entire solar system, including the outer swarm of comets. So what you're seeing here in this Bok globule is a volume of space that is comparable to our entire solar system to its furthest reaches.

In the bottom image here, what you're seeing, for comparison, is the same Bok globule, but now taken in infrared light. And in infrared light the light from those background stars is able to penetrate through the cloud, and so we see them, faintly, through the cloud. Now notice that those background stars seen through the cloud are very red in color. Now, that's not because those distant stars are all actually that red, rather, they appear red because this cloud extinguishes the light from those distant stars, and it preferentially extinguishes the bluer light from those distant stars. Astronomers say that these background stars have been reddened by the gas and dust within the Bok globule that extinguishes that light. Now, just to be clear, the stars that you see that appear to be within the cloud are not actually inside the cloud, but are background stars completely behind it, and their light is reaching us through the cloud.

Now, here's the important part: We can use the degree with which those background stars have been reddened, or their light extincted, in order to measure properties of the cloud. This is like taking core samples through different parts of the cloud, each one of those background stars giving us information about the density and the properties of the dust grains that are causing that background light to be reddened and extinctive. You'll notice that the background stars seen through the very center part of the Bok globule appear to be the reddest; they are the most extinctive; they are the reddest ones. And so what we're seeing is that, already, within this Bok globule, gravity has begun to coalesce the fine dust grains that originally came from the interstellar medium and is beginning to coalesce them into the beginnings of what will eventually be the baby star formed within.

The upshot of this type of research is that clouds like this one are entirely devoid of stars. These clouds, therefore, represent the very beginning stages of the stellar birth process. This is the stellar womb prior to conception. But gravity will continue acting upon the cloud, compressing and shaping it, making it denser still. And eventually this will lead to the central part of the cloud becoming dense and warm enough to become a stellar embryo, glowing with its own heat. When this happens, the protostar makes itself known. Even if not directly visible, you might say that the baby's kicks are definitely felt by the parent cloud.

This cloud, called GDC1, is similar to the B68 globule that we just looked at. But where B68 does not yet contain a baby star within, this cloud has already passed through the main gravitational collapse phase and has formed a baby Sun-like star in its interior. Although the protostar is still embedded within thick layers of gas and dust, it reveals its existence by spewing a two-sided supersonic jet into the surroundings.

What we are seeing in this image is the jet from that baby star within the large color cloud of gas and dust protruding from the cloud, and we can follow the direction of that jet back to the faint, red glow of the protostar still buried within the large pillar of gas and dust. You notice that there's a bright glow at the tip of that jet. That tip, that bright glow at that tip, is telling us about the heat that's generated as that jet of material rams into the surrounding gas in the stellar nursery.

You'll also notice a faint, purplish glow upstream from the glowing tip of that jet. That's caused by hydrogen atoms in the surrounding gas from the stellar nursery being heated and glowing with their signature radiation. Remember, each element in the periodic table emits a distinct fingerprint of colors. Now, if we follow that jet in the opposite direction to the far side of the pillar of gas and dust, we can just barely make out the head of another jet shooting in the other direction.

So we see that kicking baby star within the parent cloud has clear consequences that we can see even from the outside. But having seen what the manifestations of the embedded protostar look like from the outside, let's next peer into the womb by using infrared light, and compare that to the visible light with which we are more accustomed.

In this image, we have another example of a protostar that can barely be perceived in visible light, being deep within its parent cloud. However, a protruding jet reveals the kicking baby within. And in infrared light, like the obstetrician's sonogram, we can see the kicking baby directly. The jets of gas that it is spewing in opposite directions are boring through the surrounding cloud as the baby pushes to free itself and emerge from the womb. The jets that newly forming stars shoot help the newborn star to blow away the remnant cloud material from which it formed. Like a baby chick hatching from its egg, the fledgling star pokes holes in its surrounding cocoon. After a million years or so, this is how the baby star will come to be fully revealed.

Let's look at some examples of these kicking babies in closer detail, so that we can better see how these jets influence the parental cloud and so that we can begin to understand how these jets work physically. First, we're looking at an example from the Carina Nebula; we've seen this large pillar of gas and dust in the stellar nursery before, and to remind you, it has an overall extent of a few light-years. And notice, now, those jets protruding from the tips of that pillar, and these jets have structure; there are knots within it; there are gaps and small spaces within the jet, and there's also a shockwave in the front of the head of the jet, pushing out into the surrounding gas of the stellar nursery. The knots and the shockwaves formed by these jets are what we refer to as Herbig-Haro objects.

This next example is Herbig-Haro object 111. The protostar is partially visible, peeking out through a cavity that's been poked out by the jet in the surrounding disk of gas and dust around the protostar. But it's still mostly obscured by that remaining thick disk of gas and dust around the protostar. The protostar is being reddened by the dust within that thick disk of gas and dust, and so it appears very red to us. But there's also a faint blue glow that we can see within the cavity; that's caused by scattered light from the protostar reaching us as it reflects off of dust grains around the protostar, for the same reason that our sky appears blue to us.

Now, we see material being shot in a columnated jet extending far from the protostar but with a very narrow jet-like structure. And there are knots within that jet that indicate that there have been impulsive events in the past when the star has ejected clumps of material all at once. The jet appears to be twisting, and that's due to a magnetic funneling effect that we'll talk about later. And the tip of that jet is ramming into the surrounding gas in the stellar nursery, producing a bright sock.

The next example that we'll look at is Herbig-Haro number 47. In this case the protostar is completely invisible to us; it's completely enshrouded somewhere near the middle of the opposing jets that you see shooting out in opposite directions. One of those jets is pointing toward us, the other one away. And at the bright ends of those jets, are the places where those jets are ramming into the surroundings and heating that surrounding gas.

We also see a faint sheath, if you will, glowing all around along the length of those jets. Those represent previous shock waves as the jet pushed through the surrounding material in the past. And again, we see that the jet is knotted. These represent past episodes of ejection that tell us that the ejection of this jet material has occurred in bursts. And we see gaps along the way, indicating that there have been several years of time passing between subsequent bursts. Again, we see twists due to the magnetic field of the protostar confining the jet material, and we also see kinks in the jet's direction resulting from deflections of that jet material caused by the surrounding gas in the stellar nursery.

Here we have Herbig-Haro 110. Again, the protostar is completely hidden at the bottom right, and we see only one jet being shot toward the upper left. The other jet, which is being shot in the other direction, is shooting into the parent cloud and has not been revealed to us yet, at least not in visible light. The base of the jet here is extremely narrow, sort of like a nozzle from a fire hose, and that is due to the magnetic field of the protostar confining this jet material as it's being ejected, and again, we see this dramatic twisting structure of the jet. And reminiscing of, perhaps, cigar smoke rising up, this jet material is very narrow and confined at the bottom, but becomes more turbulent as it rises towards the top. Again, we see these episodic ejections in the form of knots with regular spacings between them, and overall, what we're seeing here is a jet extending some 100 times the size of our own solar system.

There are many Herbig-Haro jets to be found in a stellar nursery like this one; this is the Orion stellar nursery. Remember when we looked at this nursery before, we noticed the variegated appearance of the gas and dust in the nursery, that scalloping appearance. Well, now we can see that this is due, in part, to many, many overlapping Herbig-Haro jets coming from hundreds of forming, kicking stars shooting out their jets into the surrounding nursery that collectively sculpt the entire nursery. The individual stars are, perhaps, a light-year apart, but their jets can extend across the intervening space and can even overlap in some places.

The examples we've just seen are not the exception, but the rule. Indeed, it appears likely that all stars go through a tumultuous start. These babies don't coo so much as kick and scream. Call it a bad case of colic. But as we've seen, these colicky babies have fits and bouts, as evidenced by those knots of ejected gas that we saw in every example. What causes these colicky fits? In a word: indigestion. Let me show you what I mean.

As you can see in this actual time-lapse movie of Herbig-Haro 111, the knots in the jets steadily flow out away from the protostar. This clearly tells us that the knots represent episodes of ejection from the protostars. From these types of detailed studies we now also know that these "burp" episodes correspond directly to the pace at which the baby star ingests material from its parent cloud. In other words, when it eats too much, and too fast, it spits up. You see, at this stage of the infant star's development, it is actively feeding from the swirling disk of gas and dust that encircles it. This is how the star continues to feed and bulk itself up. This process is called accretion.

Importantly, the accretion process is mediated by the star's magnetic field, which acts to funnel the material in the disk onto the star. We'll discuss the magnetism of stars in a subsequent lecture, but for now, suffice it to say that, like the Earth's magnetic field that causes compasses to point north, stars are born with magnetic fields that can influence the motions of electrically charged material near the stars. As gravity and friction cause the material in the disk to swirl in toward the star, like water in a tub spiraling in toward the drain, the star's magnetic field directs the incoming material onto the star itself.

However, that same magnetic field also acts to launch and funnel a portion of that material away from the star, confining and focusing the material into a tight, fast stream. The star's magnetic field is stretched outward by that outflowing material and twisted around by the star's spin, so that it takes on a shape like a nozzle on a fire hose. As a result, episodes of enhanced accretion onto the star lead to episodes of enhanced ejection, as bursts of inflowing material are converted into bursts of ejected material.

Here's what happens; the material in the disk of gas and dust swirls in toward and onto the baby star at the center. That inspiraling material is clumpy, and so the baby star's feeding from that material occurs in little bursts that are then manifested in bursty clumps in the outflowing jets. Fortunately for the star, less gets ejected than is accreted, so that the star can and does grow. But in any case, it's clear that at those times when the baby feeds most heartily, it also burps most powerfully.

Occasionally, these episodic ejections can be extraordinarily dramatic, giving rise to what is called an FU Orionis event. In this type of event, the protostar suddenly and dramatically brightens by a factor of 100 times or so and can sustain this new elevated brightness for months or years. The event is named for a star in the Orion constellation that in 1937 suddenly became about 100 times brighter than it was before.

A more recent example is the so-called McNeil's Nebula, named after the amateur astronomer and astrophotographer who first spotted it. Buried deep within the gigantic Orion stellar nursery that we looked at earlier was a protostar not previously spotted, until it suddenly brightened, revealing a deeply enshrouded star, like our Sun, within an excavated cavity that was now brightly illuminated from within. Evidently, this star must have previously had such episodes that, through the action of a powerful jet, would have carved open the cavity in its surrounding cloud that we now see lit up. So the McNeil's Nebula protostar has done this before, perhaps multiple times, and we just happened to be a witness this time.

The current thinking is that these dramatic eruptions are also related to the star's magnetic field. The idea here is that, most of the time, the star's magnetic field acts like a dam with a flood gate, holding back the gas and dust in the disk swirling around the star, only letting a small amount through at a time. But every now and then, for reasons not clear, the dam breaks, and the gas and dust from the encircling disk comes crashing down onto the star, heating it and brightening it temporarily. These dam-breaks may also be what drives the knots that we see in the Herbig-Haro jets.

In fact, as you may have noticed already, those knots of ejected material in the jets very often appear to have a regular spacing, as if the star were ejecting them with a certain cadence. Measurements of the speeds of those knots expelled in the jets, and of the amount of space between them, imply that, for most stars, the knots are ejected every 50 years or so. Estimates for the often-dramatic FU Orionis eruptions is similarly every 50 years or so. So it really does appear that those dam breaks are responsible for the knot ejections.

A few years ago, one of my Ph.D. students made a startling discovery in the course of studying the details of the dramatic brightening episode of McNeil's Nebula. She found that as the protostar brightened, that brightness also had imprinted on it an oscillatory signal, modulating the total light from the star by a few percent every two hours. Based on what we know of the physical properties of protostars, two hours corresponds to the fundamental tone of the star, the frequency with which it rings if hit hard enough, like a clapper ringing a bell. In turned out that the impact of material crashing onto the star when the dam broke was hard enough to make the entire star ring for a period of a few months.

Yet, there remains much about this general picture that we don't fully understand. Why does the dam break episodically? How does it reconstitute itself? And most fundamentally, how does the star's magnetic field simultaneously funnel the material that feeds the star and expel the material that the star shoots out in those powerful jets? How does the magnetic field know how to regulate these processes so that most of the material bound for the star actually does end up on the star, and only a fraction is lost, expelled back to the cloud? I wish I could tell you we know the answers to these questions. But the role of magnetism in the stellar birth process remains one of the frontiers of stellar astrophysics research. It's questions like these that keep me busy and mostly out of trouble.

In any event, it is clear that the interrelated processes of accretion and ejection are self-regulated in such a way that the baby star feeds, grows, and through the feedback of its jets on the cloud from which it feeds, shuts off its own food supply after a few million years. So the baby star has a few million years worth of time before it is weaned from the bottle. At that point, the star has fully emerged from its cocoon, now an adolescent star ringed with just enough remnant gas and dust from its parent cloud to allow the formation of planets that will orbit it as a solar system.

These so-called protoplanetary disks can be seen in silhouette against the backdrop of a bright stellar nursery. Their structures can also be studied in detail by carefully measuring the amount of light emitted by the disk at different wavelengths; that's because the gas and dust in the disk that is closer to the star will be warmer than the material farther from the star, and consequently, that closer material will radiate most strongly at bluer wavelengths than the farther out material.

Increasingly, astronomers are discovering numerous examples of protoplanetary disks that have certain light wavelengths missing from their emissions. Because the temperature of the material in the protoplanetary disk declines with distance from the central illuminating star, we can map a specific temperature of that material to a specific distance from the star. And because we've learned that there is a direct relationship between the temperature of material and its peak wavelength of emission, we can, in turn, map a specific wavelength of emission to a specific distance from the star. So the fact that specific wavelengths of emission from the protoplanetary disk are missing indicates that the disk is devoid of gas and dust at specific distances from the star. In other words, something is carving out gaps in the protoplanetary disks. Almost certainly these are newly formed planets orbiting within the disk and sweeping out large gaps.

A beautiful example of this is the star Fomalhaut, a somewhat more mature star but which possesses a late-stage remnant of its protoplanetary disk, what we call a debris disk. That disk can be seen directly by the light from the star that reflects off of it. It has a sharp inner boundary, which appears in this image as a ring. In fact, the disk extends much farther out from the star, but its inner edge is strongly illuminated by the star, and so appears most prominently in the image. And just within that inner boundary? A planet, seen directly by Hubble Space Telescope and now observed to have moved in its orbit about the star, carving out that hole in the disk.

In our next lecture, we'll look at the complex dynamics that are at play as sibling stars birthed within a nursery jostle and interact. And in later lectures, we'll look at what happens to a family of stars as we examine the dispersal of individual stars away from these families. But what a remarkable point we've come to, to know the birth places of the stars, to be able to peer within the stellar womb, to witness firsthand the tumult of stellar infancy, to watch as these baby stars grow up and assume responsibility for their charge of planets. It seems like only 10 million years ago that little Fomalhaut was just a child, a bratty one at that. And now look, he's got a planet all his own. Where does the time go?

The Dynamics of Star Formation
Lecture 6

M‌ost stars are not born alone, but in groups. And because they're not born alone, the siblings affect one another—a sibling rivalry starting in the womb. In this lecture, you will learn about the types of stellar siblings that most stars are born with. You will learn about the latest observational evidence, and you will discover that stellar siblings undergo fascinating, and sometimes violent, dynamical interactions that influence everything from their birth weight to their future lives.

Astronomical Terminology

- Astronomers use certain terminology to describe different types of stellar siblings. First, a lone star is simply called a single. Our Sun is, to the best of our knowledge, a single star. Among all stars similar in mass to our Sun, singles are found about 40% of the time. So, while our Sun is not strange, it does represent a bit of a stellar minority.

- The most common arrangement for stars similar in mass to our Sun, about 60% of them, is a binary, or double, star system. You can think of these as twins—sometimes identical, but usually not. In fact, you can think of most binary stars as being fraternal twins; they were born together but otherwise have different characteristics. There are even binary star systems that we can think of as adopted twins; they are together now but were born from different clouds—different parents.

- Although rare, stellar siblings can also come in threesomes, foursomes, or more. There are triplets, what astronomers call trinary star systems, or triples. There are even quaternary systems, which consist of four stars, or quadruplets.

- One of the most extreme examples of multiple siblings is in fact one of the brighter objects in the sky: Castor, which is one of the two brightest objects that make up the constellation Gemini. Castor is itself six stars—sextuplets.

- The Castor sextuplets have what we call a hierarchical arrangement: The stars are not evenly separated but, rather, involve a nested sequence of tight separations within more distant orbits. Specifically, Castor contains a binary star system orbited by another binary star system, both of which are orbited by an outcast third binary star system.

- There is an important distinction that we need to make when talking about stellar siblings. The only physically meaningful types of siblings are those that are gravitationally

The constellation Gemini contains Castor, which is actually six stars.

bound to one another. Such siblings, by virtue of their mutual gravitational attraction, are most likely to have actually been born together. At the very least, it's clear that the stars are together now, so they likely share a history—past and future.

- This is in opposition to stars that at first glance might appear to be related but, in fact, are not. This can happen when two stars appear very close to one another in the sky but are actually at completely disparate distances from us.

Orbits of Star Systems
- In one type of binary star orbit, two stars orbit one another in simple circular fashion, maintaining a constant separation throughout their orbit. However, stars don't have to orbit in circles. More generally,

their orbits can be ellipses, in which case they make a close, rapid approach and then a slow, distant motion—what we call an eccentric orbit.

- With three stars, the orbits become somewhat more complex. It turns out that many, if not most, configurations of three stars are not dynamically stable for long periods of time. So, in nature we tend to find three stars in a hierarchical arrangement, with two stars close together and the third farther out. In fact, there is a fascinating, stable orbit of three stars called a figure-eight orbit.

- With four stars, while there are in principle more possible arrangements just by virtue of having more stars to arrange, there are in fact only a relatively small number of dynamically stable configurations. Again, the stable configurations have a hierarchical character. In one type of arrangement, two separate binary systems orbit one another. In a multiple hierarchical arrangement, an inner binary pair is orbited by a third star, all of which is orbited by the fourth star.

- In order to remain stable in their orbits for long periods of time, orbits generally have a hierarchical arrangement. When the orbits of multiple stellar siblings are not hierarchically arranged, some dramatic things happen. These dramatics are particularly important within the crowded, dynamical environments of stellar nurseries, the places where stars are born.

- In the relatively simple case of two stars forming out of the same dense core of cloud in a stellar nursery, which essentially represents the idealized gestation of twins within a single womb, there are multiple possibilities for the arrangements of any planets that might form from the material around the stars.

- Planets could form around one or both of the stars individually, because each star maintains a reservoir of gas and dust in a protoplanetary disk around it. Alternatively, or in addition, planets could form in the material surrounding the entire stellar binary, becoming what are known as circumbinary planets.

- In the same sort of situation, but this time where the two protostars forming within the womb get a bit too close for comfort, the possible arrangements for any planets forming in the system will be different from the binary example.

- In the case of triplet stars, where the tight binary pair has been largely stripped of its reservoir of gas and dust by the close gravitational encounter with the third sibling, planets are unlikely to be able to form around either of those stars. However, the more distant third star will be able to form planets. In addition, there may be enough material surrounding the entire triple system to form circumtriplet planets. In general, if you want to know where the planets are likely to form, just follow the protoplanetary gas and dust in the system.

- In addition, the birth of triplets within a natal cloud of gas and dust involves some amount of jostling of the stars and disruption of the system, both in the sense that protoplanetary material is disrupted and in the sense that one of the three stars is pushed out to a wider orbit compared to the other two siblings.

The Establishment of Hierarchy
- The establishment of hierarchy among stellar siblings when they are born involves a complex, highly dynamical interaction. The need for a system of three or more stars to end up in a strictly hierarchical arrangement may seem counterintuitive. After all, within our own solar system, we have one star orbited by eight planetary bodies spaced out relatively evenly in distance from the Sun.

- But the difference in a stellar triplet system is the much stronger gravitational interaction among these stellar mass bodies. The strength of that interaction is such that the only stable arrangements are hierarchical ones, so these systems naturally evolve in such a way that they either space themselves out or disrupt themselves trying.

- Indeed, more detailed studies of these interactions involving many simulations and different starting conditions reveal that the resulting stellar triplets tend to have a tightened inner binary pair and a widened outer third star. In other words, over time, two of the triplets move closer together in their orbit, while the third drifts farther and farther away from the inner pair.

- This dynamical process of at once tightening an inner binary pair of stars at the expense of expelling a third sibling turns out to help explain a long-standing mystery about binary stars. It has long been known that the most widely separated binary stars—often referred to as ultrawide binaries—could not have started out as widely separated as we see them.

- Take as an example the separation of Proxima Centauri from Alpha Centauri. Their physical separation is larger than the typical stellar womb—the clouds of gas and dust within which we know stars form. How could stellar twins be farther apart than the womb within which they were born?

- There is a third star in the Proxima Centauri and Alpha Centauri system. Most likely, the three stars actually started out their lives much more closely together, and as the two twins that make up Alpha Centauri came together, feeding from the parent cloud and growing heftier together, Proxima got pushed out, emerging as a little red dwarf star, the runt of the litter.

- Such red dwarf stars are so called because they are the lowest-mass objects that can still perform nuclear fusion in their cores, which is what defines a star. And being such lightweights, as low as one-tenth of the mass of our Sun, they have very cool surfaces, as cool as 3000 degrees Celsius, so they emit primarily long-wavelength red light.

- The most widely separated of the thousands of apparent stellar twins are separated by 3 light-years or more. However, the typical stellar womb is only about one-tenth that size. Research has shown that, in fact, one of the twins was itself a twin. What was thought to be a single star turned out to be a very tight binary pair. So, these systems are actually triplets, consisting of a tight inner binary pair and a distant third sibling that presumably was ejected during the dynamical birth process.

- In an extreme situation in which one of the triplets gets ejected hard and early, what would we see? This early ejection scenario is one of the currently favored theories to explain the existence of so-called brown dwarf stars, which are in fact not stars at all but, rather, stillborn stars—objects whose birth weights were less than the minimum amount of mass that a star needs to ignite as a full-fledged star.

- These objects, choked off during the stellar birth process from their supply of material from the parental cloud of gas and dust, emerge from that process unable to light up as a star, doomed to spend the rest of time slowly fading away, neither star nor planet.

Suggested Reading

NASA Jet Propulsion Laboratory, "Near Earth Object Program."

Wikipedia, "N-Body Simulations."

Questions to Consider

1. What might be some of the reasons that stars are so often birthed as twins, triplets, etc., whereas human twins and triplets are much more rare?

2. In what ways do stellar siblings affect one another and enhance or limit their ability to make solar systems?

The Dynamics of Star Formation
Lecture 6—Transcript

So far, we've talked about the stellar birth process from the standpoint of the nurseries within which stars are born, as well as from the standpoint of each individual star's birth. We've seen that stars are born in immense clouds of gas and dust, which, once illuminated by the embryonic stars forming within, glow as stellar nurseries that we call nebulae. Importantly, we saw how crucial gravity is as the force that initially brings together the material within the denser regions of the nursery. And it is gravity that, through its inexorable inward crush, compels the forming stars to turn on as full-fledged stars, generating the pressure and heat in their cores to push back against gravity in a life or death struggle to survive.

By studying these beautiful nebulae in different wavelengths of light, we learned about different aspects of the nurseries. We saw examples of the amazing colors of these nebulae, telling us about the elemental makeup of these nurseries, elements that become incorporated into the stars themselves as they are born from this material. This material also becomes incorporated into the systems of planets—solar systems—formed in the swirling disks of gas and dust around the newly forming stars.

We also looked at examples of massive pillars of gas and dust sculpted by the powerful radiation of the first born massive stars that begin their death throes just a million years or so after first turning on. Indeed, we saw that the death throes of massive stars within these nurseries can trigger the birth process for large numbers of smaller stars.

But an important additional aspect of the stellar birth process is this; most stars are not born alone, but in groups, or litters, if you will. In this lecture, we'll learn about the types of stellar siblings that most stars are born with. We'll learn about the latest observational evidence, and we'll look at some sophisticated computer simulations. These simulations are teaching us that stellar siblings undergo fascinating, and sometimes violent, dynamical interactions that influence everything from their birth weight to their future lives.

For the next minute or so, I'd like you to watch one of these amazing simulations. This one is from my colleague, Professor Matthew Bate from the University of Exeter. Bate developed the simulation using a powerful supercomputer designed specifically for the study of stellar birth from enormous clouds of gas and dust. This is just a teaser for now; we'll look at other related simulations a little later, at which time I hope you will more fully understand what is going on in simulations like these. But it's helpful to get a sense of how these simulations look before we jump in to understanding them and what they mean.

What we're seeing here is an initial cloud of gas and dust about three light-years across that is beginning to collapse under gravity. Gravity is causing fragments and filaments and clumps within this material to form; that's what gravity does. Now we zoom in; we see individual stars beginning to light up and emerge from the swirling gas as gravity compresses small pockets of the material to sufficient density so that individual stars can form and light up.

Now, these stars inherit the motion from the swirling gas out of which they formed, and so they continue to move through the surrounding material. Stars that were formed in proximity to one another begin to jostle and interact through gravity, and you can see some stars being kicked out even beyond the edges of the frame of the animation here, and as the swarm of stars grows, there are more and more opportunities for them to interact and jostle. As we take in the entire collection of hundreds of stars that have been newly born in the cluster, we see many of them clustered at the center, but we also see a distributed group, a halo, if you will, of stars that were ejected out to the outskirts of the cluster.

OK, we'll return to simulations in greater detail later on, and we'll talk both about the macroscopic processes affecting an entire stellar nursery and the microscopic processes affecting individual stars and siblings. But let's take a step back now to learn some fundamentals about the dynamics of star formation. And let's start by defining some terminology that astronomers use to describe different types of stellar siblings. First of all, a lone star is simply called a single. Our Sun is, to the best of our knowledge, a single star. Among all stars similar in mass to our Sun, singles are found about 40% of the time. So while our Sun is not weird, it does, in fact, represent a bit of a stellar minority, but the most common arrangement for stars similar in mass to our Sun, about 60% of them, is a double star system. But that's not all; although rare, stellar siblings can come in threesomes, foursomes, or more! Let's look at some examples of stellar siblings from some research conducted by my research team as we discuss the different types.

Two stars together make up what we call a binary, or double, star system. And in this image you can see many examples. You can think of these as twins, sometimes identical, but usually not. In fact, you can think of most binary stars as being fraternal twins; they were born together but otherwise have different characteristics. In these examples from my research team, you can tell the identical twins from the fraternal twins; the identical twins have identical colors, whereas the fraternal twins have different colors. As we'll see later, there are even binary star systems that we can think of as adopted twins. They are together now, but were born from different clouds, different parents! There are also triplets, what astronomers call trinary star systems, or triples. You can see a few examples in this image from my research team. There is even an example here of a quaternary system of four stars. Quadruplets!

One of the most extreme examples of multiple siblings is, in fact, one of the brightest stars in the sky, Castor. Castor is one of the two brightest stars that make up the constellation Gemini. In mythology, Castor and Pollux were twins. Ironically, the stars Castor and Pollux are not, in fact, twin stars; they do not orbit one another, and they don't have anything physically to do with one another. Castor, however, is itself not two, not three, not four, but six stars. Sextuplets!

Let's look more closely at the Castor sextuple system to see how the six sibling stars are arranged into one gravitationally bound system. At first glance, Castor appears to be a single bright star visible to the naked eye. With better resolution, we see what appears to be two bright stars very close together. These two are called Castor A and Castor B. Then, with a sensitive telescope, we find a third star, Castor C, in a distant orbit about the brighter Castor A and Castor B—that's three stars. But then looking even more closely with the highest angular resolution, we see that each of those three stars is, in fact, itself, a tight binary system. So Castor A is a binary, and Castor B is a binary, and those two binaries orbit about one another. Then, Castor C, also a binary, is in a distant orbit about the Castor A and Castor B binaries. In other words, the Castor sextuplets have what we call a hierarchical arrangement; the stars are not evenly separated, but rather, involve a nested sequence of tight separations within more distant orbits. Remember that idea of hierarchical arrangements because we'll come back to it shortly.

Now, there is an important distinction that we need to make when talking about stellar siblings. The only physically meaningful types of siblings are those that are gravitationally bound to one another. Such siblings, by virtue of their mutual gravitational attraction, are most likely to have actually been born together. At the very least, it's clear that the stars are together now, and so they likely share a history, past and future. This is as opposed to stars that, at first glance, might appear to be related, but in fact, are not. This can happen when two stars appear very close to one another in the sky, but in fact, are at completely disparate distances from us.

Let me give you an example; the brightest star in the constellation Capricorn, called Alpha Capricornus, is actually two stars that, from our vantage point on earth, appear to be very close together. These are easily seen through a pair of binoculars, but these two stars are not, in fact, close together at all. One of the stars that make up Alpha Capricornus, the brighter one, is about 100 light-years from us; the other is about 700 light-years from us. So though they appear side by side, in fact, one is right in front of us and the other is seven times farther away! This is what astronomers refer to as a visual double, representing a mere chance alignment of unrelated stars along a particular line of sight. We won't discuss these false siblings anymore in this lecture, as we focus, instead, on the real things.

And speaking of the real things, a nice example of a triplet in the sky is, in fact, the nearest star to us after the Sun, called Proxima Centauri. Proxima Centauri is a very faint, red dwarf star, but it is visible because it so close by, just 4.2 light-years away. Proxima is bound in a very wide orbit around two other stars that themselves comprise a binary star system, called Alpha Centauri. The Proxima and Alpha Centauri triplet system is a good example of an important principle that applies to most stellar siblings. Note that the two stars that make up the Alpha Centaurus binary pair are relatively close together; they orbit one another every 80 years or so, whereas Proxima Centauri is quite far from both, orbiting that pair on an orbit that takes nearly 500,000 years. In other words, the three stars are not arranged with equal separations, like a triangle, but rather, like a tight pair orbited by a distant outcast. This is what we call a hierarchical arrangement, and it is representative of nearly all of the triplet star systems that we know of. And not only the triplets, but the quadruplets, and so on, all hierarchical. Even the sextuple system of Castor that we looked at a moment ago shows hierarchy in the arrangement of the six stars, a binary star system orbited by another binary system, both of which are orbited by an outcast third binary star system.

Let's look at some examples of what these stellar sibling arrangements look like from the standpoint of their orbits, including how this idea of hierarchy actually looks in the orbits. First, let's look at examples of binary star orbits. Here we see one example in which the two stars orbit one another in simple circular fashion, maintaining a constant separation throughout their orbit. But stars don't have to orbit in circles; more generally, their orbits can be ellipses, in which case they can make a close, rapid approach, and then a slow distant motion, what we call an eccentric orbit.

With three stars, the orbits become somewhat more complex. It turns out that many if not most configurations of three stars are not dynamically stable for long periods of time. And so in nature we tend to find three stars in a hierarchical arrangement, with two stars close together and the third farther out, as in this example. Just for fun, there is a fascinating, stable orbit of three stars called a figure-eight orbit, which you see here. Almost like a juggler juggling three balls in the air. Figure-eight orbits almost never occur in nature, but they are fun to watch.

Now, with four stars, while there are, in principle, more possible arrangements just by virtue of having more stars to arrange, there are, in fact, again, only a relatively small number of dynamically stable configurations. And again, the stable configurations have a hierarchical character. In this example, we essentially find two separate binary systems orbiting one another, just like the two inner binaries in the Castor sextuple system that we saw earlier.

In this final example, we see a multiple hierarchical arrangement, an inner binary pair, orbited by a third, all of which is orbited by a fourth star.

The important take-away points from these examples are that, first, stellar siblings can come in twins, triplets, etc., with examples known all the way up to sextuplets. And second, in order to remain stable in their orbits for long periods of time, these orbits generally have a hierarchical arrangement. What happens when the orbits of multiple stellar siblings are not hierarchically arranged? The answer is that some dramatic things happen. And as we'll see, these dramatics are particularly important within the crowded, dynamic environments of stellar nurseries, the places where stars are born. Let's first look at a simulation of the relatively simple case of two stars forming out of the same dense core of cloud in a stellar nursery. This essentially represents the idealized gestation of twins within a single womb.

Here we see two stars surrounded by a cocoon of gas and dust; the material spirals into the stars, adding mass to both. One star gets a bit more mass, and so it gets fed more, and the disparity between the two grows. At the end, we have fraternal twins competing for food and growth, but each has some material around it to form planets, a solar system of its own.

In this type of system, there are multiple possibilities for the arrangements of any planets that might form from the material around the two stars. Planets could form around one or both of the stars individually, because each star maintains a reservoir of gas and dust in a protoplanetary disk around it. Alternatively, or in addition, planets could form in the material surrounding the entire stellar binary, becoming what are known as circumbinary planets.

Now, let's look at the same sort of situation, but this time, where the two protostars forming within the womb get a bit too close for comfort.

Here, again, we see two stars forming out of the cloud, gradually taking shape. One star starts out with a bit more mass. Now, we're going to see here a close encounter that disrupts one of the stars, causing it to become a binary. And so, overall what we see is a triplet system spreading out, becoming hierarchical. And we have material around one to form planets around it, and material around the binary pair to form circumbinary planets.

In this type of outcome, the possible arrangements for any planets forming in the system will be different from the binary example we saw a moment ago. In this case of triplet stars, the tight binary pair has been largely stripped of its reservoir of gas and dust by the close gravitational encounter with the third sibling, and so planets are unlikely to be able to form around either of those tight stars individually. However, the more distant third star will be able to form planets. In addition, there may be enough material surrounding the entire triple system to form circumtriplet planets. In general, if you want to know where the planets are likely to form, just follow the protoplanetary gas and dust in the system.

In addition, we see that the birth of triplets within a natal cloud of gas and dust involves some amount of jostling of the stars and disruption of the system, both in the sense that protoplanetary material is disrupted and in the sense that one of the three stars is pushed out to a wider orbit compared to the other two siblings. Let's take that to the next level and consider now what happens within a larger stellar nursery where the formation of stars within wombs is occurring many times over and in relatively close proximity.

In this simulation, we're going to start out zoomed in to a relatively small portion of the overall stellar nursery. We see individual stars begin to light up as they form within their individual wombs. Now, stars that are beginning to emerge from relatively nearby wombs can attract one another, and they begin to interact gravitationally. Small groupings of multiple stars that didn't necessarily form together begin to dance and swing around one another; they do some do-si-do and even swap partners. Generally speaking, the binaries become closer together, while the other siblings become pushed out, and some are even ejected, a kind of sibling rivalry.

This dance, this *pas de trois*, if you will, is a very important aspect of how stellar siblings form and develop together. The simulation you just watched gives you an idea of how dynamic these interactions can be as stars fly-by, and swing, and push, and sometimes entirely kick one another away entirely. The details of these three-body interactions are quite fascinating, so let's focus our gaze on just one such interaction and go into slow motion, so that we can really see how intricate this *pas de trois* actually is.

In this simulation we see three stars that have come close enough together so that they begin mutually interacting through gravity. Now, we're really watching this in extreme slow motion. Time is progressing in thousands of years, whereas stars live for billions. Now, you'll see, over and over again, that stars swap partners. Which one is the third sibling? Well, it changes, and it changes repeatedly. At the end of the simulation we see one of the siblings is going to get kicked out into a very distant orbit around the other two. And those remaining two become tightly locked together. And so we see the establishment of a hierarchical triplet arrangement, as we've discussed, is common for all triplet systems.

So, again, we see that the establishment of hierarchy among stellar siblings when they are born involves a complex, highly dynamic interaction. The need for a system of three or more stars to end up in a strictly hierarchical arrangement may seem counterintuitive. After all, within our own solar system, we have one star orbited by eight planetary bodies spaced out relatively evenly in distance from the Sun. But the difference in a stellar triplet system is the much stronger gravitational interaction among these stellar mass bodies. The strength of that interaction is such that the only stable arrangements are hierarchical ones, and so these systems naturally evolve in such a way that they either space themselves out or disrupt themselves trying.

Indeed, more detailed studies of these interactions involving many simulations and different starting conditions reveal that the resulting stellar triplets tend to have a tightened inner binary pair and a widened outer third star. In other words, over time, two of the triplets move closer together in their orbit, while the third drifts farther and farther away from the inner pair.

This dynamical process of at once tightening an inner binary pair of stars at the expense of expelling a third sibling turns out to help explain a long-standing mystery about binary stars. It has long been known that the most widely separated binary stars—often referred to as ultra-wide binaries—could not have started out as widely separated as we see them. Take as an example the separation of Proxima Centauri from Alpha Centauri. Their physical separation is larger than the typical stellar womb, the clouds of gas and dust within which we know stars form. How could stellar twins be farther apart than the womb within which they were born?

Well, as you now know, there is a third star in the Proxima Centauri and Alpha Centauri system. Most likely, the three stars actually started out their lives much more closely together, and as the two twins that now make up Alpha Centauri came together, feeding from the parent cloud and growing heftier together, poor Proxima got pushed out, emerging as a little red dwarf star, the runt of the litter. Such red dwarf stars are so called because they are the lowest mass objects that can still perform nuclear fusion in their cores, which is what defines a star. And being such lightweights, as low as one-tenth of the mass of our Sun, they have very cool surfaces, as cool as 3000 degrees Celsius, and so they emit primarily long-wavelength red light.

In my own team's research, we have looked closely at the most widely separated of the thousands of stellar twins that we have discovered. Some of these apparent stellar twins are separated by three light-years or more! Yet, the typical stellar womb is only about a tenth that size. Sure enough, when we trained our telescopes on each of the twins, more often than not we discovered that one of the twins was, in fact, itself a twin.

In this set of images from our research, you are seeing a very high resolution image of what we initially believed was just one star in a wider binary. And what you see in each of these example cases is that what we thought was a single star, in fact, turns out to be a very tight binary pair. So these systems are, in fact, triplets, consisting of the tight inner binary pair that you see in these images and a distant third sibling beyond the image that presumably was ejected during the dynamical birth process.

Now, let's think a bit more about what all of this dynamic interaction and partner swapping and jostling and kicking will do to the stars' abilities to continue feeding and growing from their parental clouds of gas and dust. Imagine an extreme situation in which one of the triplets gets ejected hard and early. What would we see then? This early-ejection scenario is, in fact, one of the currently favored theories to explain the existence of so-called brown dwarf stars, which are, in fact, not stars at all, but rather, stillborn stars, objects whose birth weights were less than the minimum amount of mass that a star needs to ignite as a full-fledged star. And so these objects, choked off during the stellar birth process from their supply of material from the parental cloud of gas and dust, emerge from that process unable to light up as a star, doomed to spend the rest of time slowly fading away, neither star nor planet. We'll learn more about these mysterious brown dwarfs in a later lecture.

Today, we've learned that the stellar birth process is not always gentle. Stars are not born alone, and because they're not born alone, these siblings affect one another, a sibling rivalry starting in the womb. This competition and sibling interactions will affect, for example, the mass, the final birth weight of each sibling—which has ramifications, as we'll see, for the entire life cycle of each sibling.

Solar Systems in the Making
Lecture 7

I n this lecture, you will learn about the part of the stellar life cycle in
which stars form their families of planets—their solar systems. You
will learn that one of the most successful explanations for our own
solar system involves a chaotic, dynamical process in which our planets
jockeyed for position, and in so doing, may have been responsible for
delivering the Earth its oceans in a hail of comets, the water upon which
our very lives now depend.

Protoplanetary Disks around Young Stars

- Protoplanetary disks are the birthplaces of planets; they are solar
  systems in the making. But these planetary birth sites are fleeting
  places. Stars like the Sun live a very long time, about 10 billion
  years. However, these protoplanetary disks disappear quickly after
  a star is born—typically lasting just a few million years—so planets
  have to be made very quickly.

- Using infrared measurements of disks around stars in stellar
  nurseries of differing ages, we can directly trace the evolution of
  protoplanetary disks. The amount of infrared light emitted by
  the protoplanetary disks tells us how much gas and dust is there,
  so by measuring the infrared brightnesses of different star-disk
  systems, we learn how quickly the disks are consumed by the stars
  themselves and by the planet-formation process.

- Based on these measurements, we find that when the stars are less
  than a million years old, nearly 100% of them have protoplanetary
  disks. But by the time these stellar families are just 5 million years
  old, only about 10% of the stars still have their disks. So, after an
  age of about 5 million years, stars simply do not have sufficient
  material remaining around them to be able to make planets.

- Why is the window of opportunity for planet making so brief? In large part, it is because the process of making planets itself uses up the disk material and, therefore, limits the star's planet-making potential.

- Measurements using the radio light emission of the gas in these disks tell us that young stars with masses like the Sun have just enough material in their disks to make a few times as many planets as those that comprise our solar system. More massive stars generally have somewhat more material, and less massive stars start out with somewhat less.

- These disks actually contain just enough mass within them to make a solar system's worth of planets and some extra. In fact, that bit of extra planet-making potential is probably gobbled up by the star.

- The formation of a planet within a protoplanetary disk of gas and dust can be broken down into 3 stages. First, the forming planet interacts through gravity with the nearby disk material and excites a spiral wave of density propagating through the disk. Next, as the mass of the planet grows from consumption of the disk material immediately around it, an empty gap within the disk at the planet's orbit is cleared out. Finally, the massive planet is able to interact with the nearby disk material strongly enough that it experiences a steady loss of orbital energy, causing it to spiral in closer and closer to the star.

- This spiral of the planet toward the star is referred to as planetary migration, and in some cases, the inward spiral can cause the planet to plummet all the way into the star. But while the idea of planetary migration as an integral part of the planet-formation process has become widely accepted, this idea was very surprising when it was first discovered.

- For most of the history of astronomy, and going all the way back to the philosopher Immanuel Kant in the mid-1700s, our basic ideas for the origin of our solar system were of an orderly process. In the solar nebula hypothesis, the Sun and its system of planets were thought to have originated from the gradual formation of individual bodies within a rotating nebula of gas.

- Impressively, this basic picture is not so far from the truth as we now understand it, but this picture did not imagine that the planets forming within the solar nebula might have moved or migrated from their original positions to where we see them now. And the Kantian picture certainly did not imagine a process whereby the planets might even be destroyed as they migrate all the way in to the Sun.

- So, in fact, the planetary migration idea is a relatively new one in the history of human understanding, and it represents one of the more surprising and significant advances in our quest to place our planetary home in the broader context of our own solar system and, indeed, in the context of how all other solar systems come to be.

The Planetary Migration Process: The Nice Model

- The planetary migration process serves to further clear out the disk material close in to the star, leading eventually to a cleared-out inner hole in the disk within the planet's orbit.

- The precise location within the protoplanetary disk where a planet starts out is important, because this affects the type of planet it will be. Think about the nature of the planets in our own solar system: Solid rocky planets like the Earth and Mars are close to the Sun while giant gaseous planets like Jupiter and Saturn are farther out. This was recognized early on as an important clue.

- Close to the Sun, where the protoplanetary disk was hot, light elements like hydrogen and helium would have been boiled off of a forming planet, leaving only the heavy rocky and metallic elements to form a small solid planet. Far from the Sun, where the protoplanetary disk was cold, the forming planet could hang on to a very large fluffy atmosphere of light elements.

- A successful model for our solar system must be able to account for the 8 major planets, from Mercury to Neptune; the fact that the inner planets are small and rocky whereas the outer ones are big and gaseous; and the fact that the outskirts of our solar system are littered with comets and other small bodies—the detritus of the planet-formation process.

- That basic model is now referred to as the Nice model, so named for the city in France where a group of astronomers met in 2004 with the goal of developing a comprehensive simulation of the formation of our solar system.

- In the Nice model, it is the gravitational jostling caused by the four massive planets that governs all of the ensuing action. The small rocky planets in the inner part of the solar system are innocent bystanders in a dramatic sequence of events dictated by these titans.

- The giant planets have enough mass to be able to gravitationally perturb the comets and asteroids in the Kuiper belt. Those perturbations send some of these objects inward, into the direct influence of the giant planets. And once the comets come closer to the massive planets, the comets' orbits become extremely unstable due to the strong gravitational slingshot effects caused by their close passages to the planets. Most of these comets and asteroids end up being ejected into interstellar space.

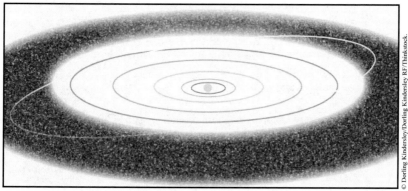

The Kuiper belt consists of hundreds of millions of objects that are supposedly leftovers from the formation of the outer planets.

- But the planets cannot kick the comets out of the Kuiper belt without consequence. Newton's third law states that for every action, there is an equal and opposite reaction. In this case, the reverse effect of kicking the comets and asteroids out of the solar system is to alter the planets' own orbits.

- In particular, the planets' orbits migrate inward or outward, and importantly, orbits that may have started off circular become elongated, what we refer to as eccentric orbits. With eccentric orbits, the giant planets can occasionally interact even more strongly because they can at times come very close together.

- At the critical moment in the Nice model, Jupiter and Saturn hit a gravitational resonance. A resonance is a kind of "sweet spot" in which the eccentricities and proximities of the planets' orbits cause them to orbit the Sun with periods that are an integer multiple of one another. In this case, it is a 2-to-1 orbital resonance, in which Jupiter goes around the Sun twice for every time Saturn goes around once. That resonance means that for a brief time, Jupiter and Saturn interacted very strongly, over and over again, and each time, Saturn's orbit got kicked so that it became more and more eccentric, more elongated.

- At that moment in the solar system's history, Saturn's orbit quickly became sufficiently elongated so that its gravitational influence could begin to affect the planets exterior to it—Uranus and Neptune. In this way, Saturn then forced Uranus and Neptune to begin migrating outward. And in the process of migrating, they actually swapped places.

- That briefly chaotic swapping motion caused Uranus and Neptune to plow through the swarm of comets and asteroids in the Kuiper belt. Their motion into that crowd caused a sudden scattering of huge numbers of comets and asteroids, splattering them all over the place.

- Some of that splash brought a very large number of comets plummeting in toward the Sun and crashing into the planets of the inner solar system, including the Earth. The geologic record of the Earth and our Moon clearly shows that an event referred to as the late heavy bombardment occurred around the time when the solar system was about 600 million years old.

- The late heavy bombardment was a brief event in which a large number of solid bodies suddenly impacted the Earth and Moon. Sure enough, the Nice model has the Neptune-Uranus swapping event occurring right around the time of the late heavy bombardment.

- An intriguing additional consequence of this scenario is its potential to explain the abundance of liquid water on Earth. Comets are often referred to colloquially as dirty snowballs—a mixture of solid material, including rocks and metals, together with ices of various compounds, including frozen water. Calculations that take into account the number of comets and their content suggest that, indeed, the Earth's oceans could have been supplied with much of their current water content by a bombardment of comets raining down from the sky.

- There is some debate in the scientific community over whether the Nice model's prediction of Neptune and Uranus swapping is correct. However, this aspect of the model is not necessarily essential to its overall success in explaining many of the features of our present-day solar system. By changing the starting conditions of the solar system's planets slightly, much of the same behavior occurs, but without the two planets swapping.

- Of course, as with any scientific model, the Nice model has its limitations. While the model seems to successfully capture many of the essential ingredients in the evolution of our solar system—with or without Neptune and Uranus swapping places—it does not attempt to explain how the planets came to be in their starting positions, only what happened to them afterward.

- Indeed, it remains an outstanding question how the planets in our and other solar systems manage to avoid falling in to their suns. The issue is that the planet-migration process within the protoplanetary disk is so effective at spiraling the newborn planets toward the star that in many calculations, the planets just keep on spiraling in until there are no planets left. The parent star in effect cannibalizes its own planets before they even really get a chance to jostle one another, as in the Nice model.

- So, many astronomers believe that the process of planet formation is sufficiently prolific that the parent stars can afford to cannibalize many of the planets, and then, once the protoplanetary disk is used up and the planet-migration process is halted, there are sufficient planets remaining to populate the solar system thereafter.

- The protoplanetary disks from which the planets form typically contain enough material for a few times as many planets as what we currently see in our solar system. Those extra planets almost certainly do get made, but we don't see them now because they ended up devoured by the Sun.

Suggested Reading

Brandner and Klahr, *Planet Formation*.

Sky & Telescope Magazine, "Video: Evolution of the Solar System."

Questions to Consider

1. How does the idea of conservation of angular momentum help explain why stars form flattened disks of gas and dust from which their planets form? Why is it a flattened disk as opposed to, for example, a bubble of gas and dust around the star?

2. How does the idea of "jockeying" planets compare to the traditional view of how our solar system came to be, and how does this dynamic view of planetary system formation help us understand the types of solar systems now being discovered around other stars?

Solar Systems in the Making
Lecture 7—Transcript

In previous lectures, we've examined the process of stellar birth in some detail. From the gravitational collapse of a gigantic cloud of gas and dust into a protostar surrounded by a dusty cocoon, and eventually, into a young star shining on its own through the power of fusion. But the process of stellar birth is also the process of planet birth, for it is from the cocoons surrounding embryonic stars that protoplanetary disks of gas and dust encircling newborn stars are made, and it is out of that protoplanetary material that the planets themselves are formed.

In this lecture, we will look at examples of the protoplanetary disks where planet formation is taking place around young stars. Importantly, because young stars gobble up, or blow away, the material in these protoplanetary disks quite quickly, we'll see that newborn stars, in fact, have a relatively short window of opportunity for making the planets that will eventually comprise their solar system. We will also look at computer simulations that show how newborn planets jockey for position close to their parent stars and how some planets can be ejected away from their parent stars and their sibling planets. Such simulations actually suggest that our own solar system might not have always looked the way it does now.

Let's start with a refresher of what the protoplanetary disks around young stars actually look like. I would like to show you six examples of very young stars in the stellar nursery in the constellation of Taurus. The six images we'll look at together were taken using the infrared camera on the Hubble Space Telescope. As we've learned, infrared light can penetrate through more intervening material than visible light can, and so, the infrared light from the central young star is acting like a light beacon, if you will, that penetrates through the protoplanetary disks silhouetted against these stars. The star is like our probe of that protoplanetary material, shining through it, revealing the nature of the surrounding disk of gas and dust. So let's have a look.

Let me start by noting that these images are particularly good ones for understanding the stellar environs within which planets form, because all of these stars happen to, by chance, be viewed nearly edge on as seen from

Earth. As a result of that fortuitous alignment, the disks of gas and dust surrounding the stars are silhouetted by the bright star in the center. This allows scientists to construct computer models involving a star plus a disk to learn in detail what the structures of these planet-forming disks are truly like. Remember that these images, like all astronomical images, are two dimensional, and the way that we can translate what we see into a complete three-dimensional understanding is by constructing three-dimensional models that correctly reproduce what we see in the two dimensional images.

Right now we're looking at a young star called IRAS 04302+2247. Don't worry, there won't be a quiz on these star names; astronomers use a variety of naming schemes to keep track of the hundreds of millions of stars in our catalogs, and many times those names involve a shorthand for the star's coordinates in the sky. Just think of it as a phone number. In any case, what we're seeing is the actual image taken by the Hubble space telescope on the left and a model intended to reproduce that image on the right. In this example, the star is completely obscured by the thick, dusty disk surrounding the star. We see light above and below that disk, scattered light from the star. This is light from the star that is reflecting and scattering off of dust grains in the surrounding material. We notice that the dark silhouette of the disk, that dark band cutting through the middle of the image, is curved at the edges, and that tells us that the surrounding protoplanetary disk is actually bowl shaped. Now, the scattered light that we see tells us that there must be some material—gas and dusk—surrounding the young star and its disk. This is a remnant of the cocoon from an earlier stage of the star's evolution.

Now, notice that the model at right is not a perfect replica of the image that you see at left; there's some messiness in the actual star-disk system, but the model gets, essentially, the key features of the image right. There's a large, spherical envelope surrounding the star and the disk, and that large envelope has been bored out by jets emerging from the young star, and we see the thick protoplanetary disk along the mid plane, along the middle, cutting through the middle of the image. And finally, the bluish or greenish hue of the scattered light that we see is indicative of relatively fine dust grains surrounding the young star and its disk.

This next example is DG Tau B. It's very similar to the previous example, except in this case, the system is tipped slightly in our direction, allowing us to just barely see the central star peeking out. Again, we see that bowl shape to the silhouette of the surrounding disk of gas and dust, and we also see a faint inner V shape to the scattered light in this case, and that's showing us directly the shape of the cavity that's been bored out by the jets shooting out from this young star.

This next example is IRAS 04016+2610. This, again, is similar to the previous example, but even more tipped in our direction, so we can more directly see the central young star. And in this case, we can see a faint glow from the other side of the envelope; that's the cavity on the other side of the disk of gas and dust. And in this case, the reddish hue of the scattered light tells us that the dust grains in this system have begun to coalesce into somewhat larger grains than those that we found in previous examples.

This example is IRAS 04248+2612. This is a very edge-on case, but notice that in this case the central dark band, the silhouette caused by the central disk of gas and dust, is very thin at the central regions. Overall, this is a slightly more evolved system; the overall envelope of gas and dust is beginning to collapse down into just the disk. The true image in this case is very complex compared to the model, and this is because of the strong jets that are shooting out of the star that stir up the surrounding protoplanetary gas and dust.

This next example is called CoKu Tau/1. This is a very unusual case because, in this case, the surrounding protoplanetary disk around the star is very small, and this suggests that there may be a stellar sibling; this is a binary star system within the protoplanetary disk that's acting like a beater and is depleting and using up the protoplanetary disk material. But we can still see the scattered light off of the remaining envelope of gas and dust.

And finally, in this last example, we have Haro 6-5B. This is the most of all, compared to the others that we've seen. In this case, the surrounding envelope of material is mostly entirely gone, and only the thick protoplanetary disk remains. Notice, though, that the disk flares out at the edges, like in the other examples, is puffy at the outskirts.

Like similar disks throughout the universe, these protoplanetary disks in the constellation Taurus are the birth places of planets. These are solar systems in the making. But these planetary birth sites are fleeting places. Remember that stars like the Sun live a very long time, about 10 billion years all told. But these protoplanetary disks disappear quickly after a star is born, typically lasting just a few million years or so. So planets have to be made very quickly.

How do we know this? Using infrared measurements of disks around stars in stellar nurseries of differing ages, we can directly trace the evolution of these protoplanetary disks. The amount of infrared light emitted by the protoplanetary disks tells us how much gas and dust is there, so by measuring the infrared brightnesses of different star-disk systems, we learn how quickly the disks are consumed by the stars themselves and by the planet formation process. Based on these measurements, we find that when the stars are less than a million years old, nearly 100% of them have protoplanetary disks. But by the time these stellar families are just 5 million years old, only about 10% of the stars still have their disks. So, after an age of about 5 million years, stars simply do not have sufficient material remaining around them to be able to keep making planets. You could say that stars have a lot of pressure to start their families very early on and assume a heavy parenting responsibility very early in their lives.

How does this happen? Why is the window of opportunity for planet making so brief? Well, in large part it is because the process of making planets itself uses up the disk material, and therefore limits the star's planet-making potential. Measurements using the radio light emission of the gas in these disks tell us that young stars with masses like the Sun have just enough material in their disks to make a few times as many planets as those which comprise our solar system. More massive stars generally have somewhat more material, and less massive stars start out with somewhat less.

But the point is that these disks actually contain just enough mass within them to make a solar system's worth of planets and a bit extra. We'll come back toward the end of this lecture to discuss where that bit of extra planet-making potential goes, but just so that you aren't held too much in suspense, the quick answer is that the star probably gobbles up the extra planets.

To better understand why the planet formation process limits the amount of time available for planets to be formed from protoplanetary disks, let's look at a movie simulation showing the basic picture for the formation of a planet like Jupiter within a protoplanetary disk of gas and dust.

In the simulation, we see a planet on a circular orbit growing in time from an initial mass of 3 Earth masses to a final mass of 10 Jupiter masses. What you see happening in the animation can be broken down into three stages. First, the forming planet interacts through gravity with the nearby disk material and excites a spiral wave of density propagating through the disk. Next, as the mass of the planet grows from consumption of the disk material immediately around it, an empty gap within the disk at the planet's orbit is cleared out. Finally, the massive planet is able to interact with the nearby disk material strongly enough that it experiences a steady loss of orbital energy, causing it to spiral in closer and closer to the star.

This in-spiral of the planet toward the star is referred to as planetary migration, and in some cases, the in-spiral can cause the planet to plummet all the way into the star. But while the idea of planetary migration as an integral part of the planet formation process has become widely accepted, let's reflect for a moment on just how surprising this idea actually was when it was first discovered.

For most of the history of astronomy, and going all the way back to the philosopher Immanuel Kant in the mid-1700s, our basic ideas for the origin of our solar system was of an orderly process. In this Solar Nebula Hypothesis, the Sun and its system of planets was thought to have originated from the gradual formation of individual bodies within a rotating nebula of gas. Now, impressively, this basic picture is not so far from the truth as we now understand it, but this picture did not imagine that the planets forming within the solar nebula might have moved, or migrated, from their original positions to where we see them now. And the Kantian picture certainly did not imagine a process whereby the planets might even be destroyed as they migrate all the way in to the Sun. So in fact, the planetary migration idea is a relatively new one in the history of human understanding and represents one of the more surprising and significant advances in our quest to place our planetary home in the broader context of our own solar system, and indeed, in the context of how all other solar systems come to be.

Let's come back now to examine the planetary migration process further. This migration serves to further clear out the disk material close in to the star, leading eventually to a cleared out inner hole in the disk within the planet's orbit.

In this image of the star Fomalhaut taken by the Hubble space telescope, we see the star at the center of a large evacuated cavity, surrounded by a disk whose inner edge is lit up by the light from the star. At the location of that inner edge of the disk, we see a planet dimly illuminated, its orbital motion now traced over a small portion of its total orbit through nearly 10 years of study with the Hubble. Now, the precise location within the protoplanetary disk where a planet starts out is important, as this affects the type of planet it will be. Think about the nature of the planets in our own solar system, solid, rocky planets like the Earth and Mars close to the Sun, giant gaseous planets like Jupiter and Saturn farther out. This was recognized early on as an important clue.

Close to the Sun where the protoplanetary disk was hot, light elements like hydrogen and helium would have been boiled off of a forming planet, leaving only the heavy rocky and metallic elements to form a small, solid planet. Far from the Sun, where the protoplanetary disk was cold, the forming planet could hang on to a fairly large, fluffy atmosphere of light elements. So you can think of massive planets like Jupiter as being like the Earth, only surrounded by an enormous, massive envelope of hydrogen and helium, its Earth-like rocky and metallic content likely forming a dense solid core deep in the planet's interior.

With these basic ingredients in mind for how planets form around stars like the Sun, let's spend some time examining in detail what has become one of the standard models for the formation of our own solar system. A successful model for our solar system must be able to account for the eight major planets from Mercury to Neptune; the fact that the inner planets are small and rocky whereas the outer ones are big and gaseous; and the fact that the outskirts of our solar system is littered with comets and other small bodies, the detritus of the planet formation process.

That basic model is now referred to as the Nice model, so named for the city in France where a group of astronomers met in 2004 with the goal of developing a comprehensive simulation of the formation of our solar system. Let's watch an animation of the model and then discuss its most important features. The animation will start out with our solar system's four giant gas planets—Jupiter, Saturn, Uranus, and Neptune—packed together somewhat more closely together than they are now.

In the animation, you are going to see something surprising happen. Neptune starts out closer to the Sun than Uranus, but then the two planets end up swapping positions. In the animation, time will be slowed down dramatically at that critical swapping moment; otherwise most of the changes would happen too quickly to notice. But the total time represented by the simulation is approximately a billion years. Finally, a quick note that in the simulation you won't see the inner planets of our solar system represented—Mercury, Venus, Earth, Mars. That's because their orbits are much smaller than the full scale of the solar system as whole, and because being very low-mass planets they don't influence the simulation very much, and so can be ignored. So let's watch this amazing simulation together.

As the simulation begins, we see the positions of the planets represented by their orbits and importantly notice that Neptune starts out closer to the Sun that Uranus, unlike the present-day configuration of these planets. And we see the surrounding Kuiper belt of comets and asteroids at the outskirts. Now, the planets begin jostling one another gravitationally and also perturbing those surrounding comets. And at the critical moment, right here, we see Jupiter and Saturn force Neptune outward, swapping positions with Uranus, and sending those planets plowing into the comets, splattering them and sending some of them into the inner solar system, and finally we have the arrangement of the planets as they are today.

Let's discuss that scenario in a bit more detail now. In the Nice model, it is the gravitational jostling caused by the four massive planets that governs all of the ensuing action. The small rocky planets in the inner part of the solar system are innocent bystanders in a dramatic sequence of events dictated by these titans. The giant planets have enough mass to be able to gravitationally perturb the comets and asteroids in the surrounding Kuiper Belt. Those perturbations send some of these objects inward into the direct influence of the giant planets. And once the comets come closer to the massive planets, the comets' orbits become extremely unstable due to the strong gravitational slingshot effects caused by their close passages to the planets. Most of these comets and asteroids end up being ejected into interstellar space.

But the planets cannot kick the comets out of the Kuiper belt without consequence. Remember Newton's Third Law, which says that for every action there is an equal and opposite reaction. In this case, the reverse effect of kicking out the comets and asteroids out of the solar system is to alter the planets' own orbits. Remember planetary migration; in particular, the planets' orbits migrate inward or outward, and importantly, the orbits that may have started off circular become elongated, what we refer to as eccentric orbits. With eccentric orbits, the giant planets can occasionally interact even more strongly because they can, at times, come very close together.

At the critical moment in the Nice model, what is happening is that Jupiter and Saturn hit a gravitational resonance. A resonance is a sort of sweet spot in which the eccentricities and proximities of the planets' orbits cause them to orbit the Sun with periods that are an integer multiple of one another. In this case, it is a two-to-one orbital resonance, in which Jupiter goes around the Sun twice for every time Saturn goes around once. That resonance means that for a brief time Jupiter and Saturn interacted very strongly, over and over again, and each time Saturn's orbit got kicked, so that it became more and more eccentric, more elongated.

So at that moment in the solar system's history, Saturn's orbit quickly became sufficiently elongated that its gravitational influence could begin to affect the planets exterior to it—Uranus and Neptune. In this way, Saturn then forced Uranus and Neptune to begin migrating outward, and in the process of migrating, they actually swapped places. That briefly chaotic swapping motion caused Uranus and Neptune to plow through the swarm of comets and asteroids in the Kuiper Belt. Their motion into that crowd caused a sudden scattering of huge numbers of comets and asteroids, splattering them all over the place.

Some of that splash brought a very large number of comets plummeting in toward the Sun and crashing into the planets of the inner solar system, including the Earth. The geologic record of the Earth and our Moon clearly show that an event referred to as the late heavy bombardment occurred around the time when the solar system was about 600 million years old. The late heavy bombardment was a brief event in which a large number of solid bodies suddenly impacted the Earth and Moon. And sure enough, the Nice model has the Neptune-Uranus swapping event occurring right around the time of the late heavy bombardment.

An intriguing additional consequence of this scenario is its potential to explain the abundance of liquid water on Earth. Comets are often referred to colloquially as dirty snowballs, a mixture of solid material, including rocks and metals, together with ices of various compounds, including frozen water. Calculations that take into account the number of comets and their content suggest that, indeed, the Earth's oceans could have been supplied with much of their current water content by a bombardment of comets raining down from the sky.

There is some debate in the scientific community over whether the Nice model's prediction of Neptune and Uranus swapping is correct. But in fact, this aspect of the model is not necessarily essential to its overall success in explaining many of the features of our present-day solar system. By changing the starting conditions of the solar system's planets slightly, much of the same behavior occurs, but without the two planets swapping. Let's look at what that might look like.

In this alternate version of the Nice model, we again see the positions of the planets starting out represented with their orbits. This time, Neptune and Uranus start out in their present configuration. The planets, again, jostle one another and perturb the surrounding comets, and at that critical moment, once again Jupiter and Saturn strongly interact. This time, however, Neptune and Uranus do not swap positions, but they still do plow outward into the comets, splattering them and sending many of them falling into the inner solar system.

Of course, as with any scientific model, the Nice model has its limitations. While the model seems to successfully capture many of the essential ingredients in the evolution of our solar system, with or without Neptune and Uranus swapping places, it does not attempt to explain how the planets came to be in their starting positions, only what happened to them afterward.

Indeed, it remains an outstanding question how the planets in our and other solar systems manage to avoid falling into their suns. The issue is that the planet migration process within the protoplanetary disk is so effective at spiraling the newborn planets toward the star, that in many calculations the planets just keep right on spiraling in until there are no planets left. The parent star, in effect, cannibalizes its own planets before they even really get a chance to jostle one another, as in the Nice model.

So, many astronomers believe that the process of planet formation is sufficiently prolific that the parent stars can afford to cannibalize many of the planets, and then, once the protoplanetary disk is used up and the planet migration process is halted, there are sufficient planets remaining to populate the solar system thereafter. Remember we said earlier that the protoplanetary disks from which the planets form typically contain enough material for a few times as many planets as what we currently see in our solar system. Those extra planets almost certainly do get made, but we don't see them now because they ended up devoured by the Sun.

Now let's look at the process of planet formation in terms of the raw building blocks of stars and planets. Are some elements more important than others? In fact, the available evidence strongly hints that in order for planets to form at all, the parent star needs to have itself formed from a cloud of gas and dust that was sufficiently enriched chemically to include a large amount of heavy elements, such as iron.

When we look at the light spectra of stars known to harbor solar systems, we can measure the abundance of different elements in those stars, which presumably reflects the chemical composition of the material from which the star itself and its planets formed. This graph summarizes the results of such measurements. It shows the percentage of known solar systems on the vertical axis with different amounts of iron on the horizontal axis. The amount of iron is measured relative to the Sun, such that zero means the same amount of iron as the Sun, positive values mean more iron than the Sun, and negative values mean less iron than the Sun. There is a clear and strong correlation, such that the vast majority of stars with planets have an amount of iron that is similar to or greater than the Sun. Stars with less iron than the Sun rarely possess planets.

Why should the availability of heavy elements in a star influence its ability to make planets? A currently favored explanation is that such heavy elements allow the formation of dense, heavy seeds of rocky and metallic material within protoplanetary disks, and those heavy seeds then either become the small rocky planets, like the Earth, or else nucleate the growth of heavier planets, like Jupiter.

In addition, the star's chemical composition may be influenced by the planet cannibalization process that we discussed earlier. To understand this, it's important to remember that when we measure the chemical composition of a star from its light spectrum, we are strictly speaking only measuring the abundance of elements in the star's outer layers, because we cannot directly see down into the opaque interior. Now imagine that a star like the Sun that gobbles up a bunch of planets of rocky or metallic material, like the Earth, and those elements that made up that rocky planet now become sprinkled on the star's surface, in effect, polluting the star with heavy elements, like iron.

This is important, because, as we've discussed in previous lectures, we can use the light spectrum of a star to measure its elemental composition, and so the presence of an iron-polluted stellar atmosphere might be used in the future as a signpost of those stars most likely to host planets like our own. The thinking is that if the star has ingested some planets in the past, that star is likely to have other planets that survived and orbit the star still.

So in truth, the chemical signature of the planet formation process on the stars that host them is twofold. First, the star needs to have inherited enough in the way of heavy elements from the previous generations of stars whose ashes now make up the new solar system. Without those heavy element ashes, the planet-formation process might not be able to nucleate at all. Secondly, like the mythical Saturn eating his own children, the stellar birth process appears to involve the cannibalization of a star's planets, such that it becomes polluted by the heavy elements that had initially been used by the nascent planets. Only the planets surviving the first few million years of the star's life then remain to jockey for position close to the parent star, sculpting the structure of the solar system in the process.

In this lecture, we've learned about the part of the stellar life cycle in which stars form their families of planets, their solar systems. We've seen that stars must start their families quickly. Their biological clocks run fast! And we've seen that one of the most successful explanations for our own solar system involves a chaotic, dynamic process in which our planets jockeyed for position, and in so doing, may have been responsible for delivering the Earth its oceans in a hail of comets, the water upon which our very lives now depend.

Telescopes—Our Eyes on the Stars
Lecture 8

E verything we know about the stars has been learned through the information carried by the light that the stars emit. But for us to make use of that light, to be able to actually learn something about the stars from it, it must also be detected and recorded. In this lecture, you will learn about the instruments astronomers use to measure starlight. Merely to detect light is not so difficult, but to record it with sensitivity and accuracy is, in fact, a remarkable feat.

Types of Telescopes

- Our eyes, in their form and function, are small telescopes with built-in light detectors. Light that reaches your eye is brought to a focus by the lens—that's the telescope part—and the focal point is the retina, which has light-sensitive chemicals to absorb the light—that's the detector part. In one way or another, that's what every astronomical telescope does: It uses an optical element to focus the incoming light and a light-sensitive detector to absorb and record the light.

- The ability of the human brain to store visual information in the form of memory is, of course, absolutely wonderful. But as a means for making accurate measurement, the human visual system leaves much to be desired. First, most of the information that we see is recorded only temporarily and then lost forever. Second, even if it could be recorded forever, there are limitations to what our eyes can do as telescopes. Our eyes are only sensitive to a small portion of the entire electromagnetic spectrum, what we call the visible part.

- In addition, the response of our retinas to the amount of light received is what we call nonlinear. That means that there is not a simple linear relationship between the apparent relative brightness of different objects and their *true* relative brightness.

- In fact, our eyes' response to light is logarithmic. A star that appears to our eyes to be 5 times as bright as another is in fact 100 times brighter. This is a helpful property for our needs in everyday life, because it means that we can handle looking at things that have a very large range of relative brightness, but for accurately measuring quantities of light, a linear response is much more desirable.

- The digital detectors attached to most telescopes have this property: They respond in a linear way to light, so a star that is twice as bright as another is faithfully recorded as being twice as bright. In addition, our eyes are relatively small—in particular, the lenses of our eyes are small—and so are only sensitive to the brightest stars. To see fainter stars requires a larger light-gathering device—in other words, a larger telescope.

- There are two basic types of telescopes that astronomers use to observe stars in visible light, and they differ only in the nature of the optical element that focuses the incoming light. The first type is called a refracting telescope, because it uses lenses to refract, or bend, the incoming light and thereby bring the light to a focus. This is the type of telescope that most people associate with astronomy, and for much of history, this is the type of telescope that astronomers used.

- The second main type of telescope is called a reflecting telescope, which is so named because, rather than using a lens to bend the light to a focus, it uses mirrors to reflect the light to a focus. Today, all major astronomical telescopes are of the reflecting variety.

- A reflecting telescope can be built with a much larger mirror than it is possible to fabricate a lens. A lens is a solid piece of glass, so a larger lens is necessarily heavier—much heavier. In contrast, a mirror can be made large but thin.

- Another advantage of a reflecting telescope is that, in contrast to the long tube of a refracting telescope, a reflecting telescope can be built relatively compact. That is because the long focus can be accommodated with multiple reflections off of a series of mirrors, in essence folding the light beam.

Properties of Telescopes

- The two most important properties of telescopes are light-gathering power and angular resolution. Light-gathering power refers simply to the ability of a telescope to collect light. The more the better, because the more light that a telescope can capture from a star, the fainter the star the telescope can see and the more sensitively it can measure the properties of that light.

- The single most important property of a telescope that determines its light-gathering power is the area of the primary optical element. In other words, the bigger the lens or mirror, the better. The bigger the primary mirror of the telescope, the more photons that can be gathered quickly. And because the area of the mirror is the square of the diameter, by making the mirror twice the diameter, you get four times the area—four times the light-gathering power—and you can see a star that is four times fainter or measure the same star four times more sensitively.

- The second important property of any telescope is its angular resolution, which refers to the crispness with which a telescope can focus the incoming light. The better the angular resolution, the crisper the focus that can be achieved, and this allows ever-finer details in the stellar images to be discerned.

- The size of the telescope's primary mirror determines its angular resolution. A small telescope has poor angular resolution, so a star appears fuzzier. The formula that describes the angular resolution of a telescope is as follows: $\theta = \lambda/D$.

- This formula says that the best focus a telescope can achieve, the smallest angle a telescope can discern, the finest detail that the telescope can make out—call that angle θ—is the wavelength of light divided by the diameter of the telescope's primary mirror. So, for a given wavelength of light, λ, this formula says that a larger-

© Daniel Stein/iStock/Thinkstock.

The Mauna Kea Observatory, an astronomical observatory in Hawaii, lies on top of the peak of the dormant volcano named Mauna Kea.

diameter mirror will achieve a smaller angle of focus. That is, with a finer focus, finer details can be discerned.

- Both of the most important properties of a telescope—light gathering power and angular resolution—depend directly on the size of the telescope's primary mirror.

Digital Detectors and Filters

- The basic component of any modern astronomical detector—the "retina"—is a digital detector similar to what you find in your digital camera or phone. That detector is essentially just a chip of silicon that is subdivided into tiny squares, called pixels.

- Each pixel is a tiny photoreceptor that responds to incoming light by what is known as the photoelectric effect, which states that when a material absorbs the energy of a photon of light, that energy causes an electron within the material to be freed, and that electron then joins other electrons as part of an electric current that can be measured.

- Each absorbed photon therefore liberates one electron, so by measuring the strength of the electric current produced in the material, we can measure accurately and linearly the number of photons that struck the detector. With many pixels in the detector linked together, each one acting as a photon counter, we can build up an image, each pixel measuring a tiny part of the overall image.

- Measuring the color of a star is one of the basic ways that we can determine its physical characteristics, and the photoelectric effect by which the digital detector measures incoming photons doesn't care about the wavelength of the incoming photons. In essence, the digital detector only tells us whether light was detected, and how much, but not its color.

- To get color information, astronomers use filters. A given filter lets through light of only a particular wavelength—for example, red, or blue, or green. By combining images taken through discrete filters, we are then able to create images that convey the true colors of the stars, and we can use the color measurements to determine things like the temperatures of the stars.

- Much of the true power of light to convey physical information about the stars comes not from mere pictures, but from the spectrum of light. Spreading the light we receive into its constituent rainbow and then dissecting those wavelengths is ultimately how we learn the star's temperature, chemical makeup, motion, and many other properties.

- To spread the light gathered by a telescope into a spectrum, the light has to be focused onto a light dispersing element, which is in general one of two different types of optical devices: a prism or a grating.

Beyond Visible and Radio Light

- Visible light is one of the only types of light that can penetrate through the Earth's atmosphere and reach telescopes on the ground. The Earth's atmosphere acts like a screen that absorbs almost all wavelengths of light. From the ground, the only wavelengths of light that can be seen coming from the outside are visible light and radio light.

- A small portion of the infrared part of the spectrum, called the near-infrared part, can be seen from high mountaintops or from instruments flown at very high altitude. At these heights, the atmosphere above is thinner, so enough of the incoming light is able to penetrate and be detected. Even for visible light, placing our telescopes on high places is an advantage.

- Beyond visible and radio light, it's necessary to put telescopes above the filtering of the Earth's atmosphere. All of our ability to study the stars at ultraviolet, X-ray, and gamma-ray wavelengths—and most of our ability to study the stars in infrared light—comes from space-borne telescopes.

- Because these telescopes gather starlight from above the Earth's atmosphere, they additionally are able to obtain images with a crispness that is normally not possible from the ground. Putting telescopes in space is an extremely expensive thing to do, but major technological efforts are underway to develop new ground-based telescopes, operating at visible and radio wavelengths, that can achieve the same degree of precision focus—the same crispness—as is currently possible only from space.

- There are two methods currently being developed that show great promise: adaptive optics and interferometry. With adaptive optics, the telescope's primary mirror is connected to a large set of precise motorized controls that can push and pull on pressure points beneath the mirror's surface, so as to intentionally deform the mirror.

- With interferometry, multiple telescopes are used in unison so as to mimic the performance of a single monolithic telescope as big as the smaller telescopes are apart. An interferometer doesn't have the same light-gathering power as a telescope, so it's not as sensitive to light from fainter stars, but in cases where angular resolution is more important than sensitivity, interferometers are the way to go. And no other approach—not even a telescope in space—can beat the fantastic fineness of detail that can be achieved.

Suggested Reading

Giant Magellan Telescope Observatory, "Giant Magellan Telescope (GMT)."

Large Synoptic Survey Telescope, "Large Synoptic Survey Telescope (LSST)."

Lowell Observatory, "Navy Precision Optical Interferometer."

Questions to Consider

1. In what ways do the basic tools of the astronomer—telescopes and cameras and spectrographs—compare to or differ from those used by scientists in other disciplines?

2. How have technical advances in astronomical instrumentation enhanced our everyday lives?

Telescopes—Our Eyes on the Stars
Lecture 8—Transcript

Everything we know about the stars, everything that we've talked about in our lectures so far, and everything that we'll discuss in our remaining lectures—everything—has been learned through the information carried by the light that the stars emit. But for us to make use of that light, to be able to actually learn something about the stars from it, it must also be detected and recorded. If light is the messenger, our telescopes and instruments are the receivers of those messages.

In this lecture, we'll take a little bit of an intermission from talking about the stars themselves to talk about the instruments astronomers use to measure their light. Even as a professional astronomer, I continue to be amazed by the fantastic technological advances that enable us to measure and record starlight. As we'll see, merely to detect light is not so difficult, but to record it with sensitivity and accuracy, that, is in fact, a remarkable feat.

Think for a moment about what detecting light from a distant star really signifies. When we see light from a distant star, we are reading a message written long ago. Consider the bright red star Aldebaran, the fiery red eye of the constellation Taurus. Aldebaran is 60 light-years away. That means that if you are 60 years old, then the light you see from Aldebaran tonight actually left the star around the time of your birth. It traveled silently, undisturbed, and undeterred across the cold, quiet vacuum of space, only to be at long last absorbed by your eye. That's a little bit of star energy, created deep within the star over ten thousand years ago, emitting from its glowing surface 60 years ago, and finally deposited into your retina. In other words, seeing the stars is not a passive thing. When we stare at the stars on a clear night, a real physical transaction occurs, a literal and direct transfer of energy from the stars to our eyes.

Indeed, our eyes are a good place to start talking about what telescopes are and what they do. Our eyes, in their form and function, are in fact, small telescopes with built-in light detectors. Here's what I mean. Light that reaches your eye is brought to a focus by the lens; that's the telescope part. And the focal point is the retina, which has light-sensitive chemicals to absorb the light; that's the detector part. In one way or another, that's what every astronomical telescope does; it uses an optical element to focus the incoming light, and a light-sensitive detector to absorb and record the light.

The ability of the human brain to store visual information in the form of memory is, of course, absolutely wonderful. But as a means for making accurate measurement, the human visual system leaves much to be desired. First of all, most of the information that we see is recorded only temporarily and then lost forever. Second, even if it could be recorded forever, there are limitations to what our eyes can do as telescopes. Our eyes are only sensitive to a small portion of the entire electromagnetic spectrum, what we call the visible part.

In addition, and this is important, the response of our retinas to the amount of light received is what we call non-linear. That means that there is not a simple linear relationship between the apparent relative brightness of different objects and their true relative brightness. In fact, our eyes' response to light is logarithmic; a star that appears to our eyes to be five times as bright as another is, in fact, 100 times brighter. This is a helpful property for our needs in everyday life, because it means that we can handle looking at things that have a very large range of relative brightness. But for accurately measuring quantities of light, a linear response is much more desirable.

The digital detectors attached to most telescopes have this property; they respond in a linear way to light. And so a star that is twice as bright as another is faithfully recorded as being twice as bright. In addition, our eyes are relatively small, in particular, the lenses of our eyes are small, and so are only sensitive to the brightest stars. To see fainter stars requires a larger light-gathering device, in other words, a larger telescope.

So with our eye as an analogy, let's now look at some examples of telescopes that astronomers use to observe stars in visible light. There are two basic types of telescopes, differing only in the nature of the optical element that focuses the incoming light. The first type is called a refracting telescope, because it uses lenses to refract, or bend, the incoming light, and thereby bring the light to a focus. This is the type of telescope that most folks associate with astronomy, and indeed, for much of history, this is the type of telescope that astronomers used. One of the largest refracting telescopes ever built is the Yerkes Telescope, in southern Wisconsin. Its primary light-gathering optical element is a lens 40 inches in diameter.

The long tube of the telescope doesn't serve a function, per se, other than to hold that massive lens up to the incoming light and at a sufficient height so that the focal point can be positioned within reach of the detector. You see, in general, the bigger the lens, the farther away from the lens the focal point is. So with a lens this large, the focal point would be beneath the observatory's floor if the lens weren't positioned high up on the long tube. At the bottom of the tube, an instrument can be attached to detect and record the light focused by the lens. That could be an astronomer's eye or a plate of photographic film, as was used decades ago, or now, a digital detector like what you have in your digital camera. We'll talk about those digital detectors and how they work in a moment.

But first, let's talk about the second main type of telescope, what we call a reflecting telescope. A reflecting telescope is so named because, rather than using a lens to bend the light to a focus, it uses mirrors to reflect the light to a focus. These days, all major astronomical telescopes are of the reflecting variety. Here I am at the Gemini Observatory in the Chilean Andes. Comparing me in the foreground, as well as the parked vehicles in the background, you get a sense of just how large these observatory structures are, quite a sight, out and up in the middle of nowhere! You can also see me in this image posing across from the eight-meter-diameter primary mirror of the telescope, and you can even see my reflection in the mirror.

A reflecting telescope can be built with a much larger mirror than it is possible to fabricate a lens. You see, a lens is a solid piece of glass, and so a larger lens is necessarily heavier, much heavier. So heavy, in fact, that a lens any bigger than the 40-inch Yerkes lens would deform itself under its own weight, and that deformation would cause a loss of focus; the images would not be as crisp and the measurement of the light less accurate.

In contrast, a mirror can be made large but thin. Think about your bathroom mirror at home. It can be made as large as you like and still the glass can be quite thin. In an astronomical reflecting telescope, the mirror can be made out of an even lighter substrate than glass. That substrate can be shaped into just the right bowl shape, like the curved surface of a lens, only in reverse, to be able to focus the incoming light. Then, an ultra-thin coating of a highly reflective material—usually aluminum or silver—is applied to the surface. That coating is typically extremely thin, perhaps just a micrometer in thickness. For comparison purposes, by the way, a human hair is about 100 micrometers thick.

Another advantage of a reflecting telescope is that, in contrast to the long tube of a refracting telescope, a reflecting telescope can be built relatively compact. That's because the long focus can be accommodated with multiple reflections off of a series of mirrors, in essence, folding the light beam. A common design in astronomy is one in which a smaller mirror, called the secondary mirror, is positioned directly above the main mirror, called the primary mirror.

The primary mirror sits at the bottom of the telescope; the telescope is really just a big truss. The incoming light rains down into the telescope, like a bucket, and the primary mirror reflects that light back up toward the sky. Up at the top of the telescope, the light is then reflected off of a smaller secondary mirror, which, in turn, reflects the light down through a hole in the center of the primary mirror. Down beneath the primary mirror, an instrument to detect and record the light can be installed. Having the instrument down there is advantageous also because it can be stationary on the floor, instead of having to ride around with the telescope as it moves.

Another innovation with reflecting telescopes is the ability to form extremely large mirrors in segments. The segments, often hexagonal in shape, can each be relatively small, and therefore, can be machined very precisely. In addition, if anything happens to one small segment, it can be more easily be repaired or replaced. Finally, and most importantly, these segments can be individually adjusted so that the overall mirror's shape maintains exactly the right shape to correctly focus the incoming light, even as the telescope bends or expands due to shifting in its position or changes in the ambient temperature during the night.

Now that we've seen examples of the two major types of telescopes—reflecting and refracting—let's talk about the two most important properties of a telescope. They are (1) light-gathering power, and (2) angular resolution. Light-gathering power refers simply to the ability of a telescope to collect light. The more the better, because the more light that a telescope can capture from a star, the fainter the star the telescope can see and the more sensitively it can measure the properties of that light. The single, most important property of a telescope that determines its light gathering power is the area of the primary optical element. In other words, the bigger the lens or mirror, the better.

To understand light-gathering power, a useful analogy is that of a bucket collecting rain water. If you want to collect as much rain water as possible, you would want to use a bucket with the largest possible diameter. That's because the rain drops are spread out uniformly over a certain area, and so the more area that you cover with the bucket, the more drops you'll catch. You could use a bucket with a smaller diameter, but then you'd have to let it sit for a longer period of time to catch the same number of raindrops.

With a telescope, we're catching photons—essentially, raindrops of light. And indeed, a telescope is often referred to colloquially by astronomers as a light bucket. The bigger the primary mirror of the telescope, the bigger the light bucket, and the more raindrops of light that can be gathered quickly. And because the area of the mirror is the square of its diameter, bigger really means better. Make the mirror twice the diameter, and you have four times the area, four times the light-gathering power, and you can see a star that is four times fainter, or measure the same star four times more sensitively.

The second important property of any telescope is its angular resolution. Angular resolution refers to the crispness with which a telescope can focus the incoming light. The better the angular resolution, the crisper the focus that can be achieved, and this allows ever finer details in the stellar images to be discerned. As an analogy, consider a faraway car approaching on a lonely stretch of road. It's dark out, and you see the headlights. Or is it a motorcycle with a single headlight? It's hard to know, because all you see at first is a single light. Only as the car gets closer, do the two headlights separate to become discernible as two separate lights. That is the limited angular resolution of your eye; the human eye has an angular resolution of about one sixtieth of a degree. When the car is far off in the distance, the two headlights subtend an angle smaller than that, and so appear to the eye as a single blended point of light. It's the same with a telescope. The size of the telescope's primary mirror determines its angular resolution. A small telescope has poor angular resolution, and so a star appears fuzzier. There is a formula that describes the angular resolution of a telescope, and it goes as follows: $\theta = \lambda/D$.

This formula says that the best focus a telescope can achieve, the smallest angle a telescope can discern, the finest detail that the telescope can make out—call that angle θ—is just the wavelength of light divided by the diameter of the telescope's primary mirror. So for a given wavelength of light (λ), this formula says that a larger diameter mirror will achieve a smaller angle of focus. That is, a finer focus, and finer details can be discerned.

So you see why it's so imperative to build ever bigger telescopes. Both of the important properties of a telescope, its light-gathering power and its angular resolution, depend directly on the size of the telescope's primary mirror. As they say, when it comes to telescopes, bigger is better. Notice, by the way, that magnification is not a property of a telescope that astronomers care about. Magnification just refers to how much a telescope zooms in on an image. But zooming in on a blurry image won't make it any crisper; that's why astronomers care first and foremost about the telescope's light-gathering power and its angular resolution. Only if the sensitivity is good and if the fineness of detail is excellent does it make sense to zoom in on that detail.

So far, we've talked about telescopes, the light buckets that gather the light from the stars. Now let's talk a bit about the detectors that we attach to the backs of those telescopes to detect and record the light. The basic component of any modern astronomical detector, the retina of the telescope, if you will, is a digital detector similar to what you find in your digital camera or phone. That detector is essentially just a wafer of silicon, with some electronics attached. That chip of silicon is subdivided into tiny squares, each one about 20 micrometers on a side, called a pixel. Each pixel is a tiny photo receptor that responds to incoming light by what is known as the photoelectric effect. The photoelectric effect is the discovery for which Albert Einstein was awarded the 1921 Nobel Prize. The photoelectric effect states that when a material absorbs the energy of a photon of light, that energy causes an electron within the material to be freed, and that electron then joins other electrons as part of an electrical current that can be measured.

Each absorbed photon therefore liberates one electron, and so by measuring the strength of the electric current produced in the material, we can measure accurately and linearly the number of photons that struck the detector. With many pixels in the detector linked together, each one acting as a photon counter, we can build up an image, each pixel measuring a tiny part of the overall image.

Now, what about color? As we've seen in previous lectures, measuring the color of a star is one of the basic ways that we can determine its physical characteristics. Well, the photoelectric effect by which the digital detector measures incoming photons doesn't care about the wavelength of the incoming photons. The pixels simply count photons whatever their wavelength might be. In essence, the digital detector is fundamentally a black-and-white detector. It only tells us whether light was detected or not, and how much, but not its color.

To get color information, astronomers, therefore, use filters. A given filter lets through light of only a particular wavelength, say red, or blue, or green. By combining images taken through discrete filters, we are then able to create images that convey the true colors of the stars, and we can use the color measurements to determine things like the temperatures of the stars, as we've talked about before. For example, this image of the IC1318 star-forming region was created by combining separate images taken through red, green, and blue filters, so that when combined, we get an image revealing which parts of the nebula emit most strongly in the different colors, and we also can see the true colors of the stars.

As we've discussed, much of the true power of light to convey physical information about the stars comes not from mere pictures, but from the spectrum of light. Spreading the light we receive into its constituent rainbow and then dissecting those wavelengths is ultimately how we learn the star's temperature, its chemical makeup, its motion, and many other properties. As one of my former professors used to be fond of saying, "A picture may be worth a thousand words, but a spectrum, that's worth a thousand pictures!"

To spread the light gathered by a telescope into a spectrum, the light has to be focused onto a light-dispersing element. That dispersing element is, in general, one of two different types of optical devices, a prism or a grating. Let's look at how a prism works as an example. A prism is usually a large piece of glass that refracts, or bends, the incoming light. The amount by which the incoming light is bent depends on the wavelength of the light, such that blue light is bent more than red light.

Consequently, white light coming in to the prism gets spread into its constituent rainbow, because where all of the wavelengths included in the incoming light were initially all focused on top of one another, the prism bends the blue wavelengths in one direction, the green into another, and the red into yet another. Then, by placing a digital detector, as we discussed before, at the position of the outgoing rainbow, we can record a picture of that rainbow, and then that digital rendering of the spectrum can be dissected and analyzed in detail. A rainbow that you see in the sky is formed in precisely this way. Raindrops act as tiny prisms, spreading the sunlight into its constituent spectrum of colors, which are then focused and detected by the little telescopes and detectors that are our eyes and retinas.

So far, we've talked about telescopes and instruments that operate at visible light wavelengths. One reason for that is obviously that, as humans, we have a predilection for those particular wavelengths of light, since those are the wavelengths we can see with our own eyes. But another reason is that visible light is one of the only types of light that can penetrate through the Earth's atmosphere and reach telescopes on the ground.

The Earth's atmosphere acts like a screen that absorbs almost all wavelengths of light. From the ground, the only wavelengths of light that can be seen coming from the outside are visible light and radio light. A small portion of the infrared part of the spectrum, called the near-infrared, can be seen from high mountaintops or from instruments flown at very high altitudes. At these heights, the atmosphere above is thinner, and so enough of the incoming light is able to penetrate and be detected. The SOFIA telescope is an infrared telescope that operates from an airplane flying at high altitude. Even for visible light, placing our telescopes on high places is an advantage. With less atmosphere above the telescope to distort and blur the incoming light, the images obtained are crisper. And less moisture in the air above the telescope means less of the incoming light is absorbed by water molecules. So when it comes to telescopes, being high and dry is a good thing.

Radio telescopes operate very effectively from the ground. But remember that the ability of a telescope to resolve light into a fine focus depends on the ratio of the wavelength divided by the mirror's diameter. You may recall from an earlier lecture that radio light typically has a wavelength of 10 centimeters or longer. Given such long wavelengths, the mirrors on radio telescopes require an enormous diameter to achieve fine focus. As an example, the largest radio telescope in the world, indeed, the largest telescope in the world of any kind, is the Arecibo telescope in Puerto Rico, 1000 feet in diameter. You may remember seeing this telescope featured in the movie *Contact*.

Beyond visible and radio light, it's necessary to put telescopes above the filtering of the Earth's atmosphere. All of our ability to study the stars at ultraviolet, X-ray, and gamma-ray wavelengths, and most of our ability to study the stars in infrared light, comes from space-borne telescopes. These include the Hubble space telescope's ultraviolet detectors, the Chandra telescope's X-ray detectors, the Fermi telescope's gamma-ray detectors, and the Spitzer telescope's infrared detectors.

Because these telescopes gather starlight from above the Earth's atmosphere, they additionally are able to obtain images with a crispness that is normally not possible from the ground. Turbulent motions in the air produce distortions of the incoming light that result in the shimmering and twinkling of the stars that we see. That twinkling is pretty, and makes for nice children's songs, but it's terrible for making the most accurate measurements.

Of course, putting telescopes in space is an extremely expensive thing to do. Where building the largest telescopes on the ground may cost 250 million dollars, building and then putting a telescope, like the Hubble, into orbit costs about ten times as much, more like a few billion dollars. Again, if we want to measure light from the stars at wavelengths other than visible and radio, we simply have no choice. But major technological efforts are underway to develop new ground-based telescopes operating at visible and radio wavelengths, of course, that can achieve the same degree of precision focus, the same crispness, as is currently possible only from space.

There are two methods currently being developed that show great promise, and over the next few years you are surely going to see these innovations coming online with new telescopes making the latest ground-breaking discoveries about the lives and deaths of stars. The two methods being developed are called adaptive optics and interferometry. With adaptive optics, the telescope's primary mirror is connected to a large set of precise motorized controls that can push and pull on pressure points beneath the mirror's surface, so as to intentionally deform the mirror.

You heard me right; the idea is to bend the mirror out of shape, on purpose. The idea is that the Earth's atmosphere has distorted the light waves that reach the telescope, and so by warping the mirror just so, that atmospheric distortion can be undone. The mirror has to be warped and re-warped very rapidly, many times every second, in order to keep up with the rapid, tiny fluctuations in the atmosphere's distortion pattern. It's the same idea conceptually as in noise-canceling headphones, which generate sound of just the right ever-changing pitch in order make the unwanted sounds around you disappear.

There are different approaches being tried for adaptive optics telescopes, but one of the most promising utilizes what is called a laser guide star. In such a system, a laser pulse is sent up into the air in the direction that the telescope is pointing. Some of that laser light is bounced back to the telescope by reflective particles in the air. Any distortions in the laser beam then tell the adaptive optics system precisely how the mirror needs to be deformed at that instant.

With interferometry, multiple telescopes are used in unison so as to mimic the performance of a single monolithic telescope as big as the smaller telescopes are apart. For example, with the new Atacama Large Millimeter Array, or ALMA, telescope, the individual small telescopes can be spaced as far apart as one kilometer. The light received from the individual telescopes is then combined in just the right way so that the angular resolution of the combined light is as good as if a single telescope dish one kilometer in diameter had been used.

Of course, there is no free lunch; such an interferometer does not have the same light-gathering power as a one-kilometer-diameter telescope. In other words, it's not as sensitive to light from fainter stars. But in cases where angular resolution is more important than sensitivity, interferometers are the way to go. And no other approach, not even a telescope in space, can beat the fantastic fineness of detail that can be achieved with an interferometer.

In this lecture, we've learned about the basic telescopic tools that astronomers use to study the stars. Light is the stars' information messenger, and the telescope is the receiver of those messages. It is the telescope that serves as our eyes on the sky, the shore onto which those distant cosmic signals making their lonely journey across the vastness of space and time wash up, asking to be seen and understood.

Mass—The DNA of Stars

Lecture 9

In this lecture, you will learn how astronomers use the information encoded in the light from binary star systems to measure the properties of the stars, to weigh the stars, and to thereby establish that mass is the DNA of the stars—determining all of its other physical properties. In order to fully appreciate how astronomers arrived at this understanding of mass as the stellar DNA, you will embark on a train of logic building on several topics that were presented in previous lectures.

Stellar Mass

- The most important property of a star is its mass, which determines all of its other characteristics. Indeed, a star's mass determines everything about its life—how long it will live, for example, and even the manner in which it will die. In a sense, mass is a star's DNA, specifying all of its physical characteristics.

- But while our own DNA governs certain things about us physically and otherwise predisposes us to certain personality characteristics and behaviors, our ultimate life paths are also influenced by many factors that have nothing to do with our DNA. But for a star, mass is totally deterministic. We would not know this fact if not for the existence of binary stars—systems in which two stars orbit one another.

- The only way to weigh anything is to measure its gravitational effect on something else, or vice versa. For example, we know the Sun's mass because we can measure its gravitational influence over the planets that orbit it, such as the Earth. We can measure how far the Earth is from the Sun in its orbit, and we can measure how long the Earth takes to orbit the Sun. These are fundamentally measures of the Sun's gravitational influence and, therefore, are measures of the Sun's mass.

- When we want to study stars other than the Sun, we need a tracer that we can easily detect and measure in order to determine the strength of that star's gravitational influence, or mass. The most convenient thing is another star, bright enough to see, itself massive enough to exert its own gravitational influence that we can measure. By studying binary stars, we not only have the requirement of an orbiting body to reveal the strength of the other star's gravity, but we are also able to measure the characteristics of two stars at once.

- There are three main types of binary star systems: visual binaries, spectroscopic binaries, and eclipsing binaries. Visual binaries are pairs of stars where we can actually see the two stars separately. Because of the great distances that most stars are from us, the fact that we can see the two stars as separate points of light means that in reality, they must be separated by a very large physical distance.

- Consequently, the orbital periods of most visual binary star systems are extremely long, too long in most cases for us to be able to see the orbital motion. However, in some cases, the stars are close enough to us that we can actually see the orbital motion of the two stars directly.

- Although visual binaries for which we can directly observe the orbit are relatively rare, they are wonderful systems to study because of the relative simplicity of making the necessary measurements. Just a series of photographs over time allows us to measure the two key properties of the orbit that we'll need: the time it takes for the stars to orbit, what we call the orbital period; and the size of the orbit, or the physical separation of the two stars.

- The second type of binary star system is a spectroscopic binary. These get their name from the fact that, visually, they appear as a single star, a single point of light. However, when we examine their spectrum, the fact that it is actually two stars becomes revealed.

- An advantage of spectroscopic binary stars is that, because they are so close together that in images they appear as a single star, we learn that they orbit one another quickly. Most spectroscopic binary star systems have orbital periods of less than a year, and many have periods of just days. So, in a short period of time, we can collect spectra at different points in the stars' orbit and measure from the spectra the information that we need to put into Newton's law to determine the stars' masses.

- Finally, there is a third type of binary system, which is arguably the most valuable of all: eclipsing binary stars. These are usually spectroscopic binary stars also, but they have a rare but special property. From our perspective on Earth, their orbit happens to be oriented such that the two stars periodically pass in front of one another, temporarily blocking—or eclipsing—the light of the other star. When this happens, the total light that we receive from the two stars together becomes temporarily diminished, and the pattern repeats.

- Eclipsing binary stars are extremely valuable and important. That's because not only can we analyze their spectra in the same way as for other spectroscopic binary stars, but we also have the eclipses, which tell us an additional crucial piece of information: the diameters of the stars.

- With knowledge of how fast the stars are orbiting one another from the Doppler measurements of the spectra, together with a measurement of how long the eclipses last, we can calculate physically how large the stars must be. This is one of the only ways that we have of directly measuring the sizes of stars.

The Hertzsprung–Russell Diagram
- Almost 100 years ago, two astronomers, Ejnar Hertzsprung and Henry Russell, used a set of hundreds of stars for which some measurements had recently become available. Importantly, this included stars that were nearby enough to have all of the basic

properties measured by which their masses and diameters could be determined.

- First, their distances could be measured via parallax. Second, their temperatures could be measured from their colors and spectra. Third, their luminosities could be measured from their apparent brightnesses together with the parallax distances. Finally, the stars' masses could be determined from binary-star measurements and diameters determined from the eclipsing binary stars in the sample.

- Hertzsprung and Russell made a diagram from these measurements that we still use today as one of the most fundamental astrophysical tools for understanding the properties of stars. It's called the Hertzsprung–Russell diagram, or H–R diagram, which is simply a graph of luminosity on the vertical axis and temperature on the horizontal axis.

- For historical reasons, the diagram is always represented with luminosity increasing on the vertical axis but with temperature decreasing on the horizontal axis. Stars farther up are more luminous, but stars farther to the right are cooler. Also, for reference, the Sun has a temperature of about 6000 degrees Kelvin and a luminosity of 1 solar luminosity.

- Temperature and luminosity are two of the fundamental properties of stars that we can directly measure for most stars. Luminosity we can measure from a star's apparent brightness and its distance. Temperature we can measure from a star's color, using Wien's law.

- There is also a second way of measuring a star's temperature, using details of its spectrum. A star's spectral type can be denoted by one of the following letters: O B A F G K M. Each letter represents stars whose spectra have a certain appearance and, therefore, a certain range of temperatures.

- In that order, O B A F G K M, the number of chemical absorption fingerprints increases. And in that order, there is a temperature

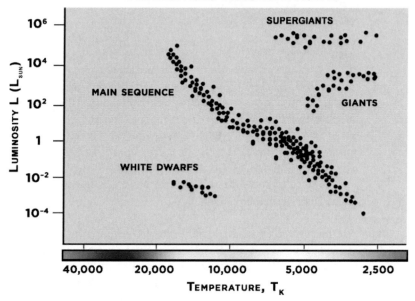

HERTZSPRUNG-RUSSELL DIAGRAM

sequence, with the O stars being the hottest and the M stars the coolest. On the H–R diagram, we can represent the stars' temperatures on the horizontal axis either as physical temperature in Kelvins or in terms of the stars' spectral types.

- Graphing the stars that Hertzsprung and Russell knew about, essentially a random assortment of stars having nothing in common other than being relatively close to us, we find that the stars are not scattered randomly in the graph. Instead, the stars form a narrow diagonal band across the diagram. About 90% of all stars reside on this diagonal swath, which is now referred to as the stellar main sequence.

- At the most basic level, the main sequence tells us that stars are made in such a way that they have only certain particular combinations of temperatures and luminosities. The hottest, bluest stars—the O-type stars—are extremely luminous, and the coolest, reddest stars—the M-type stars—are extremely dim.

- Once we know the main sequence, we can infer the luminosity of any star just by measuring its temperature. Temperature is a relatively straightforward property to determine, because we need only measure a star's color or its spectrum. Luminosity is more difficult. By "luminosity" we mean the intrinsic energy output of the star, not simply how bright it appears to us in the sky. A star can appear bright because it's simply nearby, not necessarily because it is intrinsically luminous. So, to determine a star's true luminosity, we also need to know the star's distance, such as measured using the parallax triangulation method.

- But the real power of the main sequence is what it reveals about the underlying stellar DNA. When we add the information that we have regarding the stars' masses, as measured from the binary stars, we find a clear and unmistakable correlation: The hottest, most luminous stars are also the most massive, while the coolest, least luminous stars are also the least massive.

- In other words, the main sequence of the H–R diagram reveals that what fundamentally drives the physical properties of the stars is one thing: mass. And it's a prescriptive relationship. A star born with a mass equal to that of our Sun always has the temperature of our Sun and its same luminosity.

- A few percent of stars appear at the upper right of the diagram, and a few percent appear at the lower left. A star at the upper right in the H–R diagram has a very cool temperature but an extremely large luminosity. We can infer that those stars are physically very large, and we call them giants and supergiants. These are stars at an elderly stage of life, approaching their deaths.

- At the lower left, the opposite reasoning applies. These stars are hotter than the Sun yet extremely dim. That's because they have tiny sizes. We call these white dwarfs. These are not stars at all, or at least not living stars. They are the corpses of stars like our Sun that have ended their lives and are now just fading embers made of nearly pure carbon—diamonds in the sky.

- About 90% of stars are on the main sequence, and the rest are giants, supergiants, and white dwarfs. This tells us that stars spend the vast majority of their lifetimes—about 90%—as members of the main sequence, and then a small portion of their lives, the end of their lives, as giants and white dwarfs.

- In other words, stars spend most of their lives in a happy, stable configuration with a particular temperature and luminosity, and then end their lives relatively fast. When the end comes for a star, it evidently comes quickly. And all of this is dictated by the star's mass.

- When we perform a census of the stars by their masses, we find that there are relatively few very massive stars, more mid-weight stars, and many runts. Mass is a star's DNA, and where the star gets that DNA is the stellar birth process, which establishes the star's birth weight. The stellar birth process does not birth stars of all masses equally. Rather, the process of gravitational collapse and the subsequent feeding of the baby stars in their cocoons of gas and dust prefers by an enormously large margin to make lots of little stars and only rarely makes a massive star.

Suggested Reading

Kallrath and Milone, *Eclipsing Binary Stars*.

University of Nebraska at Lincoln, "Eclipsing Binaries Simulator."

1. In what ways are binary star systems essential tools for our general understanding of the physical properties of all stars?

2. If the pattern of stellar masses involves a power law, whereby many more lightweight stars are formed for every massive star, do we expect that there will be an even larger number of brown dwarfs, stillborn stars with masses less than the minimum mass required for a star? If not, how does nature know during the stellar birth process to principally make only objects that will eventually light up as full-fledged stars?

Mass—The DNA of Stars

Lecture 9—Transcript

Already in this course, we've learned about the amazing way in which stars begin their lives, the stellar birth process. As we've seen, most, if not all stars go through essentially the same process. The birthing process starts with a collapse under gravity within a giant cloud of gas and dust. Then the embryonic stars feed from an enveloping cocoon of gas and dust, until eventually they ignite as a full-fledged star. But we've also seen that stars during their lives differ from one another in their basic properties—different temperatures, and colors, and luminosities. What determines what a star's characteristics will be during its lifetime?

The answer, in one word, is mass. Indeed, the most important property of a star is its mass, which determines all of its other characteristics. Indeed, a star's mass determines everything about its life, how long it will live, for example, and even the manner in which it will die. In a sense, mass is a star's DNA, specifying all of its physical characteristics.

But while our own DNA governs certain things about us physically and otherwise predisposes us to certain personality characteristics and behaviors, our ultimate life paths are also influenced by many factors that have nothing to do with our DNA, but it's different for a star. For a star, mass is totally deterministic; mass is not only destiny, it is fate.

This is obviously a crucially important piece of information in our understanding of the stars. But how did we come to know this vital fact? The truth is, we would not know this fact if not for the existence of binary stars, which, as you'll recall, are systems in which two stars orbit one another.

In this lecture, we'll see how astronomers use the information encoded in the light from binary star systems to measure the properties of the stars, to weigh the stars, and to therefore establish that mass is the stars' DNA. To fully appreciate how we've arrived at our understanding of mass as the stellar DNA, we need to go through a series of steps, a train of logic building on several of the topics that we've discussed in previous lectures. So let's jump in.

Why are binary stars so important for learning about stellar masses? Well, it comes down to the fact that, fundamentally, the only way to weigh anything is to measure its gravitational effect on something else, or vice versa. The same is true when we weigh ourselves. When you weigh yourself on your bathroom scale, what that scale is actually measuring is how hard the Earth is pulling down on you, which in turn reveals how much you weigh, or rather, how much mass makes up your body.

When we talk about the Sun and its basic properties, one of the things we invariably include is how much it weighs. We know the Sun's mass because we can measure its gravitational influence over the planets that orbit it, such as the Earth. We can measure how far the Earth is from the Sun in its orbit, and we can measure how long the Earth takes to orbit the Sun; these are fundamentally measures of the Sun's gravitational influence, and therefore, are measures of the Sun's heft, its mass.

When we want to study stars other than the Sun, we need a tracer that we can easily detect and measure in order to determine the strength of that star's gravitational influence, its mass. And the most convenient thing is another star, bright enough to see, itself massive enough to exert its own gravitational influence that we can measure. So by studying binary stars, we not only have the requirement of an orbiting body to reveal the strength of the other star's gravity, we are able to measure the characteristics of two stars at once.

In a previous lecture, we introduced binary star systems in the context of the dynamical nature of the stellar birth process. We learned that many, if not most, stars are born with one or more sibling stars and that these siblings have complex gravitational interactions with each other. Let's reintroduce binary stars now, this time, with an eye toward how they are used as tools for weighing stars. There are three main types of binary star systems, which we call visual binaries, spectroscopic binaries, and eclipsing binaries.

Visual binaries are, as the name suggests, pairs of stars where we can actually see the two stars separately. Now, because of the great distances that most stars are away from us, the fact that we can see the two stars as separate points of light means that in reality they must be separated by a very large physical distance. And consequently, the orbital periods of most visual binary star systems are extremely long, too long in most cases for us to be able to see the orbital motion.

However, in some cases the stars are close enough to us that we can actually see the orbital motion of the two stars directly. For instance, one of the brightest stars in the sky is Sirius, and it is, in fact, a visual binary. Sirius is close enough to us that the physical separation of the two stars corresponds to an orbital period of about 50 years. And so images of the relative positions of the two stars taken over the past hundred years or so clearly show them orbiting. For example, in this sequence of images of the two stars that make up Sirius, we can see the stars shifting their positions in a clockwise pattern from one image to the next.

Although visual binaries like Sirius, for which we can directly observe the orbit, are relatively rare, they are wonderful systems to study, because of the relative simplicity of making the necessary measurements. Just a series of photographs over time allows us to measure the two key properties of the orbit that we'll need: (1) the time it takes for the stars to orbit, what we call the orbital period, and (2) the size of the orbit, or the physical separation of the two stars.

The way this works is through application of Newton's law of gravity. We arrive at a mathematical formula that relates three quantities: the binary stars' orbital period, P; the size of the orbit, a; and the total masses of the two stars. There's the total mass of the stars that we're after, as well as Newton's gravitational constant. And so all we need to calculate the mass is to know P and a, which for a visual binary, we can see directly.

Now, there is one subtle but important trick here. That a in the formula has to be the actual physical size of the orbit in real physical units of length, such as kilometers. But, when we look at a visual binary on the sky, we don't measure the size of the orbit directly in kilometers. We measure it directly as an angle on the sky, say one one-thousandth of a degree. To convert that angular size of the orbit into a physical size of the orbit, we need to know one more thing, which is the distance from us to the binary. But as we've discussed in a previous lecture, the distance is something we can measure through the technique of parallax, by which we triangulate on the distance to a star by observing it from opposite sides of the Earth's orbit about the Sun. So now we have all of the ingredients to measure the masses of the stars in a binary system.

But do we have to rely on these visual binaries whose orbits take decades? Thankfully, no, we have some other options. The second type of binary star system is a spectroscopic binary. These get their name from the fact that, visually, they appear as a single star, a single point of light. However, when we examine their spectrum, the fact that it is actually two stars becomes revealed.

To understand how this works, recall what we discussed in a previous lecture about the Doppler effect. When a star moves toward us, its light spectrum is shifted to shorter, or bluer, wavelengths; when a star moves away from us, its light spectrum is shifted to longer, or redder, wavelengths. When two stars orbit one another as a binary system, at any given moment one star will be moving toward us, while the other, on the opposite side of the orbit, will be moving away from us. Therefore, at every time in the orbit, one of the stars' spectra will be blueshifted and the other redshifted relative to the other.

An advantage of spectroscopic binary stars is that, because they are so close together that in images they appear as a single star, this means they orbit one another quickly. Most spectroscopic binary star systems have orbital periods of less than one year, and many have periods of just days. So in a short period of time, we can collect spectra at different points in the stars' orbit and measure from the spectra the information that we need to put into Newton's law to determine the stars' masses.

Finally, there's a third type of binary system, and arguably, the most valuable of all—eclipsing binary stars. These are usually spectroscopic binary stars also, but they have a rare but special property. From our perspective on earth, their orbit happens to be oriented such that the two stars periodically pass in front of one another, temporarily blocking—or eclipsing—the light of the other star. When this happens, the total light that we receive from the two stars together becomes temporarily diminished, and the pattern repeats.

Eclipsing binary stars are extremely valuable and important. That's because not only can we analyze their spectra in the same way as for other spectroscopic binary stars, we also have the eclipses, which tell us an additional crucial piece of information: the diameters of the stars. That's because with knowledge of how fast the stars are orbiting one another from the Doppler measurements of the spectra, together with a measurement of how long the eclipses last, we can calculate physically how large the stars must be. This is one of the only ways that we have of directly measuring the sizes of stars, and so this is a very important technique.

As an aside, but since we're on the subject, there is another important method for measuring the sizes of stars directly that is becoming increasingly technically possible. Recall in a previous lecture we discussed the technique of interferometry, in which multiple telescopes are spaced far apart and used in concert to achieve a fineness of detail comparable to a single enormous telescope. Well, this technique has now been used to directly measure the diameters of a few dozen stars, and in the coming years, we expect this technique to be developed even further. This is an important development, because it allows us to measure these fundamental stellar characteristics for single stars, so that we don't have to rely only on eclipsing binary stars to obtain this important information.

Even with these exciting technological advances coming up, binary stars remain the gold standard for our understanding of these basic stellar characteristics. So now that we've reviewed the different types of binary stars and have reviewed how we go about making measurements of their basic characteristics, including their masses, let's look at what insight emerges from combining these measurements.

Almost 100 years ago, two astronomers, Ejnar Hertzsprung and Henry Russell, used a set of hundreds of stars for which the types of measurements we've discussed had recently become available. Importantly, this included stars that were nearby enough to have all of the basic properties measured by which their masses and diameters could be determined. First, their distances could be measured via parallax. Second, their temperatures could be measured from their colors and spectra. Third, their luminosities could be measured from their apparent brightnesses, together with the parallax distances. And finally, the stars' masses could be determined from the binary star measurements we've just discussed and diameters determined from the eclipsing binary stars in the sample.

What Hertzsprung and Russell did next was to make a diagram from these measurements that we still use today as one of our most fundamental astrophysical tools for understanding the properties of stars. It's called the Hertzsprung-Russell diagram, or H–R diagram for short. We'll be using this diagram quite a bit in upcoming lectures to interpret how stars evolve during their lives. So let's get to know the H–R diagram.

The H–R diagram is simply a graph of luminosity on the vertical axis and temperature on the horizontal axis. Note that for historical reasons, the H–R diagram is always represented as shown here with luminosity increasing on the vertical axis but with Temperature decreasing on the horizontal axis. So keep that in mind as we use this diagram; stars farther up are more luminous, but stars farther to the right are cooler. Also, for reference, you see the position of the Sun. It has a temperature of about 6000 degrees Kelvin and a luminosity of one solar luminosity. For comparison, Sirius is up and to the left, meaning it is hotter, at about 10,000 degrees Kelvin, and it is more luminous, at about 50 times the Sun's luminosity. As we've discussed, temperature and luminosity are two of the fundamental properties of stars that we can directly measure for most stars. Luminosity we can measure from a star's apparent brightness and its distance. Temperature we can measure from a star's color, using Wien's Law.

There is also a second way of measuring a star's temperature, using details of its spectrum that we haven't discussed yet but need to mention now. As we've discussed, a star's spectrum has imprinted on it the chemical absorption signature of the various elements that make up its atmosphere. But in addition, that chemical absorption signature also conveys the star's temperature. The chemical absorption imprint is always the same in terms of which wavelengths are absorbed, as we talked about before. But depending on the temperature of the star, those absorbed wavelengths will appear more or less dark. In general, the cooler stars have more of these dark absorbed wavelengths, and we can use that information to make a very precise measurement of the star's temperature, complementing the more crude measurement that we can make just from the star's overall color using Wien's Law.

The relationship between the number of dark absorption bands and temperature gives rise to a taxonomy that has traditionally been designated by a series of letters. Each letter represents stars whose spectra have a certain appearance, and therefore, a certain range of temperatures. That sequence is, O B A F G K M. A traditional mnemonic for remembering that order is "Oh, Be A Fine Girl. Kiss Me." That one's obviously a bit sexist, so an alternative is, "Only Bad Astronomers Forget Generally Known Mnemonics." Remember, "mnemonics" is spelled with an M. It can be fun to try to come up with your own.

Anyway, in that order, O B A F G K M, the number of chemical absorption fingerprints increases. And in that order, we have a temperature sequence, with the O stars being the hottest and the M stars the coolest. O-type stars, having temperatures exceeding 30,000 degrees Kelvin, are very blue in color, and in fact, their peak wavelength of emission is in the ultraviolet. Near the middle of the sequence are the G-type stars with temperatures around 5000 to 6000 Kelvin. Our Sun is a G-type star. Finally, at the bottom of the sequence, the M-type stars are also sometimes referred to as red dwarf stars. Their temperatures are less than 3500 Kelvin, and their peak wavelengths are in the infrared, so we see them as having a deep red color.

If you're wondering why astronomers would choose a classification scheme based on letters and not put the sequence in alphabetical order, well, the reason is purely historical. When the spectra of stars were originally being classified in the early twentieth century, it was not yet understood that the appearance of the spectra was related to the stars' temperatures. Only later was this connection made, at which point the original alphabetically ordered sequence was re-ordered to be in temperature order, and also, some of the letters were consolidated with others so that, thankfully, we only have to remember seven.

Coming back to the Hertzsprung-Russell diagram, we can represent the stars' temperatures on the horizontal axis either as physical temperature in Kelvins or in terms of the stars' spectral types that we've just discussed. So, for example, when we locate the Sun along the horizontal axis, we see that it can be represented either by its temperature of about 6000 degrees Kelvin, or by its G spectral type.

And here now is the amazing thing about the H–R diagram. Graphing the stars that Hertzsprung and Russell knew about, essentially a random assortment of stars having nothing in common other than being relatively close to us, we find that the stars are not scattered randomly in the graph. Instead, the stars form a narrow diagonal band across the diagram. Some 90% of all stars reside on this diagonal swath, which is now referred to as the stellar main sequence.

What does it mean, this main sequence? Well, at the most basic level, it tells us that stars are made in such a way that they have only certain particular combinations of temperatures and luminosities. The hottest, bluest stars—the O-type stars—are extremely luminous, and the coolest, reddest stars—the M-type stars—are extremely dim. This by itself is very interesting and useful. It's useful because, once knowing the main sequence, we can infer the luminosity of any star just by measuring its temperature. Temperature is a relatively straightforward property to determine, since we need only measure a star's color or its spectrum. Luminosity is more difficult. Remember that by luminosity we mean the intrinsic energy output of the star, not simply how bright it appears to us in the sky. A star can appear bright because it's simply nearby, not necessarily because it is intrinsically luminous. So to determine a star's true luminosity, we also need to know the star's distance, such as measured using the parallax triangulation method that we've discussed before.

But the real power of the main sequence is what it reveals about the underlying stellar DNA. When we add the information that we have regarding the stars' masses, as measured from the binary stars, we find a clear and unmistakable correlation; the hottest, most luminous stars, are also the most massive, while the coolest, least luminous stars are also the least massive. In other words, the main sequence of the Hertzsprung-Russell diagram reveals that what fundamentally drives the physical properties of the stars is one thing—mass. And it's a prescriptive relationship. A star born with a mass equal to that of our Sun, always has the temperature of our Sun and its same luminosity. A massive star weighing in at ten times the mass of our Sun, like Spica, will always have Spica's same temperature and color and luminosity. It will also have Spica's same diameter. And a star born a runt, weighing just one-tenth of the mass of the Sun, will always share Proxima Centauri's hue, temperature, dimness, and tiny size.

Now, I mentioned that some 90% of all stars in the sky are part of the main sequence. What about the rest? Well, when we represent their temperatures and luminosities and diameters on the H–R diagram, we find that they occupy just a couple of other regions of the graph. A few percent of stars appear at the upper right of the diagram, and a few percent appear at the lower left. What do these represent?

Well let's think about what those combinations of temperatures and luminosities imply. A star at the upper right in the H–R diagram has a very cool temperature but an extremely large luminosity. How can a star be as cool as Proxima Centauri but as luminous as Spica? Just make it really big. Its cool temperature means that each patch of its surface will be dim, but if its total surface area is large enough, it can still radiate an enormous overall luminosity. So we can infer that those stars at the upper right are physically very large, and we call them giants and supergiants. As we'll see later, these are stars at an elderly stage of life, approaching their deaths.

At the lower left, the opposite reasoning applies. These stars are hotter than the Sun, yet extremely dim. That's because they have tiny sizes. We call these white dwarfs. As we'll see later, these are not stars at all, or at least not living stars. They are the corpses of stars like our Sun that have ended their lives and are now just fading embers made of nearly pure carbon, diamonds in the sky.

In a later lecture, we'll focus more on these interesting giants and white dwarfs. But for now, what can we infer about the stellar life cycle from the relative numbers of stars in these different parts of the H–R diagram? Ninety percent of stars are on the main sequence, and the rest are giants, supergiants, and white dwarfs. This tells us that stars spend the vast majority of their lifetimes—about 90%—as members of the main sequence and then a small portion of their lives, the end of their lives to be specific, as giants and white dwarfs. In other words, stars spend most of their lives in a happy, stable configuration with a particular temperature and luminosity and then end their lives relatively fast. When the end comes for a star, it evidently comes quickly, and all of this is dictated by the star's mass.

There is another very important message in the H–R diagram. When we perform a census of the stars by their masses, that most important of properties, we find that stellar masses follow a particular pattern. There are relatively few very massive stars, more midweight stars, and lots and lots of the runts. In fact, a detailed accounting finds that the number of stars of different masses follows a so called power-law relationship, meaning that the number of stars increases as a certain exponential power of mass. Our current best estimates of that exponent put it at around put. That means that for every 100 stars with a mass of our Sun's mass, there will be one star with a mass ten times the Sun, but ten thousand stars with a mass of one-tenth of the Sun.

This exponential law governing the number of stars of different masses is fundamentally telling us that the stellar birth process does not birth stars of all masses equally. Rather, the process of gravitational collapse and the subsequent feeding of the baby stars in their cocoons of gas and dust prefers by an enormously large margin to make lots of little stars and only rarely makes a massive star. Mass is a star's DNA. And where the star gets that DNA is the stellar birth process which, as we've seen in previous lectures, establishes the star's birth weight.

A question for some time was whether the exponentially increasing number of stars with the smallest birth weights might continue to masses even smaller than the minimum mass for a star. For a star to be a star, which is to say, for a star to be able to ignite and perform nuclear reactions to sustain its light and heat output, it needs to have a mass of at least 8% of the mass of our Sun. A forming star that does not reach that minimum mass would essentially be stillborn, what we call a brown dwarf.

A few years ago, my research team had the thrill of discovering the first eclipsing binary star system in which the two siblings are brown dwarfs. Only because they orbit one another and eclipse one another, we were able to directly measure their masses and their diameters, allowing us to show for the first time that such a pair of stillborn stars could exist. In a future lecture we'll talk all about brown dwarfs, and we'll discuss this fascinating eclipsing binary system in more detail.

In this lecture, we've learned that the mass of a star is its DNA, determining all of its other physical properties—its temperature, its color, its luminosity, its size. As we'll see in later lectures, the star's mass also determines the manner in which it will end its life. It's probably a good thing that our own birth weights are not nearly as deterministic of our eventual course in life. And I, for one, am glad that the eventual endpoint of my life is not known in advance. Or if it is, perhaps it's written only in the stars.

Eclipses of Stars—Truth in the Shadows
Lecture 10

In this lecture, you will learn how eclipses can have tremendous applications to many aspects of the stars and the systems of planetary worlds that orbit them. Eclipses help us directly measure the properties of stars from afar and are, therefore, invaluable tools for telling us what stars are like, physically, at different stages of their life cycle—from eclipses of our own Sun by our Moon, to eclipses of more distant stars by our Moon, to the shadows cast by tiny worlds orbiting other suns and whose properties we can measure by virtue of the light we don't see.

Solar Eclipses

- Solar eclipses by the Moon have been observed by people for as long as humans have existed. A total eclipse of the Sun by the Moon occurs somewhere on the Earth every 18 months or so. But any particular location will be able to see a total eclipse only once every few hundred years. Even so, over the course of human history, this is frequent enough that many, many eclipses have been observed, documented, and discussed over the centuries.

- Solar eclipses are astounding cosmic coincidences. First, the Moon orbits the Earth in very nearly the same plane as the Earth orbits the Sun. This allows the Moon to pass directly between the Earth and the Sun. But it's not perfectly aligned—it's tipped by a few degrees—so the Moon's path does not intersect the line between the Earth and Sun every time around. If it did, solar eclipses would occur every month, and they probably wouldn't be regarded as any more special than the new Moon once a month.

- The other coincidence is that the relative sizes of the Sun and Moon are almost exactly the same as their relative distances from the Earth. The Moon is a lot smaller than the Sun, but it's also a lot closer. And it's closer by just the right amount so that the Moon appears to cover the same angle in the sky as does the Sun. So, when the Moon's orbit brings it directly between the Earth and Sun, it can almost exactly cover the Sun entirely. That is what we call a total solar eclipse.

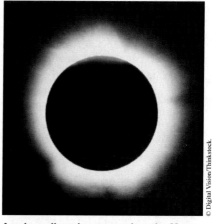

© Digital Vision/Thinkstock.

A solar eclipse happens when the Moon comes between the Sun and Earth, from the perspective of a person on Earth.

- A total eclipse can only be seen from a relatively small region on the surface of the Earth. That's because the shadow cast by the Moon on the Earth is quite small, owing to the Moon's small size compared to the Earth. Pictures taken from space of the Moon's shadow on the Earth show this beautifully. And because the Earth is 70% ocean, most total eclipses occur over the ocean and must be viewed from ships, which adds to the rarity of their sightings.

- Not all eclipses of the Sun by the Moon are total. There are also partial eclipses, when the Moon's path brings it close to but not perfectly in a line between the Earth and Sun. And there are also annular eclipses. These beautiful events occur when the Moon passes directly in front of the Sun as in a total eclipse, but the size of the Moon appears to be just a bit smaller than the Sun.

- This happens because the Moon's orbit around the Earth is not a perfect circle—it is slightly elongated, or elliptical—so sometimes the Moon is a little farther away from the Earth than at other times so that it does not fully block the full face of the Sun. It leaves a ring of sunlight showing, an annulus.

- With total eclipses, the fortuitous coincidence of the alignment and the relative sizes of the Moon and Sun means that we can observe phenomena on the Sun's surface that we would not be able to observe otherwise. As the totality of an eclipse occurs, we see a phenomenon known as Baily's beads, which are the result of bright sunlight reaching our eyes through craters and valleys on the Moon. Finally, the red glow of the Sun's thin, hot chromosphere becomes visible as a ring around the Moon's silhouette, and then the total eclipse reveals the faint, shimmering glow of the Sun's superhot corona.

- It was through eclipses of our Sun that these hot layers extending above the Sun's surface first became known and, by extension, how astronomers came to suspect that other stars might exhibit these phenomena also.

- In addition, during a total solar eclipse, phenomena such as prominences and flares on the Sun's surface can be observed directly. Although these energetic magnetic events are now routinely seen with ultraviolet and X-ray telescopes, normally these are not visible to the eye because of the intense glare of the Sun's photosphere. But during an eclipse, they can be seen directly, and before the development of modern telescope technologies, these prominences and flares provided some of the first clues about the Sun as a magnet.

- Eclipses can and do happen also with other stars besides our own Sun. The most common version of this is when two stars orbiting one another, a binary star system, have their orbit viewed edge on from the perspective of Earth. When that happens, the stars periodically eclipse one another, and we temporarily receive less total light from the system because the light of one of the stars is blocked.

- One of the most important uses of such eclipsing binary star systems is that they allow us to directly measure the diameters of stars by using the following formula: $d = rt$, or distance equals rate times time. With stellar eclipses, we can determine the speed of the stars' motions from the Doppler shift of their light spectra. Then, we can time how long the dimming of the total light from the system lasts, as one star passes directly in front of and across the other. The speed of the stars' motion, times the duration of that crossing, tells us directly the distance across the star that was traveled—its diameter.

Lunar Occultation

- Amazingly, there are a few bright stars in the sky that can be eclipsed by our Moon. For technical reasons, this is referred to as lunar occultation. Lunar occultations are another tool for measuring the diameters of distant stars. That's because the edge of the Moon is so sharp that when it passes in front of a distant light source, we can actually measure the brief but finite amount of time that it takes the Moon's edge to traverse that distant light source.

- If the light source were extremely distant, or very small, then it would effectively be a point of light, so the Moon's edge would traverse it instantaneously. The light from the star would just instantly wink out. However, if the star has some measurable dimension, then the Moon's edge will take a measurable amount of time to traverse it.

Coronagraphs

- Nature is not always kind enough to provide an ideal object for eclipsing something that we might wish to eclipse. In those cases, we can create our own eclipses using a device known as a coronagraph to act like an eclipsing body.

- One of the most common uses of coronagraphs is to observe and measure coronal mass ejections from our Sun. Coronal mass ejections are energetic, magnetic blasts that fling material from the Sun's surface out into interplanetary space. But like other solar phenomena that we can only see when the Sun's glare doesn't drown them out, these coronal mass ejections are very faint because they are diffuse.

- However, by placing a metal disc in a satellite and holding it out in front of the satellite's camera at just the right distance, the face of the Sun can be obscured, and the wispy coronal mass ejections can be seen faintly emerging from the edges of the coronagraph.

- With these types of coronagraphic studies of solar coronal mass ejections, we can directly measure the amount of material being ejected from the Sun, as well as its direction and speed. This helps astronomers improve predictions of bad space weather that might cause disruptions in our power grids or knock out communications satellites. This same type of coronagraphic technique can also be used to image faint material or objects around other stars.

Applications of Eclipses

- One of the most important uses of eclipses today for the study of planets around other stars is the transit method, which has revolutionized our ability to discover solar systems by the hundreds and provides astronomers with a powerful way to understand the nature of these other worlds, to ascertain their characteristics, and to use them as fine probes of the surfaces of their host stars.

- When a planet orbiting its star passes directly in line between its star and our sightline, the planet blocks a tiny portion of the star as seen by us. We can't see the star's surface per se; the star just appears as a point of light. Nonetheless, the small blocking of a bit of the star by the planet casts a tiny shadow on us from that enormous distance. And, as a result, we briefly see a tiny bit less light from the star. The star is being eclipsed in a very small way by its planet.

- Mathematically, the fractional amount by which the light of the star is briefly dimmed during the transit of the planet is the same as the fraction of the planet's size to the star's size. More precisely, it is the ratio of the planet's apparent area to the star's apparent area.

- Simply by measuring the fractional amount of light dimming during a planetary transit, we have a measurement of the planet's physical size in relation to the star's physical size. However, we cannot know the actual physical size of a transiting planet without first knowing the size of the star that it eclipses.

- This is where eclipsing binary star systems come to be so valuable and important. By learning what the diameters of stars of different temperatures are from the study of eclipsing binary star systems, we can confidently infer the diameters of other stars from their temperatures.

- When studying these transiting planets, astronomers can use a technique known as transmission spectroscopy to make direct measurements of the chemical compositions of their atmospheres. The measurements reveal that some of those planets appear to have strong temperature inversions in which the outer layers of the planet are actually hotter than the inner layers, whereas others do not. This phenomenon is not yet well understood.

- Another application of eclipses is microlensing, which refers to the brief but dramatic brightening of a star when another star passes directly between it and us, causing the gravitational bending of the more distant star's light to be focused toward us. If the foreground star, the one doing the focusing of the starlight, should possess a planet, the brightening signal will exhibit a bump.

- It is as though the foreground object is a lens, and the planet orbiting the lensing star is a blemish on the lens. This is an amazing application of Einstein's theory of relativity, which was originally vetted using the bending of starlight by the Sun, to the eclipse of distant stars by more nearby ones. And because the microlensing brightening is so dramatic, it can be used to detect planets among stars at great distances.

Suggested Reading

Mobberley, *Total Solar Eclipses and How to Observe Them.*

Steel and Davies, *Eclipse.*

1. What types of information about the Sun are possible to gain only during solar eclipses? Would we be able to ascertain the same information, or as accurate information, if not for the happy accident of the Moon's angular size being the same as the Sun's?

2. What are some of the challenges involved in detecting planets around other stars through their eclipses? What types of stars might be best suited for detecting such planetary transits? What are the pros and cons of searching for planetary transits around different types of stars or around stars at different stages of their lives?

Eclipses of Stars—Truth in the Shadows
Lecture 10—Transcript

Eclipses of the Sun by our Moon have long fascinated humans. Eclipses are as powerful as they are beautiful. Through these eclipses we are able to study phenomena that cannot be studied any other way, as the intense glare from the Sun is temporarily blocked out, allowing us to directly see processes that are faint, but important, to understanding how the sun works. Astronomers also use the power of eclipses to study other stars and to study the planets that encircle those stars. By providing astronomers with the means to measure the sizes and other properties of stars, eclipses are one of the most powerful tools that we have for understanding what stars are physically like at different stages of their life cycles.

In this lecture, we'll look at several examples of phenomena that can only be observed and studied when the Moon blocks out the Sun's intense glare. We will also look at how astronomers use eclipses of other stars, either by our Moon or by those stars' own binary star siblings, to measure the diameters of stars and to identify orbiting companions to those stars. We'll see how astronomers can create the conditions for an eclipse even without the Moon in order to temporarily block out the glare of the Sun and other stars at will. And we'll see how eclipses of stars by their planets allow us to study these other worlds and the potential for planetary atmospheres capable of supporting life.

So let's start closest to home and look at eclipses of our own Sun by our Moon. Solar eclipses by the Moon have been observed by people for as long as humans have existed. A total eclipse of the Sun by the Moon occurs somewhere on the Earth every 18 months or so. But any particular location will be able to see a total eclipse only once every few hundred years. Even so, over the course of human history, this is frequent enough that many, many eclipses have been observed, documented, and discussed over the centuries.

Why do solar eclipses happen? The answer is that it is an astounding cosmic coincidence. First, the Moon orbits the Earth in very nearly the same plane as the Earth orbits the Sun. This allows the Moon to pass directly between the Earth and the Sun. But it's not perfectly aligned; it's tipped by a few degrees, and so the Moon's path does not intersect the line between the Earth and Sun every time around. If it did, solar eclipses would occur every month, and they probably wouldn't be regarded as any more special than the new moon once a month.

The other coincidence is that the relative sizes of the Sun and Moon are almost exactly the same as their relative distances from the Earth. The Moon is a lot smaller than the Sun, but it's also a lot closer. And it's closer by just the right amount so that the Moon appears to cover the same angle in the sky as does the Sun. So when the Moon's orbit brings it directly between the Earth and Sun, it can almost exactly cover the Sun entirely. That is what we call a total solar eclipse.

A total eclipse can only be seen from a relatively small region on the surface of the Earth. That's because the shadow cast by the Moon on the Earth is quite small, owing to the Moon's small size compared to the Earth. Pictures taken from space of the Moon's shadow on the Earth show this beautifully. And because the Earth is 70% oceans, most total eclipses occur over the ocean and must be viewed from ships. Perhaps you've even participated in a so-called eclipse cruise. In any case, this adds to their rarity.

Now, not all eclipses of the Sun by the Moon are total. There are also partial eclipses—when the Moon's path brings it close to but not perfectly in a line between the Earth and Sun. And there are also annular eclipses. These beautiful events occur when the Moon passes directly in front of the Sun, as in a total eclipse, but the size of the Moon appears to be just a bit smaller than the Sun. This happens because the Moon's orbit around the Earth is not a perfect circle; it is slightly elongated, or elliptical, and so sometimes the Moon is a little farther away from the Earth than at other times, so that it does not fully block the full face of the Sun. It leaves a ring of sunlight showing, an annulus.

Coming back to those total eclipses, the fortuitous coincidence of the alignment and the relative sizes of the Moon and Sun means that we can observe phenomena on the Sun's surface that we would otherwise not be able to observe. As the totality of an eclipse occurs, we see a phenomenon known as Baily's beads, which are the result of bright sunlight reaching our eyes through craters and valleys on the Moon. Finally, the red glow of the Sun's thin, hot chromosphere becomes visible as a ring around the Moon's silhouette, and then the total eclipse reveals the faint shimmering glow of the sun's super-hot corona.

It was through eclipses of our Sun that these hot layers extending above the Sun's surface first became known, and by extension, how astronomers came to suspect that other stars might exhibit these phenomena also. In addition, during a total solar eclipse, phenomena such as prominences and flares on the Sun's surface can be observed directly. Although these energetic magnetic events are now routinely seen with ultraviolet and X-ray telescopes, normally these are not visible to the eye because of the intense glare of the Sun's photosphere. But during an eclipse, they can be seen directly, and before the development of modern telescope technologies, these prominences and flares provided some of the first clues about the Sun as a magnet, which we'll discuss in detail in another lecture.

The total solar eclipse of 1919 was a particularly important one in the annals of astrophysics. That's because this eclipse was used by one of the most important astronomers of the 20th century, Sir Arthur Eddington, to verify one of the key predictions of Einstein's theory of relativity, which was still quite new. Einstein's theory predicted that light can be bent in the presence of a massive object. This could be verified by observing the slight deflection of the light of a star passing right near the edge of the Sun. This would have been impossible to do under normal circumstances, with the Sun's glare drowning out the light of any stars along its sightline. But during a total eclipse, this could be done. So Eddington mounted an expedition to an island off the coast of Africa for just this purpose.

Eclipses can and do happen also with other stars besides our own Sun. The most common version of this is when two stars orbiting one another, a binary star system, have their orbit viewed edge on from the perspective of Earth. When that happens, the stars periodically eclipse one another, and so we temporarily receive less total light from the system, because the light of one of the stars is blocked.

As I mentioned in our last lecture, one of the most important uses of such eclipsing binary star systems is that they allow us to directly measure the diameters of stars. The way this works is conceptually quite straightforward. It's really just the old distance equals rate times time. For example, if I tell you that I'm traveling down a straight road at 50 miles per hour, and I tell you that I drove a particular stretch for one hour, you would immediately know that I covered 50 miles of distance. Similarly, with stellar eclipses, we can determine the speed of the stars' motions from the Doppler shift of their light spectra, as we've already discussed. And then, we can time how long the dimming of the total light from the system lasts, as one star passes directly in front of and across the other. The speed of the stars' motion, times the duration of that crossing, tells us directly the distance across the star that was traveled; that's its diameter.

This is important, because there is no other direct way to accurately measure the diameters of most stars. There are some very few stars nearby enough that with modern telescopes we can directly image their surfaces and measure their diameters that way. But we would simply not know nearly as much as we do about the fundamental properties of stars if not for the ability to directly measure their diameters hundreds of times over using the eclipse technique.

But let's not give up on our own Moon just yet. Amazingly, there are a few bright stars in the sky that can be eclipsed by our, yes our, Moon. For technical reasons, this is actually referred to as lunar occultation. The best example of this is of the bright star Aldebaran. Now, you might be thinking, so what, the Moon happens to pass in front of a bright star. But these lunar occultations actually are another tool for measuring the diameters of distant stars. That's because the edge of the Moon is so sharp, like a knife's edge, that when it passes in front of a distant light source, we can actually measure the brief, but finite, amount of time that it takes the Moon's edge to traverse that distant light source. If the light source were extremely distant, or very small, then it would, effectively, be a point of light, and so the Moon's edge would traverse it instantaneously. The light from the star would just instantly wink out. But, if the star has some measurable dimension, then the Moon's edge will take a measurable amount of time to traverse it.

In the case of Aldebaran, when we look at a graph of the light of the star as it re-emerges from the back side of the Moon after the occultation, we can see that the light doesn't go from zero to full tilt instantly, rather, it takes about 100 milliseconds. Combining that travel time with the known speed of the motion of the Moon in its orbit, we get the diameter of the distant star. Aldebaran, being a red giant star, turns out to have a diameter that is about 45 times the diameter of the Sun. And it can be measured through a total eclipse by our own Moon. Now, nature is not always kind enough to provide an ideal object for eclipsing something that we might wish to eclipse. In those cases, we can create our own eclipses using a device known as a coronagraph to act like an eclipsing body. Let's look at a few examples of this modern, innovative technique.

One of the most common uses of coronagraphs is to observe and measure coronal mass ejections from our Sun. Coronal mass ejections are energetic, magnetic blasts that fling material from the Sun's surface out into interplanetary space. But like other solar phenomena that we can only see when the Sun's glare doesn't drown them out, these coronal mass ejections are very faint, because they are diffuse. But by placing a metal disc in a satellite and holding it out in front of the satellite's camera at just the right distance, the face of the Sun can be obscured, and the wispy coronal mass ejections can be seen faintly emerging from the edges of the coronagraph.

With these types of coronagraphic studies of solar coronal mass ejections, we can directly measure the amount of material being ejected from the Sun, as well as its direction and speed. This helps astronomers improve predictions of bad space weather that might cause disruptions in our power grids or knock out our communications satellites. The subject of space weather is a fascinating and important one that we will discuss in detail in a separate lecture. This same type of coronagraphic technique can also be used to image faint material or objects around other stars. For example, debris disks around stars, which are the wispy remnants of the gas and dust left behind from the stage in the star's life when it had a protoplanetary disk. The Hubble space telescope has a coronagraph on board that can be used in just this way.

Here are two examples of nearby stars whose debris disks have been imaged with the Hubble Space Telescope's coronagraphic camera. At the center of each image, there is a dark spot; that's where the light of the central star has been blocked out by the coronagraph held out in front of the camera. This then allows the extremely faint debris disks to be seen, as they reflect a tiny fraction, perhaps one ten-thousandth of the star's light. In one case, we happen to be seeing the debris disk nearly edge on; and in the other case, we happen to be seeing it nearly face on. By directly imaging these whiffs of leftover material from the planet formation process, we actually have a tool for inferring the presence of planets around these stars.

Indeed, there have now been several planets imaged around nearby stars using coronagraphic telescopes. This is, in fact, one of the most promising and cutting edge areas of technological development in the exploration for and study of other solar systems. Over the past 20 years, other techniques have allowed hundreds of planets around other stars to be discovered. Those techniques are the Doppler wobble technique, in which a planet causes its host star to move back and forth due to its gravitational tug; and the transit technique in which a planet passes directly in front of its star. We'll return to the transit technique in a moment.

But the coronagraphic method for studying planets is particularly exciting, because it permits astronomers to directly see the planets, to collect photons from the planets directly. And this, in turn, allows us to measure important physical properties of these other worlds in a manner similar to how we study the light emitted by stars. We can use the colors and the chemical signatures in the light spectra of the planets to determine the planets' temperature and atmospheric compositions.

Perhaps the best example of a planet directly imaged with a coronagraphic camera is the planet around the star Fomalhaut, using the coronagraph on the Hubble Space Telescope. Here, Fomalhaut itself is not visible because its glare has been blocked out by the coronagraph in front of the camera, leaving a dark spot in the center of the image. There is some residual glow from the star that even the coronagraph cannot prevent. But the faint planet orbiting at the edge of a debris disk suddenly becomes visible, if only barely, and Hubble has been able to watch the planet orbit Fomalhaut over the past several years.

This brings us to one of the most important uses of eclipses today for the study of planets around other stars. The so-called transit method has revolutionized our ability to discover solar systems by the hundreds, and provides astronomers with a powerful way to understand the nature of these other worlds, to ascertain their characteristics, and to use them as fine probes of the surfaces of their host stars.

Remember that image from space of the Sun casting our Moon's shadow on the Earth's surface? You can use that same image to imagine the transit method. Here, a planet orbiting its star passes directly in line between its star and our sightline. When that happens, the planet blocks a tiny portion of the star, as seen by us. Mind you, we can't see the star's surface per se; the star just appears as a point of light. Nonetheless, the small blocking of a bit of the star by the planet casts a tiny shadow on us from that enormous distance. And as a result, we briefly see a tiny bit less light from the star. The star is being eclipsed in a very small way by its planet.

Mathematically, the fractional amount by which the light of the star is briefly dimmed during the transit of the planet is the same as the fraction of the planet's size to the star's size. More precisely, it is the ratio of the planet's apparent area to the star's apparent area. Imagine the planet as a small, circular disc blocking a portion of the star's bright, large disc. The area of a circle is just pi times its radius squared. So dividing these circular areas for the planet and the star, we find that the fractional blocking factor of the planet on the star is just the planet's radius squared, divided by the star's radius squared.

That means that simply by measuring the fractional amount of light dimming during a planetary transit, we have a measurement of the planet's physical size in relation to the star's physical size. A dimming of one part in ten thousand implies a planet that is 1% as large as its star. If the star is the size of the Sun, then that planet would be the same size as Jupiter. But the other side of this coin is that we cannot know the actual, physical size of a transiting planet without first knowing the size of the star that it eclipses. And this is where those eclipsing binary star systems that we discussed earlier come to be so valuable and so important.

By learning what the diameters of stars of different temperatures are from the study of eclipsing binary star systems, we can confidently infer the diameters of other stars from their temperatures. So the name of the game in the study of transiting planets is to carefully measure the temperature of the host star and thereby infer its diameter. And that, in turn, from the fractional dimming caused by the planet, allows the planet's diameter to be measured.

Why should we care about the diameters of these planets? The main reason is that the size of a planet is our best way of guessing at its overall makeup. A planet that is the size of Jupiter or larger, as are most of the planets so far discovered, must, according to most calculations, be a giant gaseous planet like our own Jupiter. However, a much smaller planet must in contrast be made of something denser, such as liquid water or rock.

Of course the holy grail of these searches and investigations of other solar systems is to find a tiny rocky world like our own Earth. This is very difficult, because a planet such as ours is extremely tiny compared to a star such as our Sun, and consequently, the fractional dimming of light caused by the tiny planet's eclipse of the star is one part per million, an extremely challenging measurement.

So astronomers have increasingly turned their attention to red dwarf stars. These are certainly legitimate stars, but by virtue of being less massive than our Sun, they are both cooler and, most important for the transit method, smaller. That means that the light dimming of an Earth-like planet might be as large as one part in ten thousand. Still challenging, but this measurement is much more doable than one part per million.

Importantly, with these transiting planets, astronomers can use a technique known as transmission spectroscopy to make direct measurements of the chemical compositions of the atmospheres of these alien worlds. So now let's talk about that works. Most of the planets so far discovered through the technique of planetary transits are so-called hot Jupiters. These are massive, gas giant planets. But unlike our own Jupiter, which orbits relatively far out from the Sun, these Jupiters orbit extremely close to their suns. As a result, their temperatures can approach those of cool stars, and their atmospheric properties are correspondingly exotic.

Now, remember that atoms and molecules can absorb light at specific wavelengths, depending on the elements involved. So when one of these hot Jupiters passes in front of its star, the molecules in the planet's atmosphere will absorb light at certain specific wavelengths, in addition to the light blocked by the planet just due to its geometric size. At other wavelengths, the atmosphere is transparent, because the molecules aren't capable of absorbing those wavelengths, and so the planet just blocks the light based on its geometric size. In other words, at wavelengths where the atmosphere can absorb light, the planet effectively appears a bit larger and so blocks a little bit of additional light. This wavelength dependence of the fraction of light blocked is very tiny, typically only a few parts in 10,000. But it has been successfully measured for about a dozen planets now. Those spectra indicate that the hot-Jupiter atmospheres contain water, methane, carbon monoxide, and possibly carbon dioxide, in addition to hydrogen and helium, not unlike our own Jupiter.

The abundances of those molecules appear to vary from planet to planet for reasons that are not yet fully understood. One possible explanation is that gas giants form with different amounts of hydrogen, carbon, and oxygen, depending on their location in the protoplanetary disks from which they form. Another possible explanation is that the close proximity of these hot Jupiters to their host stars results in the selective destruction of some molecules from the intense radiation they receive from the stars.

Importantly, these planets also pass behind their host stars as they orbit around the stars, causing events known as secondary eclipses. As the planets circle around, from Earth we see the side of the planet that is being directly irradiated by the star, and so we see the thermal infrared emission from the planet caused by the heating from the star. By measuring the wavelength-dependent decrease in that infrared radiation during those secondary eclipses, we can learn about the shape of the planet's emission spectrum, which, in turn, provides information about the planet's atmospheric composition and how temperature varies as a function of altitude within the planet's atmosphere. This type of infrared eclipse measurement has now been successfully done for nearly 50 of these planets. And the measurements reveal that some of those planets appear to have strong temperature inversions in which the outer layers of the planet are actually hotter than the inner layers, whereas others do not. This phenomenon is not yet well understood, but a few hypotheses have been proffered.

Because these hot Jupiters orbit very close to their suns, it is generally believed that their rotation and orbits should be tidally locked, meaning that the same side of the planet always faces the host star. This is the same phenomenon by which tides on the Earth have caused the Moon to always present the same face to us. One open question is whether heat absorbed on the dayside of those planets is efficiently transported around to the nightside of the planet. By measuring changes in the brightness of the planet as it orbits its host star, we can determine the amplitude of the day-night temperature gradient in the planet's atmosphere. Such measurements obtained by the Spitzer Space Telescope for a handful of hot Jupiters indicate that most appear to have strong winds that help to keep their nightsides warm by transporting heat from the dayside. In addition, the weather on these hot Jupiters appears to be relatively unchanging, as repeated observations of the dayside brightness of these planets over several years have failed to detect any significant variability.

We have one more application of eclipses to discuss; it is what we call microlensing. Microlensing refers to the brief but dramatic brightening of a star when another star passes directly between it and us, causing the gravitational bending of the more distant star's light to be focused toward us. If the foreground star, the one doing the focusing of the starlight, should possess a planet, the brightening signal will exhibit a bump. It is as though the foreground object is a lens and the planet orbiting the lensing star is a blemish on the lens. This is an amazing application of Einstein's theory of relativity, which itself was originally vetted using the bending of starlight by our Sun, to the eclipse of distant stars by more nearby ones. And because the microlensing brightening is so dramatic, it can be used to detect planets among stars at great distances. A future space mission called WFIRST is planned to utilize this technique to dramatically increase the number of known solar systems throughout our galaxy.

In this lecture, we've seen how the power of a single concept—eclipses—can have tremendous application to many aspects of the stars and the systems of planetary worlds that orbit them. Eclipses help us to directly measure the properties of stars from afar and are, therefore, invaluable tools for telling us what stars are like, physically, at different stages of their life cycle. From eclipses of our own Sun by our Moon, to eclipses of more distant stars by our Moon, to the shadows cast by tiny worlds orbiting other suns and whose properties we can measure by virtue of the light we don't see, sometimes, the truth really is in the shadows.

Stellar Families

Lecture 11

In this lecture, you will learn about the different types of star clusters—the stellar families into which most stars are born and spend the early stages of their lives. You will learn that these stellar families come in two basic varieties, representing what might be considered modern versus old-fashioned families. Like human families, most modern stellar families disperse over time, in contrast to the old-fashioned stellar families, which were and remain extremely tight knit.

Types of Star Clusters

- There are two main types of star clusters: open clusters and globular clusters. Globular clusters are highly organized, dense, shapely clusters with a spherical or globular appearance, like a swarm of bees around a hive. Open clusters are looser, generally somewhat less ordered in their structure. They appear more "open" as opposed to the dense, compact appearance of the globulars.

- The two types of clusters represent two very different kinds of entities—in terms of their compositions, where they are found, and the evolutionary stage of our galaxy when they were formed. The globular clusters are ancient, old-fashioned families. The open clusters are younger, more modern families.

- The stars forming in the giant clouds of gas and dust of stellar nurseries tend to have a certain organization. In particular, the most massive stars are generally found at the centers of these nurseries, surrounded by a larger collection of less massive stars. In addition, when the stars are still in their nurseries, they are embedded in the gas and dust from which they formed.

Open Clusters

- Looking at the youngest open clusters that we know, we can see the vestiges of these beginnings. We can see remnants of the gas and

The two general types of star clusters are open clusters and globular clusters.

dust from which the stars were formed, wisps still floating within the cluster. Think of these wisps as the placental material left over from the stellar birth process.

- Another aspect of the stars in the youngest open clusters, representing stellar families soon after birth, is that we see in them the full span of stellar types—from the most massive to the least massive and everything in between.

- When we measure the masses of all of the hundreds of stars in a young open cluster, we see that the pattern of stellar masses is created in the birth process, and the stars carry that pattern forward through the rest of their lives.

- As we look at increasingly older open clusters, we find three important characteristics that reveal what happens to the stars in these families over time. First, we find that the older clusters have a looser distribution with more space on average between the stars. Second, we find that the older clusters tend to have fewer members. Third, the pattern of stellar masses changes, with fewer and fewer massive stars.

Globular Clusters

- Globular clusters are much more extended stellar families. Whereas typical open clusters have hundreds to perhaps a couple thousand members, globular clusters have tens of thousands to a million members. As a result of their sheer heft, they are very tight knit, and they are much better able to remain bound together for a very long time—at least for as long as the galaxy and the universe are old.

- In fact, we only know of globular clusters that are extremely old, as old as our galaxy. These relic clusters tell us about the conditions of our galaxy when it first formed. In fact, by studying the chemical makeups of these stars, we find some of the least chemically endowed stars in our galaxy. In other words, these stars are made up almost purely of hydrogen and helium, in the proportions that hydrogen and helium were produced in the big bang.

- These stars did not benefit from the production of heavier elements by previous generations of stars. These stars were formed from pristine material from the birth of our universe. But, interestingly, there are trace amounts of heavy elements in these otherwise pristine ancient stars.

- This harkens back to a time when our galaxy was taking shape from an immense protogalactic cloud of gas produced in the big bang. That protogalactic cloud had not taken the flattened pancake shape that we have now. Instead, the galaxy initially had a more amorphous shape, so the very first generations of stars to form would have formed in clusters distributed about the center of the galaxy and bearing that amorphous distribution.

- As the galaxy evolved into the spiral disk galaxy that we have now, the globular clusters retained their distribution, continuing to swarm about the center of the galaxy. The motions of these globular clusters keep them far from the disk of the galaxy, where they might be disturbed by other stars or clouds of gas and dust. Consequently, these ancient stellar families have remained tight knit from the very beginning.

- But that's not to say that they have not changed since their formation over 10 billion years ago. In fact, they have changed steadily over time. As these stellar families have aged, individual stars have died, with the most massive stars dying first and then successively lower-mass stars dying off. As a result, only the lowest-mass stars remain in these families.

- In a typical globular cluster, all of the remaining stars have masses similar to our Sun or less. Interestingly, though, in many globular clusters, we see some stars that appear to be too massive to still be alive. These are called blue straggler stars.

Dynamical Interactions

- The stars in globular clusters have undergone dynamical interactions over the eons, and these interactions have shaped the clusters in important ways. Stars born as triplets undergo a dance in which two of the siblings are brought closer together, and a third star is pushed out into a distant orbit. A similar process occurs within the globular cluster as a whole.

- In this process, many individual gravitational flybys and kicks among the stars lead to a population of the stars coming closer together at the core of the cluster, while another group is pushed out into wider orbits about the cluster. So, the cluster ends up with what we call a core-halo structure, dense at the center and surrounded by a somewhat looser swarm of stars that have been kicked to the outer limits of the cluster.

- These dynamical interactions also produce other important effects for globular clusters. Some of the stellar corpses of stars that have already died—the white dwarfs—will be members of binary star systems. And when those binary systems come closer and closer together as a result of the dynamical interactions in the cluster, mass from the stellar sibling of the white dwarf can spill onto the white dwarf and cause it to explode as a supernova. That supernova explosion can flash forge heavy elements, which then can lightly pollute the otherwise pristine stars in the cluster.

- Another consequence of these dynamical interactions in the cluster is that, occasionally, two stars can be brought close enough together to merge. When that happens, the resulting star will be more massive, having the sum of the two original stars' masses. So, the cluster can, for a short time, appear to have a few massive stars that should have died long ago—as if they had been resuscitated or reborn—which explains those blue stragglers.

- These types of dynamical interactions can take place within open clusters as well, and although the consequences are generally somewhat less dramatic, the byproducts are nonetheless very interesting. When an open cluster is undergoing the slow process of disintegrating, its members slowly drifting apart, there can be occasional flybys of stars that come close to stars within the cluster. As a result, the stars receive a strong gravitational kick that ejects them from the cluster.

- While most of these types of ejections happen to individual stars randomly, it can happen on occasion that two stars in different parts of the cluster undergo such an ejection at nearly the same time and in nearly the same direction. When that happens, the two ejected stars, which initially had nothing to do with one another, become joined by gravity as an ultrawide binary system.

- The widest of these ultrawide binary systems is much, much wider than the stellar wombs in which stars are normally born, and instead have separations similar to the typical sizes of open clusters. This is not a coincidence. It is a direct consequence of these binary systems having been brought together through the mutual-ejection mechanism.

Prospects for Habitability

- Planets are much more likely to arise in open clusters than in globular clusters. That's because the chemical composition is more complex—and, therefore, more supportive of life—in the open clusters. And that's because the open clusters have formed more recently from chemically enriched material, which includes the ashes of previous stellar generations.

- At the same time, even open clusters represent relatively densely populated environments. Certainly, they are more densely populated environments than a random isolated point in space, such as where our own solar system currently resides. So, planets that find themselves in solar systems within open clusters will be more likely to be disrupted by jostling from other stars in the cluster. Any planets that become stripped from their parent stars would then float freely through the cluster—what we refer to as solivagant planets.

- Imagine a civilization on a planet happily in orbit about a sun in an open cluster. The night sky would be filled with extremely bright stars, similar to our full Moon. Now imagine the havoc wreaked if that planet then became stripped away from its sun. Deprived of life-giving light and heat, surely this would spell disaster for the planet, doomed to drift through the rest of the cluster, a lifeless barren world that was.

- Imagine what life might be like on a planet in a globular cluster. While the elemental compositions of stars in a globular cluster don't favor the formation of rocky worlds or life on them, it might still be possible.

- Imagine a sky filled with not a single sun but hundreds or thousands all around. Astronomers have, in fact, found a planet in a globular cluster that orbits a binary star system and is 13 billion years old. In this case, the two suns are a white dwarf and a neutron star, one the corpse of a low-mass star and the other the corpse of a massive star.

Archinal, *Star Clusters*.

HubbleSite, "Star Clusters."

Questions to Consider

1. In what ways are star clusters uniquely important for piecing together the time line of the life cycles of stars?

2. What might be some of the reasons that globular star clusters of the type seen in the halo of our galaxy are no longer formed? Why does the galaxy now make only open clusters?

Stellar Families
Lecture 11—Transcript

So far in this course, we've talked about the stellar life cycle, and in particular the stellar birth process, in terms of the experience of individual stars and of siblings born together. But individual stars and their siblings are generally not born in isolation. Rather, they are born in large extended families of hundreds to thousands of stars that we call star clusters.

In this lecture, we'll see examples of different types of star clusters, the stellar families into which most stars are born and spend the early stages of their lives. We'll also see that these stellar families come in two basic varieties, representing what you might call modern versus old-fashioned families. Like human families, most modern stellar families disperse over time, in contrast to the old-fashioned stellar families, which were and remain extremely tight knit.

Let's start by describing the two main types of star clusters. One type are called open clusters, the other type are called globular clusters. These names originate historically as simple visual descriptors of the appearances of these two types of clusters. The globular clusters are highly organized, dense, shapely clusters, with a spherical or globular appearance, like a swarm of bees around a hive. One such globular cluster is Messier 80, or M80 for short. It is at a distance of 33,000 light-years from us and has a very large diameter of about 100 light-years. There are tens of thousands of stars in this rich cluster.

Open clusters are looser, generally somewhat less ordered in their structure. They appear more open, as opposed to the dense and compact appearance of the globulars. An example of an open cluster is the Pleiades cluster, also known as the Seven Sisters. The Pleiades cluster is easily visible in the autumn sky and is sometimes confused for the Little Dipper because of the dipper-like arrangement of the bright Seven Sister stars. Compared to the M80 globular cluster that is situated far away and far above the plane of our Galaxy, the Pleiades cluster is very nearby, only about 400 light-years away, and is in the disk plane of our Galaxy. It is also much looser than the M80 globular cluster, and it is overall smaller, about five light-years across.

Moreover, the Pleiades cluster has fewer stars overall, a few hundred, as opposed to tens of thousands.

So we see that the two types of clusters represent two very different kinds of entities, in terms of their compositions, where they are found, and more fundamentally, the evolutionary stage of our galaxy when they were formed. The globular clusters are ancient, the old-fashioned families. The open clusters are younger, the modern families.

Open clusters are found exclusively in the disk of our galaxy, where there are ample numbers of the massive gas and dust clouds within which stellar nurseries reside and the majority of new stars are formed. This is as opposed to globular clusters, which are found, instead, in the older, surrounding halo of the galaxy. So already there is a hint that open clusters represent a relatively youthful population.

In addition, detailed analysis of the spectra of stars in open clusters shows that their chemical makeups tend to be relatively enriched in heavy elements, the types of elements that stars have created over many generations in the galaxy. This tells us that the stars in open clusters formed relatively recently, because they must have formed from material that has been processed and re-processed by previous generations of stars.

Finally, the stars in open clusters tend to spin fast. The spins can be measured by measuring the brightness variations of the stars as spots on their surfaces, like sunspots, rotate with the star in and out of view. Since stars wind down over time, spinning ever more slowly, these fast spinners indicate that the stars in the open clusters are relatively young.

Let's look again at some of the examples of stellar nurseries that we've looked at before, but now with an eye toward how these become the open clusters. Recall from our discussion of stellar nurseries that the stars forming in these giant clouds of gas and dust tend to have a certain organization. In particular, remember that the most massive stars are generally found at the centers of these nurseries surrounded by a larger collection of less massive stars. In addition, when the stars are still in their nurseries, they are embedded in the gas and dust from which they formed.

Looking at the youngest open clusters that we know, we can see that the vestiges of these beginnings. For example, in the Pleiades, we immediately see the so-called seven sisters, which are the most massive stars in the family that formed at the center of the stellar nursery. We also can see remnants of the gas and dust from which the stars were formed, wisps still floating within the cluster. Think of these wisps as the placental material left over from the stellar birth process.

Another aspect of the stars in the youngest open clusters, representing stellar families soon after birth, is that we see in them the full span of stellar types, from the most massive to the least massive and everything in between. In a previous lecture we learned that stellar masses follow a particular pattern, which reveals an exponential law. Specifically, we learned that for every 100 stars with a mass like our Sun's, there will be only one star with a mass ten times the Sun's. When we measure the masses of all of the hundreds of stars in a young open cluster, that so-called initial mass function is clearly present. This tells us that the pattern of stellar masses is created in the birth process and the stars carry that pattern forward through the rest of their lives.

As we look at increasingly older open clusters, we find three important characteristics that reveal what happens to the stars in these families over time. First, we find that the older clusters have a looser distribution with more space on average between the stars. Second, we find that the older clusters tend to have fewer members. Third, the pattern of stellar masses changes, with fewer and fewer massive stars.

Let's examine three example clusters over a range of ages to illustrate these three features. First, let's look at the open cluster known as NGC 265. By the way, NGC stands for the New Galactic Catalog, so this cluster is number 265 in that catalog. NGC 265 is a relatively young cluster with an age of about 300 million years, making it just a bit older than the very young Pleiades cluster. Note how rich and dense the stars are in this cluster. We see mainly many bright blue young stars, and just a couple of red stars, which are the nearly dead red giant stars in the cluster.

Next, NGC 2266 is an open cluster with a medium age of about one billion years. This cluster has fewer stars overall; the stars that are there are somewhat more loosely spaced, and we see a larger proportion of nearly dead red giant stars, indicating that this is an aging stellar population.

Finally, let's look at open cluster, Messier 67. This is a very old cluster, as open clusters go, with an age of about four billion years. In other words, the stars in this cluster are similar in age to our own Sun. Here we see even fewer stars. The stars are much more spread apart, and there is an even larger proportion of old, red giant stars, indicating that this is an aged population indeed. One of the reasons that the older clusters have fewer members is simply that, as the stars in the cluster age, some of them will die. In particular, the most massive stars will die first, and so that also explains why the pattern of stellar masses in the cluster increasingly includes only the lowest mass stars.

As for the loosening of the cluster, there are two factors at work here. First, as the more massive stars in the cluster die off, their mass is ejected from the cluster in supernova explosions. This causes an overall reduction in the gravitational binding of the cluster, because the overall mass is less, and therefore, the overall gravity of the cluster is less. So, the remaining stars find themselves less strongly bound, and they begin to drift apart. Secondly, as the cluster moves through the galaxy, it is gently jostled by other stars, as well as those large clouds of gas and dust that we've discussed. These perturbations also cause the stars within the cluster to be disturbed and to further separate. Indeed, little by little, the stars in these clusters drift apart sufficiently that as the cluster ages, more and more of the stars will have left the cluster. As a result, as we look to older and older open clusters, we find fewer and fewer of the stars. In fact, there are relatively few known examples of open clusters older than about five billion years. Despite the fact that our galaxy is some 12 billion years old, some of these modern stellar families evidently become totally dispersed significantly faster than that. These are families in which the kids all grow up and move away.

So, how about globular clusters; how do they differ? Well the first thing to say is that these are much more extended stellar families. Whereas typical open clusters have hundreds to perhaps a couple thousand members, globular clusters have tens of thousands to a million members. As a result of their sheer heft, they are very tight-knit, and they are much better able to remain bound together for a very long time. How long? At least for as long as the galaxy and the universe are old. So, essentially, forever. In fact, we only know of globular clusters that are extremely old, as old as our galaxy. So these relic clusters tell us about the conditions of our Galaxy when it first formed. In fact, studying the chemical makeups of these stars, we find some of the least chemically endowed stars in our galaxy. That's to say, these stars are made up almost purely of hydrogen and helium in the proportions that hydrogen and helium were produced in the big bang itself.

So these stars did not benefit from the production of heavier elements by previous generations of stars. These stars were formed from pristine material from the birth of our universe. But interestingly, there are trace amounts of heavy elements in these otherwise pristine ancient stars. Remember that as a clue for now, and we'll come back to it later.

This harkens back to a time when our galaxy was taking shape itself from an immense protogalactic cloud of gas produced in the big bang. That protogalactic cloud had not taken the flattened pancake shape that we have now. Instead, the galaxy initially had a more amorphous shape, and so the very first generations of stars to form would have formed in clusters distributed about the center of the galaxy and bearing that amorphous distribution.

As the galaxy evolved into the spiral disk galaxy that we have now, the globular clusters retained their distribution, continuing to swarm about the center of the galaxy. The motions of these globular clusters keep them far from the disk of the galaxy where they might be disturbed by other stars or by clouds of gas and dust. Consequently, these ancient stellar families have remained tight knit from the very beginning. But that's not to say that they have not changed since their formation over ten billion years ago. In fact, they have changed, steadily, over time. First of all, as these stellar families have aged, individual stars have died, with the most massive stars dying first, and then successively lower mass stars dying off. As a result, in these families, only the lowest mass stars remain. In a typical globular cluster, all of the remaining stars have masses similar to our Sun or less.

Interestingly, though, in many globular clusters, we see some stars that appear to be too massive to still be alive. These are the so-called blue straggler stars. In this image of the globular cluster NGC 6397, we are looking at the very center of what is a much larger cluster. Whereas the stars toward the edges of this image are mainly old, red stars, representative of the stars in the cluster generally, in the center we see a clutch of blue stars. Those are the blue stragglers. We'll come back to them shortly.

It's important to note that the stars in globular clusters have undergone dynamical interactions over the eons, and these interactions have shaped the clusters in important ways. Think back to the types of interactions that we discussed in the context of stellar nurseries. Recall that stars born as triplets undergo a dance in which two of the siblings are brought closer together, and a third star is pushed out into a distant orbit.

A similar process occurs within the globular cluster as a whole. In this process, many individual gravitational flybys and kicks among the stars leads to a population of the stars coming closer together at the core of the cluster, while another group is pushed out into wider orbits about the cluster. So the cluster ends up with what we call a core-halo structure, dense at the center and surrounded by a somewhat looser swarm of stars that have been kicked to the outer limits of the cluster.

We can see an example of that core-halo structure in the Omega Centauri globular cluster. The image at left in visible light shows the broader halo distribution of old red stars, whereas the image at right in ultraviolet wavelengths shows the tighter core of hotter blue straggler stars concentrated toward the center. Those blue stragglers were brought together through the dynamical interactions that I mentioned, making them a tight-knit group at the expense of the old red stars that increasingly have gotten kicked out toward the outskirts of the cluster.

These dynamical interactions also produce other important effects for globular clusters. First of all, some of the stellar corpses of stars that have already died—the white dwarfs—will be members of binary star systems. And when those binary systems come closer and closer together as a result of the dynamical interactions in the cluster, mass from the stellar sibling of the white dwarf can spill onto the white dwarf and cause it to explode as a supernova. That supernova explosion can flash forge heavy elements, which then can lightly pollute the otherwise pristine stars in the cluster. Remember I told you earlier the clue that these stars do possess trace amounts of heavy elements.

Another consequence of these dynamical interactions in the cluster is that occasionally two stars can be brought close enough together to merge. When that happens, the resulting star will be more massive, having the sum of the two original stars' masses. And so the cluster can, for a short time, appear to have a few massive stars that should have died long ago, as if they had been resuscitated or reborn. So that explains those blue stragglers.

These types of dynamical interactions can take place within open clusters as well. And although the consequences are generally somewhat less dramatic, the byproducts are, nonetheless, very interesting. When an open cluster is undergoing the slow process of disintegrating, its members slowly drifting apart, there can be occasional flybys of stars that come close to stars within the cluster. As a result, the stars receive a strong gravitational kick that ejects them from the cluster.

Now, here's the interesting part. While most of these types of ejections happen to individual stars randomly, it can happen on occasion that two stars in different parts of the cluster will happen to undergo such an ejection at nearly the same time and in nearly the same direction. When that happens, the two ejected stars, which initially had nothing to do with one another, become joined by gravity as an ultra-wide binary system.

My own research group has helped to identify thousands of these ultra-wide binary systems. In this gallery of thumbnail photos we see just a handful out of the thousands of examples we have so far discovered. Each of these thumbnail images is about five light-years across, and the two stars in each physical binary system are indicated by arrows. The most widely separated of these are about three light-years apart. Indeed, one of the interesting findings from our work is that the widest of these ultra-wide binary systems are much, much wider than the stellar wombs in which stars are normally born, and instead, have separations similar to the typical sizes of entire open clusters. This is not a coincidence; it is a direct consequence of these binary systems having been brought together through the mutual ejection mechanism.

So now that we've described the different types of star clusters and related them to the stellar birth process and to the history of our galaxy, let's now look at how we can use these stellar families to understand the manner in which all stars live out their lives, as well as a tool for determining the age and distance to the stars.

To do this, we'll need to make use of the stellar astrophysicist's favorite tool: The Hertzsprung-Russell diagram, or the H–R diagram for short. To remind you, the H–R diagram is a graph in which stars are represented by their temperatures on the horizontal axis and their luminosities on the vertical axis. We talked previously about how the majority of stars are found in this diagram along a narrow diagonal swath that we call the main sequence. Individual stars spend most of their lives as part of that main sequence, their positions along the sequence determined by one thing—their mass. Massive stars are hot and luminous and so are at the upper left; whereas low mass stars are cool and dim, and so are at the lower right.

Let's now use the H–R diagram to represent the stars in a young open cluster, such as the Pleiades. These stars all have different masses, and so they are found at different places along the main sequence. But they do have one thing in common: they are all the same age, because they were all born from the same stellar nursery at the same time. For reference, the position of a star like our Sun is indicated. Now let's represent a different cluster in the H–R diagram, a globular cluster this time. We see something pretty different. We see a group of stars that trace the lower portion of the main sequence. These are the low-mass stars. But the higher mass stars that populate the upper part of the main sequence in the open cluster are missing in the globular cluster.

In actuality, those missing stars are still there, they're just in different parts of the H–R diagram. In other words, the higher mass stars that were in the globular cluster at the beginning, have evolved, changed their temperatures and luminosities, so that they are now in different parts of the graph. Other stars are just now in the process of evolving, and we can use these stars to accurately measure the age of the cluster.

Let's look at how this works. We can depict how an individual star of a particular mass starts out as part of the main sequence then changes its temperature and luminosity as it evolves. We begin with a group of stars of different masses but with the same age, just as we have in a cluster. Then running the clock forward in time, we see the stars begin to peel away from the main sequence, shifting to the right and up as they become red giants and subsequently die.

First the most massive stars at the top of the main sequence peel away; then, successively lower mass stars farther and farther down the main sequence do the same thing. After about 10 billion years, the stars that are peeling away from the main sequence are the stars that weigh the same as our Sun. That's how long a Sun-like star lives before ending its life—10 billion years.

We can also depict the same thing, but this time using a line in the H–R diagram to represent where the group of stars will be in the H–R diagram after a given amount of time. At the beginning, the line traces out the main sequence, because that's where stars of all masses in the cluster are at the beginning of the cluster's life. But as the cluster ages, and the more massive stars peel away from the main sequence, the line at the top also bends over. As progressively lower mass stars peel away from the main sequence, the point where the line bends away from the main sequence shifts to progressively lower and lower positions.

With this tool, we can now compare the stars in any cluster with this line representation in the H–R diagram, and read off the age of the cluster. For example, applying this to the Pleiades open cluster, we infer an age of about 100 million years. Applied to a globular cluster in which stars like our Sun are just beginning to end their lives, we infer an age for the cluster of about 10 billion years.

Now that we've discussed the two major types of stellar clusters and their properties, as well as the ways in which we can use clusters as tools to measure the ages of stars, let's close by considering what these different stellar families are like from the standpoint of planets and the prospects for habitability. The first thing to say is that planets are much more likely to arise in open clusters than in globular clusters. That's because the chemical composition is more complex, and therefore, more supportive of life in the open clusters. And that's because the open clusters have formed more recently from chemically enriched material that includes the ashes of previous stellar generations.

At the same time, even open clusters represent relatively densely populated environments. Certainly, they are more densely populated environments than a random isolated point in space, such as where our own solar system currently resides. So planets that find themselves in solar systems within open clusters will be more likely to be disrupted by jostling from other stars in the cluster. Any planets that become stripped from their parent stars would then float freely through the cluster, what we refer to as solivagant planets.

Just imagine a civilization on a planet happily in orbit about a Sun in an open cluster. The night sky would be filled with extremely bright stars, similar to our full moon. Parents might sing lullabies to their children about their stars, but it probably wouldn't be "Twinkle, Twinkle, *Little* Star"! Now, imagine the havoc wreaked if that planet then became stripped away from its sun. Deprived of life-giving light and heat, surely this would spell disaster for the planet, doomed to drift through the rest of the cluster, a lifeless barren world that was.

Finally, just for kicks, let's imagine what life might be like on a planet in a globular cluster. While the elemental compositions of stars in a globular cluster don't favor the formation of rocky worlds or life on them, it might still be possible. Just imagine a sky filled with not a single sun but hundreds or thousands all around. Would there be any notion of night versus day? And if this world were located near the center of the globular cluster, with a dense forest of 100 thousand stars in every direction, would you even know there was a larger universe beyond out there? Just imagine what such a civilization's notions of cosmology might be. How might they answer big questions such as "where do we come from"?

Just science fiction, right? Not so fast! Astronomers have, in fact, found a planet in the globular cluster Messier 4. Amazingly, this 13-billion-year-old planet orbits a binary star system; and not just any ordinary binary star system, in this case, the two suns are a white dwarf and a neutron star, one the corpse of a low-mass star and the other the corpse of a massive star. If there is life on that world, my goodness the stories they must have to tell! Truly, when it comes to the life cycle of stars, real life can be much stranger than fiction.

In this lecture, we've seen that open clusters represent modern families, formed in stellar nurseries in the disk of our galaxy, and made from the enriched chemical material that previous generations of stars produced in their lives. Among the most ancient of previous stellar families are the globular clusters, still swarming about the center of the galaxy, a vestige of the time when the galaxy as a whole was forming. The modern stellar families are less tight knit, dispersing over time. But inheriting their enriched chemical material from stars that originally lived and died in the globular clusters, the modern families in a real way owe their enriched lives to those ancient families that came before them.

A Portrait of Our Star, the Sun
Lecture 12

In this lecture, you will take an imaginary trip into the Sun—a star firmly in the midlife stage—as if plunging in at the top and diving down to its core, and you will learn about the different layers of the Sun's interior as you do so. Then, you will take a step back out to explore the phenomena occurring on the Sun's surface and the effects of these phenomena on humans on Earth.

Diving into the Sun

- Obviously, we can't actually dive into the Sun, but suppose that we could pull the feat off without the unfortunate consequence of incineration. The Sun is made entirely of gas, not liquid. But, even so, it is a sufficiently dense gas that it behaves like a liquid. For the purposes of our imaginary dive, thinking of the Sun as a liquid can be a helpful way of conceptualizing its fluid nature.

- The visible surface of the Sun is what we refer to as the photosphere. It is the glowing yellow surface that we see as the shining Sun in the sky. Like the surface of a swimming pool, the Sun's surface shimmers and undulates. However, unlike the water in a pool, the Sun's surface is opaque. It appears as a wall of light, and you can't see into it, let alone see down to the bottom.

- As you prepare to dive in, your first sensation would be that of an intense blast of heat. The Sun's temperature at its surface is about 6000 degrees Celsius—hot enough to melt any solid. But unlike a fluid, the Sun at its surface is extremely rarefied; it is comparable in density to air. And because the density at the Sun's surface is so low, you would experience no buoyancy, nothing to float you back up to the top. Instead, you'd sink faster and faster.

- In fact, just like falling through a cloud, you would simply free-fall. If you could continue to fall at this pace, it would take about 30 minutes to reach the center of the Sun 700,000 kilometers down. However, you wouldn't free-fall all the way down, because the Sun's density increases steadily toward its center.

- By the time you got about halfway down, the density would be comparable to water. And here, like being in a pool, it would require effort to push farther down against the buoyancy pushing you back up.

- If you could continue plunging downward toward the Sun's core, you'd push through increasingly dense fluid. Eventually, you'd be pushing through a fluid so dense that it would be like swimming through tar. Then, at the bottom, you'd be in fluid more than 100 times the density of water but that was still gas. This is because the physics of matter are such that a gas can remain a gas without solidifying, even at extremely high densities, as long as the temperature is high enough.

- Finally, down in the Sun's core, the temperature would be about 15 million degrees Celsius. This is the central furnace of the Sun, the place where the Sun's energy is ultimately generated through the process of nuclear fusion. And the nuclear fusion in the cores of stars like the Sun is the most important thing that stars in this stage of the life cycle do.

The Sun's Energy
- Floating back up to the surface, the first region of the Sun's interior that you'd pass through is the hot, dense core itself. The core comprises approximately 20% of the Sun's interior. This is the region where the energy that percolates its way to the surface is generated.

- From the core, the energy has to work its way toward the surface. But it can't just stream freely to the surface because it has to pass through layer after layer of dense, opaque gas. So, we say that the energy from the core is transported or carried from the core toward the surface by a few different mechanisms. It takes a long time for the energy to be carried all the way to the surface.

- Emerging from the dense core, for the next 50% or so of the Sun's interior, we move through a region known as the radiation zone. In this region, the energy from the nuclear furnace at the core is carried principally in the form of light—or radiation.

- The gas comprising the Sun's interior in the radiation zone is essentially a static medium through which light energy passes. As we pass through this radiation zone, the Sun's density drops from about 10 times the density of water at the bottom to about a tenth the density of water closer to the surface.

- Next, we transition from the radiation zone into a region known as the convection zone. This is an important transition because here the energy percolating up from the bottom is no longer transported principally in the form of light. Rather, it is transported by bulk motion of the fluid. The gas roils and boils, producing large-scale

Sunlight is solar radiation that is visible at the Earth's surface.

currents of movement similar to convective motions of magma beneath the Earth's surface.

- In fact, the upper 25% or so of the Sun's interior is dominated by these convective motions. Consequently, if buoyancy weren't strong enough to return us all the way to the surface in our imaginary journey, the convective motions would propel us the rest of the way.

- Finally, as we push back up through the photosphere and look back down, we see that the Sun's surface is not smooth and uniform but, rather, roiling all over, much like the surface of a pot of boiling water. These boiling motions of the Sun's surface are called granulation, because of how the churning surface appears from Earth.

- Now that we've reemerged from the Sun's interior in our imaginary swim from the center to the surface, let's consider how the energy generated in the Sun's core percolates its way out. The energy that we ultimately see as the Sun's bright yellow glow at its surface is generated down at the bottom, in the Sun's core, through the process of nuclear fusion.

- The energy rising from the Sun's core takes a while to percolate to the surface because the energy, in the form of light, cannot simply stream out through the dense interior. Rather, it slowly diffuses its way out. The light energy generated in the core travels a very small distance before being absorbed in the surrounding layer, then reemitted, then reabsorbed, and so on.

- Finally, that energy emerges from the surface, about 17,000 years after it was initially produced. If the Sun were to suddenly stop generating energy right now, it would continue to shine for 17,000 years before the photons produced just now finally worked their way to the surface.

- The surface of the Sun that we see glowing a bright yellow—the photosphere—is not the end of the story. In fact, the Sun's influence extends far beyond the photosphere, as the roiling of the Sun's surface causes heat to be steadily injected into the Sun's hot, tenuous (meaning "low density") outer layers called the chromosphere and corona.

- The chromosphere is located immediately above the Sun's photosphere. It emits a distinctly red pigment caused by the fact that, whereas the photosphere emits a continuous rainbow of colors, the chromosphere's tenuous nature causes it to emit only certain specific colors of light. One of the most prominent colors that it emits in visible light is a bright red color emerging from highly heated hydrogen atoms.

- We don't ordinarily see this red light from the chromosphere because the chromosphere is very tenuous compared to the photosphere and emits very faintly in comparison. Its light is normally drowned out by the glare of the photosphere, but during a total solar eclipse, the chromosphere can become visible to the naked eye.

- The chromosphere is estimated to be approximately 20,000 degrees— so hot that this tenuous layer of gas emits primarily ultraviolet light. In fact, this is where most of the Sun's ultraviolet light is produced. Thankfully for us, most of it is absorbed by the Earth's atmosphere, limiting the damage it can do to our skin and eyes.

- Above the chromosphere is an even hotter layer called the corona, which is extremely hot—about 2 million degrees Celsius. At such an extremely hot temperature, it emits primarily X-ray light. Even so, the corona can be seen in visible light, but because it's extremely tenuous, it glows dimly and can be seen only during a total solar eclipse.

The Sun's Magnetic Nature
- The Sun's surface is frequently pocked with dark spots that we call sunspots. These spots are darker than the surrounding surface because they represent cool regions—typically a few hundred degrees cooler than the rest of the photosphere—so, by virtue of being cooler, they emit less light.

- Sunspots represent the footpoints of the Sun's magnetic field—the points on the Sun's surface where the magnetic field pokes out and pokes back in. Magnets always have a north pole and a south pole. Sure enough, sunspots almost always appear in pairs. The sunspot pairs correspond to the north and south poles of a strong localized magnetic field on the Sun. The Sun as a whole has a magnetic field that behaves similarly to the Earth's magnetic field.

- However, because the Sun behaves like a fluid, its global magnetic field is constantly twisted and distorted by the Sun's rotation. This twisting causes the Sun's magnetic field to become kinked, and the places where these magnetic kinks occur are the places where the magnetic field pokes up through the surface, creating one sunspot, and back down into the surface, creating the paired sunspot.

- The kinked magnetic field poking up and back down through the surface causes the upwelling heat from below to be impeded at those points. Consequently, the surface at these spots becomes cooler than the surrounding surface, and we end up with a pair of cool—hence, dark—sunspots.

- The same magnetic field, if it becomes sufficiently kinked, can protrude far above the surface. This protruding magnetic field can be seen in the form of what we call prominences, gigantic loops that carry hot gas from the photosphere up into the surrounding layers. In extreme cases, these prominences can be quite large, elevating hot gas to heights of a tenth or more of the Sun's radius above the surface.

- When these magnetic loops erupt, not only does an erupting prominence inject gas and heat into the corona, but it can also release its energy violently enough to expel the heated gas far out into space. We call such an event a coronal mass ejection. These coronal mass ejections can fling the ejected gas as far as the Earth and beyond, producing a gust that we call a solar storm, which we see on Earth as aurorae and which can even knock out communications satellites in Earth's orbit.

- These erupting magnetic fields on the Sun generate an enormous amount of heat, producing flashes of X-ray radiation called flares. When these extremely powerful X-ray flares erupt, we often see particularly strong "gusts" of gas launched out from the Sun. Fortunately for us, such extreme flares and coronal mass ejections are rare. Much more common are the more modest "puffs" of gas ejected from the Sun.

- Together, the strong gusts and the weaker puffs lead to a steady stream of gas flowing from the Sun's surface out into the solar system. We refer to this stream as the solar wind. This wind, which permeates interplanetary space, flows fastest from the Sun's north and south poles, with a speed of some 1000 kilometers per second. This wind extends the Sun's influence far from its surface to the farthest reaches of the solar system.

Suggested Reading

Big Bear Solar Observatory, "Solar Movies."

Simon, *The Sun*.

Questions to Consider

1. In what ways does the Sun represent our best opportunity for understanding the inner workings of all stars?

2. Try to imagine what the core of the Sun would "feel" like if touched. Consider its density, yet the fact that it is a gas. Would it be hard or soft? How would it compare to the hardest solids?

A Portrait of Our Star, the Sun
Lecture 12—Transcript

Congratulations! You've reached the middle of this course on the life cycle of stars. And so appropriately, we're now going to shift our attention to stars in the middle of their lives. Fortunately, for stars there are no midlife crises to worry about. In fact, stars at midlife are in the most stable, secure, and productive stage of their lives.

Now, as we've already learned, all stars are not all the same. Like studying a population of people, when we study the stars we find a variety of characteristics. But just like people, especially when you consider adults in midlife, there are certain essential features that are common to all stars. So we can learn a great deal about the stars generally by choosing one representative star and studying it in as much detail as possible. The nearest star to us, the one that we can study in the most exquisite detail, is our Sun, and it happens to be a star firmly in the midlife stage. So we'll start our exploration of midlife stars here, right close to home.

In this lecture, we'll take an imaginary trip into the Sun, a Fantastic Voyage of sorts, as if plunging in at the top and diving down to its core, and we'll talk about the different layers of the Sun's interior as we go. Then we'll take a step back out to explore the phenomena occurring on the Sun's surface and the effects of these phenomena on us here at Earth. So let's dive in.

Literally, imagine yourself about to dive in. Imagine yourself about to dive into a pool of water. You're standing at the edge looking down at the water. You see the shimmering surface, smooth but undulating lightly. Through the clear surface you can see down to the bottom of the pool. You aim your head down, give a little jump, and plunge in. As you dive in, your body immediately feels the cool water enveloping you. You glide through the water effortlessly, but its density does offer resistance, and so your downward motion quickly slows down. In fact, the tendency to float is quite strong, and so if you wanted to reach something at the bottom of the deep end of the pool, you'd have to propel yourself downward with force. When you stop forcing yourself downward, you rapidly bob back up to the top.

How would diving in to the Sun compare? Well, obviously we can't literally dive into the Sun, but let's take a leap of imagination here and suppose that we could pull the feat off without the unfortunate consequences of incineration. How would the experience compare to diving into a swimming pool? The first thing to say is that the Sun is made entirely of gas, not liquid. But even so, it is a sufficiently dense gas that it behaves like a liquid. So for the purposes of our imaginary dive, thinking of the Sun as a liquid can be a helpful way of conceptualizing its fluid nature.

As you look down at the surface of the Sun from above, preparing to dive in, what do you see? The visible surface of the Sun is what we refer to as the photosphere. It is the glowing yellow surface that we see as the shining Sun in the sky. Like the surface of the swimming pool, the Sun's surface shimmers and undulates. However, unlike the water in the pool, the Sun's surface is opaque, so it appears as a wall of light, and you can't see into it, let alone see down to the bottom.

As you prepare to dive in, your first sensation would be that of an intense blast of heat. The Sun's temperature at its surface is about 6000 degrees Celsius. That's hot enough to melt any solid. For example, the temperature of molten lava is about 1000 degrees Celsius. So this dive is not going to be nearly as cool and refreshing as that dive into the swimming pool. But unlike a fluid, such as lava, which has the density of rock, the Sun at its surface is extremely rarefied. It is comparable in density to air. So plunging in to the Sun's surface, we can imagine diving into a cloud. A glowing, hotter-than-lava cloud. And because the density at the Sun's surface is so low, you would experience no buoyancy, nothing to float you back up to the top. Instead, you'd sink faster and faster. In fact, just like falling through a cloud, you would simply free fall, more of a swan dive from a great height!

If you could continue to fall at this pace, it would take about 30 minutes to reach the center of the Sun 700,000 kilometers down. But in fact, you wouldn't free fall all the way down, because the Sun's density increases steadily toward its center. By the time you got about halfway down, the density would be comparable to water. So finally you'd be in surroundings similar to the swimming pool, at least as far as the density is concerned.

And here, like being in the pool, it would require effort to push further down against the buoyancy pushing you back up.

If you could keep yourself plunging downward toward the Sun's core, you'd push through increasingly dense fluid. Eventually, you'd be pushing through a fluid so dense it would be like swimming through tar. And then, at the bottom, you'd be in fluid more than 100 times the density of water. Still gas, but more than 10 times denser than solid steel! This is because the physics of matter are such that a gas can remain a gas without solidifying, even at extremely high densities, as long as the temperature is high enough.

Finally down in the Sun's core, the temperature would be some 15 million degrees Celsius. This is the central furnace of the Sun, the place where the Sun's energy is ultimately generated through the process of nuclear fusion, which we'll discuss in our next lecture. And as we'll see throughout the remainder of our course, the nuclear fusion in the cores of stars, like the Sun, is the most important thing that stars in this stage of the life cycle do. This is their life's work.

But now let's imagine floating back up to the surface. What are the different regions of the Sun's interior that we'd pass through? Well, the first region is the hot, dense core itself. The core comprises approximately 20% of the Sun's interior. This is the region where the energy that percolates its way to the surface is generated. From the core, the energy has to work its way toward the surface. But it can't just stream freely to the surface, because it has to pass through layer after layer of dense, opaque gas. So we say that the energy from the core is transported, or carried, from the core toward the surface by a couple of different mechanisms. And as we'll see in a moment, it takes a good, long while for the energy to be carried all the way to the surface.

So then, emerging from the dense core, for the next 50% or so of the Sun's interior, we move through a region known as the radiation zone. In this region, the energy from the nuclear furnace at the core is carried principally in the form of light, or radiation. The gas comprising the Sun's interior in the radiation zone is essentially a static medium through which light energy passes. As we pass through this radiation zone, the Sun's density drops from about 10 times the density of water at the bottom to about a tenth the density of water closer to the surface.

Next we transition from the radiation zone into a region known as the convection zone. This is an important transition, because here, the energy percolating up from the bottom is no longer transported principally in the form of light. Rather, it is transported by bulk motion of the fluid. The gas roils and boils, producing large-scale currents of movement similar to convective motions of magma beneath the Earth's surface. In fact, the upper 25% or so of the Sun's interior is dominated by these convective motions. Consequently, if buoyancy weren't strong enough to return us all the way to the surface in our imaginary journey, the convective motions would propel us the rest of the way.

And finally, as we push back up through the photosphere and look back down, we see that the Sun's surface is not smooth and uniform, but rather, it's roiling all over, much like the surface of a pot of boiling water. These boiling motions of the Sun's surface are called granulation because of how the churning surface appears from Earth.

As seen in this movie of the Sun, the surface appears to be composed of a large number of bright granules separated by darker inter-granular lanes. Let's watch for a moment as these granules churn and bubble. The brightness of the granules is due to the upwelling motion of the underlying hot gas. The intergranular lanes appear dark, because they represent the cooler gas falling back down, only to be heated again and forced back up, producing a steady, churning, roiling motion.

Now that we've reemerged from the Sun's interior in our imaginary swim from the center to the surface, let's consider how the energy generated in the Sun's core percolates its way out. The energy that we ultimately see as the Sun's bright yellow glow at its surface is generated down at the bottom, in the Sun's core, through the process of nuclear fusion. This is similar to the way that the heat escaping from a boiling pot of water as rising steam has its origin in the burner beneath the pot.

But the energy rising from the Sun's core takes a good, long while to percolate to the surface. This is because the energy, in the form of light, cannot simply stream out through the dense interior. Rather, it slowly diffuses its way out. The light energy generated in the core travels a very small distance before being absorbed in the surrounding layer, then reemitted, then reabsorbed, so on and so on. Finally, that energy emerges from the surface, about 17,000 years after it was initially produced. If the Sun were to suddenly stop generating energy right now, it would continue to shine for 17,000 years before the photons produced just now finally worked their way to the surface.

The surface of the Sun that we see glowing a bright yellow, the photosphere, is not the end of the story. In fact, the Sun's influence extends far beyond the photosphere, as the roiling of the Sun's surface causes heat to be steadily injected into the Sun's hot, tenuous outer layers called the chromosphere and the corona. When I use the word tenuous, by the way, I mean that these layers have very low density.

The chromosphere is located immediately above the Sun's photosphere. The word chromosphere literally means sphere of color, in reference to the distinctly red pigment that it emits. That red glow is caused by the fact that, whereas the photosphere emits a continuous rainbow of colors, the chromosphere's tenuous nature causes it to emit only certain specific colors of light. One of the most prominent colors that it emits in visible light is a bright red color emerging from highly heated hydrogen atoms. We don't ordinarily see this red light from the chromosphere, because the chromosphere is very tenuous compared to the photosphere, and so emits very faintly in comparison. Its light is normally drowned out by the glare of the photosphere. But during a total solar eclipse, the chromosphere can become visible to the naked eye.

The chromosphere is quite hot. Remember I said that the photosphere has a temperature of about 6000 degrees Celsius. Well, the temperature of the chromosphere is estimated to be approximately 20,000 degrees. That's so hot that this tenuous layer of gas emits primarily ultraviolet light. In fact, this is where most of the Sun's UV light is produced. Thankfully for us, most of this UV light is absorbed by the Earth's atmosphere, limiting the damage it can do to our skin and eyes.

Above the chromosphere is an even hotter layer called the corona. The corona is extremely hot, about 2 million degrees Celsius. Now, at such an extremely hot temperature, it emits primarily X-ray light. Even so, the corona can be seen in visible light, but because it's extremely tenuous, it glows dimly and can be seen only during a total solar eclipse. In an eclipse, the Moon blocks out the intense glare of the Sun's photosphere, allowing us to see the chromosphere and the faint, eerie glow of the corona. In fact, it was through eclipses of the Sun that its corona was first identified, appearing clearly in some of the earliest drawings of the Sun's appearance during total eclipse, long before we had the means to study the Sun in X-ray light.

How do the chromosphere and the corona come to be so hot? To understand that, we need to look at the Sun's photosphere once again, but this time, paying special attention to the manifestations of the Sun's magnetic nature. The Sun's surface is frequently pocked with dark spots that we call sunspots. These spots are darker than the surrounding surface, because they represent cool regions, typically a few hundred degrees cooler than the rest of the photosphere. And so by virtue of being cooler, they emit less light.

An important clue to the physical cause of sunspots lies in their detailed appearance. Look at a sunspot, and you'll notice a radial pattern at the edges, like spokes on a wheel. Perhaps you've played with or seen a toy in which small metal filings can be moved around with a small magnet. Where the magnet makes contact with the toy, the metal filings align in a radial pattern. If the pattern of the metal filings around the magnet and the pattern at the edge of a sunspot seem similar to you, well, that's not a coincidence!

In fact, sunspots represent the footpoints of the Sun's magnetic field, the points on the Sun's surface where the magnetic field pokes out and pokes back in. As you probably know, magnets always have a north pole and a south pole. Well, sure enough, sunspots almost always appear in pairs. The sunspot pairs correspond to the north and south poles of a strong, localized magnetic field on the Sun. The Sun as a whole has a magnetic field that behaves similarly to the Earth's magnetic field. However, because the Sun is a fluid, its global magnetic field is constantly being twisted and distorted by the Sun's rotation. This twisting causes the Sun's magnetic field to become kinked, sort of like twisting up a rubber band. And the places where these magnetic kinks occur are the places where the magnetic field pokes up through the surface, creating one sunspot, and back down into the surface, creating the paired sunspot. The kinked magnetic field poking up and back down through the surface causes the upwelling heat from below to be impeded at those points. Consequently, the surface at these spots becomes cooler than the surrounding surface, and voila, we end up with a pair of cool, hence dark, sunspots.

So, what does this all have to do with how the chromospheres and corona come to be so hot, especially if the action of the Sun's magnetic field is to cause regions of the photosphere to be cooler? Well, the same magnetic field, if it becomes sufficiently kinked, can protrude far above the surface. This protruding magnetic field can be seen in the form of what we call prominences, gigantic loops that carry hot gas from the photosphere up into the surrounding layers. In extreme cases, these prominences can be quite large indeed, elevating hot gas to heights of a tenth or more of the Sun's radius above the surface.

For example, in this image, we see a prominence with a total length of about half a million miles, which would dwarf the Earth by comparison. When such large prominences occur, the kinked magnetic field is eventually forced to snap, releasing the contained gas and depositing the energy of the field into the chromosphere and the corona. Think of that analogy of the kinked rubber band again, only now imagine the rubber band becoming so kinked that it snaps. The tension of the stretched rubber band suddenly releases, causing the rubber band to quickly untwist and release a large amount of energy that can sting your hand!

The rubber band analogy is a good one for understanding the ways in which the twisting and kinking of the Sun's magnetic field causes it to snap. But what this analogy doesn't convey is that these twisted magnetic field lines also form tubes that can contain and transport gas from the Sun's surface. Here it is helpful to think of a straw; the straw is long, thin, and flexible. And fluid can flow along within it from one end to the other.

We can literally see these straw-like structures in detailed movies of the Sun's magnetic field. In these images we see that these magnetic tubes, in fact, often come in large numbers of thin tendrils that together form a canopy-like structure that we call arcades. Why does the gas from the Sun's surface flow within and along these magnetic tubes? That's because the gas is hot enough that many of the atoms within it have had one or more of their electrons stripped away, leaving them electrically charged. An electrically charged particle, in the presence of a magnetic field, will be forced to move along that magnetic field. So the hot gas from the Sun's surface can flow along these magnetic tubes but not across them.

Now, coming back to what happens when these magnetic loops erupt, not only does an erupting prominence inject gas and heat into the corona, it can also release its energy violently enough to expel the heated gas far out into space. We call such an event a coronal mass ejection. These coronal mass ejections can fling the ejected gas as far as the Earth and beyond, producing a gust that we call a solar storm, which we see on Earth as aurorae, and which can even knock out communications satellites in Earth orbit. We'll discuss this space weather in a future lecture.

These erupting magnetic fields on the Sun generate an enormous amount of heat, producing flashes of X-ray radiation called flares. These flares come in a large range of energies, reflecting the range of energies that the magnetic fields producing them possess when they snap. The least energetic X-ray flares are called A-class, and the most energetic are called X-class. An X-10 flare is 10 times more energetic than an X-1.

One of the strongest X-ray flares observed on the Sun was the so-called Halloween flare of 2003; actually, it occurred a couple of days before Halloween. This flare was classified an X-40. When these extremely powerful X-ray flares erupt, we often see particularly strong gusts of gas launched out from the Sun. Fortunately for us, such extreme flares and their associated coronal mass ejections are rare. Much more common are the more modest A-class flares, which lead to more modest puffs of gas ejected from the Sun.

All together, the strong gusts and the weaker puffs lead to a steady stream of gas flowing from the Sun's surface out into the solar system. We refer to this stream as the solar wind. This wind, which permeates all of interplanetary space, flows fastest from the Sun's north and south poles, with a speed of some 1000 kilometers per second. This wind extends the Sun's influence far from its surface to the farthest reaches of the solar system. Like a child's sprinkler, spraying water emanating from the bottom out through flailing tubes into the surrounding yard, the Sun's energy flows from where it is generated at the center, up through the surface, and out through erupting tubes into the surrounding space.

Back beneath the Sun's surface, things are calmer, but there is complexity here as well. The Sun rotates once every 26 days or so. This rotation drives coherent fluid motions just beneath the Sun's surface. Just like the Gulf Stream current that flows within the Atlantic Ocean, these fluid currents move in large-scale, circular motions. These motions, moreover, differ at different latitudes on the Sun. The Sun's surface rotates in a manner that we describe as differential rotation. The equator rotates somewhat faster than the poles. The equator rotates once every 25 days, where at the poles, the surface rotates once every 36 days.

A similar pattern of differential rotation is seen also on Jupiter and gives rise to the vortical motions of clouds churning on Jupiter's surface. On the Sun, a manifestation of differential rotation is in the so-called Butterfly diagram of sunspot motions on the Sun's surface. Sunspots initially appear at higher latitudes, and then drift toward the equator and get pulled, or stretched, forward as they do, as a consequence of their drift from the slower rotation at high latitudes to the faster rotation at the equator. Interestingly, as the Butterfly diagram motion of the sunspots shows, the Sun undergoes a kind of magnetic cycle every 11 years. At the start of this 11-year cycle, there are relatively few sunspots, mostly at higher latitudes on the Sun's surface. Then, at the peak of the activity cycle, there are many sunspots, mostly congregated near the equator.

We know that this cycle must be intimately related to the Sun's magnetic nature, both because it manifests directly in the number and position of sunspots, which as we discussed, represent the footpoints of the Sun's magnetic field, and, because the magnetic polarity of sunspot pairs reverses with each 11-year cycle. All of this must be fundamentally connected to the

Sun's rotation and to the differential nature of that rotation. But the details of what causes the 11-year cycling of sunspots are still poorly understood.

Because sunspots are relatively easy to study on the Sun, records of their numbers and of their positions on the Sun's surface date back hundreds of years. The astronomers Schwabe and Wolf began detailed studies in the mid-1800s using data collected as far back as Galileo in the early 1600s. In addition to first noticing the 11-year cyclical pattern of sunspots, the historical record also clearly shows a surprising extended period from about 1645 to about 1715 during which there were remarkably few sunspots; this is referred to as the Maunder Minimum after the husband-wife astronomer team, Edward and Annie Maunder. Interestingly, later analyses of the sunspot record, together with weather patterns on Earth, show a curious correlation between the Maunder Minimum on the Sun and an event referred to as the Little Ice Age on Earth, a time during which weather data show unusually harsh winters on Earth. Apparently, subtle changes within the Sun have palpable effects for us here on Earth.

These phenomena that we've described on the Sun manifest themselves in the variety of forms of light we receive from the Sun. The visible surface that we see with our eyes, the photosphere, principally emits the ROYGBIV visible light with which we are all so familiar. The sunspots on the Sun's surface, which appear dark in visible light, are not in fact totally black, but rather, emit principally infrared light. Meanwhile, the hotter chromosphere radiates principally ultraviolet light, and the very hot corona emits mainly X-ray radiation. Magnetic flares on the Sun produce strong flashes both in X-rays and in radio light.

I hope that through this lecture you've developed a richer portrait of our Sun than that of simply a yellow glowing ball in the sky. The Sun is a gigantic glowing ball of gas, but it has a rich inner structure with strata, ranging from the extremely hot and dense core—denser than solid lead—to the more rarefied outer layers. The heat produced deep within the Sun percolates toward the surface, first as radiation, and then as convective currents of fluid motion carrying that heat to where it is ultimately liberated at the surface as the intense glow of radiation that we see on Earth. We've also seen that the Sun is a magnetic entity, with magnetically driven eruptions that flash in X-ray light and that drive gusts of solar wind into interplanetary space.

Most importantly, the Sun is the nearest middle aged star to us. It is our midlife star. And just as we parents provide a template of adulthood to our children, through an intimate understanding of our Sun, we can understand in rich detail how other adult stars work, representing this crucially important stage of the stellar life cycle. Our next lecture will begin the amazing story of how the Sun actually generates the heat and light upon which our lives, and all life on Earth, depends.

$E = mc^2$—Energy for a Star's Life
Lecture 13

I n this lecture, you will learn about the process by which the Sun shines, and by extension, you will come to understand how all stars shine. The Sun is a "nuclear factory," alchemizing the simplest of all elements, hydrogen, into helium and carbon—the same carbon that is the basis of all life. This is the result of a mighty struggle against gravity, a matter of life or death for the Sun. It's a 10-billion-year struggle that the Sun, through the power of $E = mc^2$, is able to withstand for awhile.

The Energy of Our Sun

- For its entire existence—from its birth and throughout its life and, finally, to its death—the Sun is wholly given over to a mighty battle, and this mighty battle is to hold itself up against the unrelenting crush of gravity.

- As long as the Sun possesses mass—and the Sun has a mass equivalent to one million Earths—gravity will not relent. Unless the Sun stands up for itself, gravity will crush it into nonexistence. So, for as long as it is able, the Sun does the only thing it can do to survive, which is push back. The Sun produces heat, which creates outward pressure against the compressing force of gravity.

- The Sun shines a total of 400 trillion trillion watts. The source of the Sun's immense energy output is connected to its immense mass. Energy (E) is the same as mass (m) times a conversion factor (c^2, which is just a number): $E = mc^2$. In essence, this equation— Einstein's most famous equation—says that mass is energy.

- This is a deeply profound idea. We normally think of matter and energy as fundamentally different entities. Matter is the "stuff" of the world, whereas energy is what makes it go. However, this equation says that, in fact, matter is another form of energy. Therefore, mass, being a form of energy, can be turned into another form of energy, such as heat or light. That's what stars do; they convert mass to energy. And the way they do it is nuclear fusion.

- To understand why converting matter into energy can create so much energy as to generate 400 trillion trillion watts for billions of years, we have to appreciate the conversion factor, c^2, which is the speed of light—a fundamental constant of nature. It's a really big number: 300 million meters per second. Now, square 300 million, and you've got an enormously big number: 90,000 trillion. In other words, converting even a tiny amount of matter produces a lot of energy because the conversion factor that relates matter and energy is so big.

Nuclear Fusion
- The basic process by which the Sun taps into the power of $E = mc^2$ is the fusion of light elements into heavier elements. Fusion is the process of sticking atoms together to make heavier atoms. This is not the chemical process of attaching separate atoms to one another to make molecules, such as 2 hydrogen atoms and 1 oxygen atom making water molecules, H_2O; rather, it is the welding together of the *nuclei* of atoms to make entirely new, heavier atoms— new elements.

- Fusion requires enormous pressure, which so far is only achievable at the centers of stars. It turns out that sticking the nuclei of atoms together involves a loss of mass, which is released from the newly formed atom in the form of very energetic light.

- The simplest example of this process is the fusion of hydrogen (element number 1 on the periodic table) into helium (element number 2 on the periodic table). In fact, this is the very process that fuels the Sun now and fuels most stars for most of their lives. The Sun, like most stars, is 75% hydrogen, so it has a lot of hydrogen to work with.

- The nucleus of a hydrogen atom is very simple: a single proton. The nucleus of a helium atom is only slightly more complex: two protons and two neutrons. The masses of protons and neutrons are very, very similar, so for now, you can think of a helium nucleus with its 2 protons plus 2 neutrons as being akin to 4 protons.

- In fact, the fusion process involves sticking 4 hydrogen atoms together; 2 of the 4 hydrogen protons become neutrons in this process. The end result of the process is that the 4 hydrogen atoms have fused to form a single helium atom—that is, 1 atom with 2 protons and 2 neutrons in its nucleus.

- Let's compare the total mass of 4 hydrogen atoms to the mass of 1 helium atom. They should be the same—right?—because the 4 hydrogen atoms were used to make the 1 helium atom. A hydrogen atom has an atomic weight of a little over 1, and a helium atom has an atomic weight of a little over 4, so a helium atom weighs almost exactly 4 times as much as a hydrogen atom—but there is a slight discrepancy. In fact, a helium atom weighs about 0.7% less than what 4 hydrogen atoms weigh. It's a tiny difference, but it makes all the difference.

- The act of manufacturing helium out of hydrogen trades a small amount of mass into a lot of energy: c^2. This energy keeps the Sun hot enough to keep up the tremendous pressure required to push back against gravity, which has not stopped squeezing. In addition, now the Sun has 4 fewer hydrogen atoms and 1 more helium atom than what it started out with. The Sun is alchemizing itself, turning itself from an immense ball of hydrogen into an immense ball of helium—about 0.7% less immense, but still less immense.

- This basic process of fusing hydrogen atoms into helium atoms can supply a star's power needs for a long time. For a star like the Sun, it can go on for about 10 billion years, but once the supply of hydrogen fuel is consumed, the Sun will stop making nuclear energy. And at that instant, without missing a beat, gravity is there, already pouncing, bearing down to try to crush the Sun.

The Periodic Table of the Elements

Period \ Group	1	2	3	4	5	6	7	8	9	10	11	12	13	14	15	16	17	18
1	1 H 1.008																	2 He 4.003
2	3 Li 6.941	4 Be 9.012											5 B 10.81	6 C 12.01	7 N 14.01	8 O 16	9 F 19	10 Ne 20.18
3	11 Na 22.99	12 Mg 24.31											13 Al 26.98	14 Si 28.09	15 P 30.97	16 S 32.07	17 Cl 35.45	18 Ar 39.95
4	19 K 39.10	20 Ca 40.08	21 Sc 44.96	22 Ti 47.88	23 V 50.94	24 Cr 52	25 Mn 54.94	26 Fe 55.85	27 Co 58.47	28 Ni 58.69	29 Cu 63.55	30 Zn 65.39	31 Ga 69.72	32 Ge 72.59	33 As 74.92	34 Se 78.96	35 Br 79.9	36 Kr 83.8
5	37 Rb 85.47	38 Sr 87.62	39 Y 88.91	40 Zr 91.22	41 Nb 92.91	42 Mo 95.94	43 Tc (98)	44 Ru 101.1	45 Rh 102.9	46 Pd 106.4	47 Ag 107.9	48 Cd 112.4	49 In 114.8	50 Sn 118.7	51 Sb 121.8	52 Te 127.6	53 I 126.9	54 Xe 131.3
6	55 Cs 132.9	56 Ba 137.3	57 La 138.9	72 Hf 178.5	73 Ta 180.9	74 W 183.9	75 Re 186.2	76 Os 190.2	77 Ir 192.2	78 Pt 195.1	79 Au 197	80 Hg 200.5	81 Tl 204.4	82 Pb 207.2	83 Bi 209	84 Po (210)	85 At (210)	86 Rn (222)
7	87 Fr (223)	88 Ra (226)	89 Ac (227)	104 Rf (257)	105 Db (260)	106 Sg (263)	107 Bh (262)	108 Hs (265)	109 Mt (266)	110 Ds (271)	111 Rq (272)	112 Uub (285)	113 Uut (284)	114 Uuq (289)	115 Uup (288)	116 Uuh (292)	117 Uus 0	118 Uuo 0

6	58 Ce 140.1	59 Pr 140.9	60 Nd 144.2	61 Pm (147)	62 Sm 150.4	63 Eu 152	64 Gd 157.3	65 Tb 158.9	66 Dy 162.5	67 Ho 164.9	68 Er 167.3	69 Tm 168.9	70 Yb 173	71 Lu 175
7	90 Th 232	91 Pa (231)	92 U (238)	93 Np (237)	94 Pu (242)	95 Am (243)	96 Cm (247)	97 Bk (247)	98 Cf (249)	99 Es (254)	100 Fm (253)	101 Md (256)	102 No (254)	103 Lr (257)

- And for a little while, it will. The Sun's core will shrink under its own weight, becoming even denser than before. But then, $E = mc^2$ will come to the rescue again, as the Sun finds itself sufficiently hot and dense to be able to once again perform nuclear fusion, now sticking helium atoms together—the same helium atoms that the Sun previously fabricated from hydrogen.

- Here, the basic fusion process turns 3 helium atoms into 1 carbon atom—each helium atom contributes 2 protons and 2 neutrons, so that makes a total of 6 protons and 6 neutrons, which is what carbon is. In fact, a carbon atom weighs a tiny bit less than the combined weight of 3 helium atoms. Once again, in the process of making a heavier atom from lighter ones, a little bit of mass is traded in for pure energy.

- Once the fusion of helium into carbon has run its course, gravity takes over once again, and this will lead to the Sun's demise. It will begin to slough off its outer layers, sprinkling carbon atoms into the surrounding cosmos. Meanwhile, its core is crushed down by gravity into an unimaginably dense and inert ball of carbon atoms—a massive diamond in the sky. Those diamonds in the sky, called white dwarf stars, have their own amazing story.

Seeing into the Sun

- There are two ways that astronomers can see into the Sun: by using strange particles called neutrinos and through the science of sunquakes (also known as helioseismology).

- The same process that the Sun uses to create energy from mass via $E = mc^2$—nuclear fusion—produces a strange particle called a neutrino as a by-product. One of the basic ways in which neutrinos are created is when a proton converts into a neutron, through a process known as beta decay. This is a part of the nuclear fusion process.

- Neutrinos are strange because they have almost no mass and move at nearly the speed of light. They are very difficult to detect. In fact, to detect them requires enormous underground particle detectors involving massive amounts of ultrapure water. But they can be detected, and our neutrino detectors see a steady stream of them pouring out of the Sun.

- Because we know how much energy the Sun produces, and we know how much energy is released in each nuclear fusion reaction, we can calculate how many of those by-product neutrinos should be produced.

- The number of neutrinos we see is the same as the number we expect if nuclear fusion is occurring at the rate required. The fact that we see solar neutrinos at all, and the fact that we see the right number of them, is a powerful confirmation of our basic understanding of nuclear fusion as the engine of why the Sun shines.

- Sunquakes are another powerful probe of the Sun's interior. A good analogy for this is earthquakes on the Earth. Earthquakes are one of the most important ways that scientists are able to determine what the interior structure of the Earth is like. That is because earthquakes produce sound waves that travel through the Earth, and the manner in which those sound waves propagate tells us the physical properties—such as the density—of the material through which they travel.

- A similar thing happens on the Sun—only it happens continuously. Hot gas at the surface of the Sun is constantly roiling and bubbling. Heat from below pushes the gas up; then, the gas cools and drops back down. This type of undulation produces the roiling motions of the Sun's surface, similar to boiling water in a pot, called granulation.

- This constant undulation of the Sun's surface causes it to vibrate. It "rings" like a bell. And just as a large, thick bell vibrates with a low tone whereas a small, thin bell vibrates with a high pitch, these vibrations tell us about the properties of the Sun's interior, such as its density. You can think of these vibrations as sound waves, because that is what they are: acoustic waves that propagate through the Sun's interior.

- When we observe the oscillations of the Sun's surface, we see a specific pattern of frequencies, just as we do when we hear the ringing of a bell. And these, in turn, give us information about what the density and temperature of the Sun is throughout its interior.

- Scientists have used the Sun's vibrations to graph the density of the Sun. When they compare the density that they measure through sunquakes to the density as predicted by the fusion model, there is exquisitely good agreement. Just as neutrinos confirm that nuclear fusion reactions occur in the Sun, sunquakes confirm that the physics of nuclear fusion must be responsible for the Sun's internal structure.

Suggested Reading

Bodanis, $E = mc^2$.
IceCube Neutrino Observatory, "IceCube Neutrino Observatory."

Questions to Consider

1. What are the most significant pieces of direct evidence from the Sun that confirm our basic understanding of nuclear fusion as the source of the Sun's power and of the internal structure of the Sun?

2. Having learned that energy and mass are actually equivalent, and having discussed how fusion converts a bit of mass into a lot of energy, how might one tap into the equivalence of mass and energy the other way around? What would it mean to convert energy into mass?

$E = mc^2$—Energy for a Star's Life

Lecture 13—Transcript

One of the basic questions we can ask about the Sun is, how does it shine? What are the physical processes that drive the prodigious quantities of light and heat that the Sun generates? Of course, we denizens of Earth have many reasons to care deeply about the Sun shining, and we have a huge stake in its continuing to do so. As warm-blooded life forms, we depend on the Sun's warmth. And we depend on the Sun's light to provide plants, through photosynthesis, with their energy so that we, in turn, can have the food we need to survive. Yet it's easy to take the Sun for granted. Everyone appreciates a sunny day, but how often do we stop to think and remember that what makes it sunny is, well, the Sun. Worse, we learn from an early age to avoid looking at the Sun (for good reasons!), and so our relationship to our life-giving star can ironically be a cold and distant one.

But not only is the Sun central to our world and our existence, it is, by virtue of being the nearest star to us, our best opportunity to understand in detail how all stars work. By understanding how the Sun shines, we begin to understand the general forces at work in the Universe that compel all stars to shine. In this lecture, we'll talk about the basic forces that drive the Sun to do what it does. The Sun is right now a star in middle age, the stage of the stellar life cycle in which stars do what is, arguably, their most important work. So by understanding this critically important stage of the Sun's life, we can start to piece together the broader narrative of the life cycles of all stars.

So to begin, and for context, let me remind you just how much light the Sun produces. The Sun shines a total of 400 trillion trillion watts—that's a four, followed by 26 zeroes. If every person on our planet turned on a thousand 100-watt light bulbs, that would still be only one trillionth of the Sun's wattage. The Sun makes a lot of light. But why? Why does the Sun make such an effort to produce so much energy? Does the Sun perform the life-giving service of heat and light as an act of cosmic kindness? Of course we know that the Sun is driven by physics, not human motives. But if we were to ascribe something like a human emotion to the Sun, how would we characterize its behavior?

The truth is, we would have to describe its motives as having nothing to do with altruism, but instead as based entirely on self-preservation. You see, for its entire existence—from its birth and throughout its life and, finally, to its death—the Sun is wholly given over to a mighty battle, and this mighty battle is to hold itself up against the unrelenting crush of gravity. What do I mean when I describe gravity as trying to crush the Sun? We ordinarily think of gravity as exerting a pull, and indeed, that is essentially how gravity works. Any two masses will attract one another, pulling the one toward the other. But the Sun experiences this pull as a squeeze, because the mass at the center of the Sun continually pulls the mass near the surface down in toward the center. And so the outer layers of the Sun are constantly bearing down on the mass at the center, thus squeezing the core with the full force of the Sun's weight.

Gravity is what initially brought the raw material of the Sun together and gave it birth. Gravity is what squeezes the Sun with the full force of its own weight. And it is this squeeze of gravity that compels the Sun to begin fusing together the nuclei of atoms, generating immense amounts of heat and light. And gravity is what will inevitably break the Sun apart, causing it to die a dramatic death. Gravity, as they say, is a harsh mistress. So long as the Sun possesses mass—and the Sun has a mass equivalent to 1 million Earths—gravity will not relent. Unless the Sun stands up for itself, gravity will literally crush it into nonexistence. So for as long as it is able, the Sun does the only thing it can do to survive, which is, it pushes back.

So how exactly does the Sun push back against the crush of gravity? Well, essentially, the Sun produces heat, which creates outward pressure against the compressing force of gravity. If you've ever watched a pot boil over, you've seen that kind of outward pressure at work. The heat of the contents inside the pot actually generates enough pressure to push the lid up off the pot. So, how does the Sun produce the heat necessary to create the outward pressure required to hold itself up? And how is it able to continuously replenish that heat, for billions upon billions of years, as it spills out into space in the form of 400 trillion trillion watts of light energy? The answer is: $E = mc^2$.

You've probably seen that little equation all your life. You've probably associated it, correctly, with Einstein, and so you've probably understood it to have some deep importance. Maybe you even learned at some point that it has something to do with the Sun and nuclear bombs. But if you're like most people, you probably couldn't explain what those connections are, let alone what it has to do, ultimately, with us and all of life on Earth. However, after our discussion today, I hope you will come to join me in regarding that equation as one of the most beautiful expressions in all of science.

Let's take a moment to put Einstein's most famous equation into historical context. For a long time, the question of what powered the Sun was unanswered. Astronomers knew how much energy the Sun generated. Remember, the Sun's luminosity is measured to be about 400 trillion trillion watts. And geologists knew how old the solar system was; age dating of rocks indicated an age of some billions of years. But what physical process could generate that much energy for that long remained a mystery.

Let's briefly consider two basic processes that might be seen as the Sun's energy sources. First, suppose that the Sun is simply burning in the way that a log in a campfire burns. A campfire is just a chemical reaction. As wood burns, chemical bonds between atoms are being broken. This liberates energy, which campers around the fire see as light and feel as heat. Could this kind of chemical reaction be generating the Sun's heat and light also? Well, typical chemical reactions, like a campfire, release in one second about one ten-millionth of a trillionth of a watt per atom. One ten-millionth of a trillionth is a tiny number; imagine slicing a birthday cake into 10 million trillion pieces! That's not a lot of energy per atom. But on the other hand, the Sun has a lot of atoms, about a billion trillion trillion trillion trillion of them—that's a one followed by 57 zeroes! So how long can a process last that produces one ten-millionth of a trillionth of a watt if it happens a billion trillion trillion trillion trillion times? The answer is about 20 thousand years. That's a long time, but nowhere near long enough. If the Sun were a campfire, it would have burnt down to ash a long, long, long time ago.

A second possibility could be gravity itself. Remember we said that the crush of gravity is what drives the Sun to try to hold itself up. Well, perhaps gravity is providing heat directly through what we might call the energy of

223

falling. A good analogy for that is a roller coaster going down a track. As you sit in the coaster and begin climbing up that steep incline, your adrenaline begins pumping in anticipation of the coming drop. After the mechanics of the coaster have contributed the energy to get you to the top of the track, all of the energy associated with this initial height at the top of the track is converted into the energy of falling. This, in turn, leads to friction on the track, which creates heat. If you were to touch the track of a coaster that just went speeding by, it would feel warm. Or think about the example of dropping a glass; it would shatter when it hit the floor. The energy of your hand initially holding the glass up against the pull of gravity is converted into the energy of motion and then into the energy of sound and heat.

So, suppose that the Sun were steadily shrinking, falling into itself, and releasing gravitational energy in the process. How long would that last? The Sun is very massive—a million Earths worth—and very large, so there is a lot of weight and a long way to fall. Is it enough? The answer is that the Sun could produce its current luminosity for about 10 million years this way. That's about a thousand times better than the chemical burning idea, but still far too little. In fact, we estimate that the current age of the Sun is about 4.5 billion years old, and we anticipate that it will go on shining for a quite while yet!

If you're thinking that the source of the Sun's immense energy output must somehow be connected to its immense mass, you'd be right. And that's where $E = mc^2$ comes in. So let's get into it. This is going to be fun. Let's look at the equation $E = mc^2$ and read it as a sentence. It says, Energy (that's the E) is the same as (that's the equal sign) mass (that's the m) times a conversion factor (that's the c^2; it's just a number). Ignore that conversion factor number for a second, and the equation says—and means quite literally: Mass is Energy.

This is a deeply profound idea. We normally think of matter and energy as distinct, as fundamentally different entities. Matter is the stuff of the world, whereas energy is what makes it go. But $E = mc^2$ says that, in fact, matter is another form of energy. And so mass, being a form of energy, can be turned into another form of energy, such as light or heat. That's what stars do; they convert mass to energy, and the way they do it is nuclear fusion.

But let's not get ahead of ourselves. First, to understand why converting matter into energy can create so much energy as to generate 400 trillion trillion watts for billions of years, we've got to appreciate that conversion factor that we set aside a minute ago, the c^2. So what is c? It is none other than the speed of light, which is a fundamental constant of nature. C is just a number, but it's a really big number: 300 million meters per second. Now square 300 million and you've got an enormously big number: 90,000 trillion—that's a 9 with 16 zeroes after it! In other words, converting even a tiny amount of matter produces a lot of energy, because the conversion factor that relates matter and energy is so very big.

So let's come back now to how the Sun actually taps into the power of $E = mc^2$. The basic process is the fusion of light elements into heavier elements. Fusion is the process of sticking atoms together to make heavier atoms. This is not the chemical process of attaching separate atoms to one another to make molecules, such as two hydrogens and one oxygen making water molecules, H_2O. Rather, it is the welding together of the nuclei of atoms to make entirely new, heavier, atoms, new elements. It requires enormous pressure, which so far is only achievable at the centers of stars. It turns out that sticking the nuclei of atoms together involves a loss of mass, which is released from the newly formed atom in the form of very energetic light.

Let's look at the simplest example of this process, which is the fusion of hydrogen (element number 1) into helium (element number 2). In fact, this is the very process that fuels the Sun now and fuels most stars for most of their lives. The Sun, like most stars, is 75% hydrogen, so it has a lot of the stuff to work with. So think back to what you remember about the periodic table of elements. The nucleus of a hydrogen atom is very simple, a single proton. The nucleus of a helium atom is only slightly more complex, two protons and two neutrons. The masses of protons and neutrons are very, very similar, so for now, you can think of a helium nucleus with its two protons plus two neutrons as being akin to four protons. In fact, the fusion process involves sticking four hydrogen atoms together. Two of the four hydrogen protons become neutrons in this process, more about that in a minute. But the end result of the process is this; the four hydrogen atoms have fused to form a single helium atom, that is, one atom with two protons and two neutrons in its nucleus.

We can use any standard Periodic Table of the Elements to look up the weight of any element. So let's compare the total mass of four hydrogen atoms to the mass of one helium atom. They ought to be the same, right, since the four hydrogen atoms were used to make the one helium atom? At first glance all appears kosher; a hydrogen atom has an atomic weight of a little over one and, a helium atom has an atomic weight a little over four. So, yes, as expected, a helium atom weighs almost exactly four times as much as a hydrogen atom. But look at the atomic weights on the Periodic Table more closely, and you'll see a slight discrepancy. In fact, a helium atom weighs about 0.7% less than what four hydrogen atoms weigh. It's a tiny difference, but it makes all the difference!

The act of manufacturing helium out of hydrogen trades a small amount of mass into a lot of energy. Remember c^2! This energy keeps the Sun hot enough to keep up the tremendous pressure required to push back against gravity, which has not stopped squeezing. Not only that, but now the Sun has four fewer hydrogen atoms and one more helium atom than what it started out with. The Sun is alchemizing itself, literally turning itself from an immense ball of hydrogen into an immense ball of helium, okay, about 0.7% less immense, but immense still.

This basic process of fusing hydrogen atoms into helium atoms can supply a star's power needs for a long time. For a star like the Sun, it can go on for about 10 billion years. But then what? Well, once the supply of hydrogen fuel is consumed, the Sun will stop making nuclear energy. And at that instant, without missing a beat, gravity is there, already pouncing, bearing down to try to crush the Sun. And for a little while, it will. The Sun's core will shrink under its own weight, becoming even denser than before. But then, $E = mc^2$ will come to the rescue again, as the Sun finds itself sufficiently hot and dense to be able to, once again, perform nuclear fusion, now sticking helium atoms together, the same helium atoms that the Sun previously fabricated from hydrogen.

Here, the basic fusion process turns three helium atoms into one carbon atom; each helium atom contributes two protons and two neutrons, so that makes a total of six protons and six neutrons, which is what carbon is. And guess what, look up the weight of a carbon atom and you'll find that, sure enough, it weighs a tiny bit less than the combined weight of three helium atoms. Once again, in the process of making a heavier atom from lighter ones, a little bit of mass is traded in for pure energy.

Let's pause for a moment in the story to fully appreciate the significance of how far we've come. Carbon is the basic building block of all living things. It is quite literally the stuff of life. It is in every cell in your body. It is the basic ingredient of the photosynthesis process that produces the biomass that we ultimately consume. Why should carbon be such an important, essential element? The answer is, because it is abundant, and it is abundant because it is one of the basic steps in the nuclear fusion sequence by which stars like the Sun—for a while—generate the energy they need to survive against gravity.

For a star like the Sun, this fusion sequence of hydrogen, to helium, to carbon is the end of the line. Once the Sun uses up the helium in its core and converts it all into carbon, it will not be able to initiate another round of fusion. The reason is essentially that the Sun is not sufficiently massive to permit the squeeze of gravity to become strong enough to force the fusion of those carbon atoms into still heavier ones. Stars that are much more massive than the Sun do continue the fusion sequence to heavier elements, but more about that later.

But coming back to the Sun, what happens once the fusion of helium into carbon has run its course? Well, gravity takes over once again, and this will lead to the Sun's demise. It will begin to slough off its outer layers, sprinkling carbon atoms into the surrounding cosmos. Meanwhile, its core is crushed down by gravity into an unimaginably dense and inert ball of carbon atoms—a massive diamond in the sky. Those diamonds in the sky, called white dwarf stars, are an amazing story all their own that we'll return to in later lectures.

What an amazing story, the Sun, a self-preserving, gravity-defying, chemically alchemizing, life-giving star! But is it true? Can something so simple as $E = mc^2$ really explain all of that? If you take my word for it, I'm flattered, but I hope you're wondering to yourself how we really know. After all, we can't actually watch the nuclear reactions happening inside the

Sun. We can't stick a thermometer into the Sun's center to verify that our physical understanding of its insides is correct. We can't see into the Sun, or can we? Actually, incredibly, we can. Let's briefly talk about two ways that astronomers can see into the Sun. The first is by using a strange kind of particle called neutrinos, and the second is through the science of sunquakes, also known as helioseismology.

Let's start with neutrinos. We've already talked about the process of nuclear fusion and the way in which the Sun uses it to create energy from mass via $E = mc^2$. What I haven't told you is that the same process produces something as a by-product, a strange particle called a neutrino. One of the basic ways in which neutrinos are created is when a proton converts into a neutron, through a process known as beta decay. This is a part of the nuclear fusion process we discussed. Remember that two of the four hydrogen protons that come together to form helium become neutrons in the nucleus of the resulting helium atom, and this produces neutrinos as a by-product.

Neutrinos are strange because they have almost no mass and move at nearly the speed of light. They are very difficult to detect. In fact, to detect them requires enormous underground particle detectors involving massive amounts of ultra-pure water. But they can be detected, and our neutrino detectors see a steady stream of them pouring out of the Sun. Since we know how much energy the Sun produces, and we know how much energy is released in each nuclear fusion reaction, we can calculate how many of those by-product neutrinos should be produced. And amazingly, the number of neutrinos we see is the same as the number we expect if nuclear fusion is occurring at the rate required. The fact that we see solar neutrinos at all, and the fact that we see the right number of them, is a powerful confirmation of our basic understanding of nuclear fusion as the engine of why the Sun shines.

Sunquakes are another powerful probe of the Sun's interior. A good analogy for this is earthquakes on the Earth. You probably know that earthquakes are one of the most important ways that scientists are able to determine what the interior structure of the Earth is like. That is because earthquakes produce sound waves that travel through the Earth, and the manner in which those sound waves propagate tells us the physical properties—such as the density—of the material through which they travel.

Well, a similar thing happens on the Sun, only it happens continuously. You see, hot gas at the surface of the Sun is constantly roiling and bubbling. Heat from below pushes the gas up, then the gas cools and drops back down. This roiling motion is just like what you see in a pot of boiling water, creating the churning undulations of the surface of the water. This same type of undulation produces the roiling motions of the Sun's surface that we talked about before, called granulation.

Well, this constant undulation of the Sun's surface causes it to vibrate. It literally rings like a bell. And just as a large, thick bell vibrates with a low tone, whereas a small, thin bell vibrates with a high pitch, so the Sun's vibration tells us about the properties of the Sun's interior, such as its density. You can think of these vibrations as sound waves, because in fact, that is what they are; they are acoustic waves that propagate through the Sun's interior. When we observe the oscillations of the Sun's surface, we see a specific pattern of frequencies, just as we do when we hear the ringing of a bell. And these, in turn, give us information about what the density and temperature of the Sun is throughout its interior. So what does the Sun's pattern of vibration—its ringing—tell us?

Well, scientists have used the Sun's vibration to graph the density of the Sun throughout its interior. And what these graphs show is that the Sun's density goes from being extremely high at the core to being very low at the surface. Now, we also have very good computer models that predict the physical conditions of the Sun's interior based on the assumption that its interior heat is generated by nuclear fusion reactions. And here's the crucial point. When we compare the density that we measure through sunquakes to the density as predicted by the fusion model, we find exquisitely good agreement. And so, just as neutrinos confirm that nuclear fusion reactions occur in the Sun, so do sunquakes confirm that the physics of nuclear fusion must be responsible for the Sun's interior structure. So you don't have to take the story of nuclear fusion in the Sun on my word. Believe the neutrinos and the sunquakes.

In this lecture we've discussed the process by which the Sun shines, and by extension, we have come to understand how all stars shine. The Sun is a nuclear factory, alchemizing the simplest of all elements, hydrogen, into helium and carbon, the same carbon that is the basis of all life. This is the result of a mighty struggle against gravity, a matter of life and death for the Sun. It is a 10-billion-year struggle that the Sun, through the power of $E = mc^2$, is able for a while to withstand. Our Sun, right now, every second, is holding its own against gravity, holding strong thanks to $E = mc^2$, as simple hydrogen atoms are turned into helium. Alas, it is a struggle that the Sun must eventually lose, as we'll discuss in detail in later lectures. We've also seen how measurements of neutrinos pouring out of the Sun and of sunquakes on the Sun give us powerful confirmation of this basic picture.

This is an incredible testament to the power of science to understand one of the most profound physical processes in all of the Universe, the deep relationship between matter and energy, the manner in which stars at midlife—stars like the Sun—convert matter into energy for billions of years. Thanks to science, we have a very compelling answer to the basic question of how the stars shine.

Stars in Middle Age
Lecture 14

In this lecture, you will learn about stars during the stable, long-lived portion of their life cycle that we call the main sequence stage. During this stage, the stars are able, through the energy they generate in fusion reactions, to hold strong against gravity. However, as they use up their stores of fusionable material, they undergo a series of shorter-lived resuscitations, during which they are briefly able to withstand gravity again for a time before they finally die. Importantly, it is in the resuscitations that stars experience before death that the stars manufacture many of the elements so essential to us, including carbon and oxygen.

The Stefan–Boltzmann Law

- Recall that the Hertzsprung–Russell diagram is a graph in which the properties of stars are represented by their temperatures on the horizontal axis and their luminosities on the vertical axis. One of the most important features of this diagram is what we call the main sequence, a diagonal swath on which 90% of stars are found.

- All stars begin their lives at some place on this main sequence. Precisely where is determined by their mass—their birth weight. And because stars spend about 90% of their lifetime on the main sequence, this means that the mass of the star determines the temperature and luminosity that the star will have throughout the majority of its life.

- In addition, a star's temperature and luminosity together determine its size through a relationship called the Stefan–Boltzmann law. That relationship states that the total luminosity of a star, L, equals the amount of light radiated by each square meter of the star's surface times the total number of square meters of surface area the star has.

- The total amount of energy radiated by 1 square meter of the star's surface is given by Planck's radiation law, which says that E (the amount of light radiated each second by 1 square meter of radiating surface) is a constant, σ, times the temperature to the 4th power: $E = \sigma T^4$.

- That's the energy of 1 square meter. How many square meters are there in the surface of a star? That's just the surface area of a sphere, A, which is 4 times pi (π) times the radius squared. Altogether, the Stefan–Boltzmann law says that the total luminosity of the star is L, which is the product of the luminosity per square meter, E, times the total surface area in square meters, A. And that equals σ times temperature to the 4th power times 4π times the radius squared: $E = 4\pi r^2 \sigma T^4$.

- Using this formula, the quantities graphed in the H–R diagram—temperature and luminosity—are together connected to the radius or size of the star. For a star of any particular combination of temperature and luminosity, you can solve the equation for the star's radius. The star's temperature determines how much luminosity it can radiate per square meter, so it has to have a certain radius in order to have enough surface area to radiate the total luminosity of the star.

- The star's temperature and luminosity are themselves determined by the star's mass. Remember that mass is a star's DNA. So, then, altogether the mass of the star is what dictates all three of these basic stellar properties: temperature, luminosity, and size.

- More practically speaking, the Stefan–Boltzmann law also means that for a star at any place in the H–R diagram, representing a specific combination of temperature and luminosity, we can directly infer the size of the star.

- Fundamentally, stars in different parts of the H–R diagram represent stars at different stages of evolution. Stars start on the main sequence, then move up and to the right into red giants, and then finally swoop over the left and down into white dwarfs.

The Fusion of Helium into Carbon

- What defines the long-lived, main sequence stage of a star's life is that it has a sufficient amount of hydrogen in its core to be able to fuse helium, thereby generating the energy that it needs to create the heat and pressure required to hold itself up against gravity. That state of gravitational equilibrium is highly stable, and the star remains unchanged throughout this long middle stage of life.

- A star's mass determines where along the H–R diagram main sequence it will be. In addition, the star's mass determines how long the star will remain a main sequence star. In other words, how massive a star is determines how quickly it will use up the store of hydrogen in its core. It turns out that how quickly a star will use up the hydrogen in its core—converting it all to helium—depends on the square of the star's mass.

- To see why this is, there are two factors to consider. First, how much energy can a star of a given mass produce through fusion? Second, how quickly does that star use up its fusion energy? The first factor is dependent simply on the mass of the star: The more massive the star, the more massive its core, and the more hydrogen mass is available to convert into helium. So, all things being equal, a more massive star has more fuel to burn.

- But all things are not equal. A more massive star burns its fuel more fiercely. To be specific, the luminosity of a main sequence star is proportional to its mass to the third power. So, a more massive star has more mass to burn, but it burns through that fuel much, much more quickly. With one factor proportional to the mass and the other factor proportional to the mass cubed, we end up with a final answer of the lifetime of the star being inversely proportional to the mass squared.

- For stars of the Sun's mass and heavier, their lifetimes are—relative to the age of the universe—finite. There comes a time when these stars use up the store of hydrogen in their core, having converted it all to helium. The most massive stars do this very rapidly; the less massive stars do it more slowly. But for all of these stars, they do run out of gas, and the end does come.

- The first thing that happens then is that the star stops fusing hydrogen into helium. And at that instant, gravity, which has been there all along, bearing down on the star, begins taking advantage of the star's reduced heat output and begins choking the star, shrinking its core toward oblivion.

- But this squeezing actually acts to give the star new life, at least for a short time. By compressing the star's core, which is now made of helium, gravity heats up the star's core to an extremely high temperature. At first, this compression and heating causes the core to actually generate even more heat than it did when it was happily fusing hydrogen into helium. And that causes the core to levitate its outer layers so that they expand outward and the star as a whole swells even as its core is compressing.

- As the surface swells, it cools. So, we have a red giant star, cool at the surface but enormously luminous because of the immense surface area from which it can radiate. A relatively low-mass star like our Sun will become such a red giant. A much more massive star will undergo the same process but, by virtue of its even more swollen size and even more prodigious luminosity, will become what we call a red supergiant.

- What happens at this stage is that gravity has compressed and heated the core of the star to such a degree that a new round of nuclear fusions is able to ignite. Only this time, instead of hydrogen being fused to helium, we have 3 helium atoms being fused together to make 1 carbon. The resulting carbon atoms weigh a bit less than the 3 helium atoms that make them up, leading to a conversion of mass to energy according to $E = mc^2$.

- The moment at which the fusion of helium into carbon ignites is sudden, and like defibrillator jolting a heart back into life, the star undergoes what is known as a helium flash. Instantly, the core pulses back to life as it is once again able to generate its own energy and heat with which to hold itself up against gravity. The star is now a stable red giant or red supergiant, depending on its mass.

- But this resuscitated state does not last for long, astronomically speaking. For a star like the Sun, which had managed to support itself through fusion of hydrogen into helium for 10 billion years, this stage lasts only about 10% as long—about 1 billion years.

- Why does the fusion of helium to carbon sustain the star for such a much shorter period of time compared to the hydrogen to helium fusion? The basic answer is that fusing helium to carbon is much less efficient at producing energy than is hydrogen to helium.

The Middle-Life and Near-Death Stages of Stars' Lives

- For a star like our Sun with a mass of 1 solar mass, the star begins its life on the main sequence of the H–R diagram. It will sit there happily for about 10 billion years. Then, as hydrogen in the star's core is exhausted and completely converted to helium, the star evolves to the right and up in the diagram, as it becomes a red giant star, its surface cooling but its luminosity increasing dramatically.

- When the red giant's core compresses enough to become hot enough to start fusing helium to carbon, the star shifts a bit to the left as its surface becomes a bit hotter, and the star stably fuses helium to carbon for a while.

- Finally, once the helium is fully depleted and converted to carbon in the core, the star again shifts up and to the right, becoming a cool but extremely luminous red giant, and then at last sheds its outer layers as a planetary nebula, revealing the inert, hot carbon core, which becomes a white dwarf at the far lower left in the diagram.

- In contrast, for a much more massive star—for example, a star 10 times the mass of our Sun—the star again starts out on the main sequence. However, as it completes each stage of nuclear fusion, it swings far to the right, its surface much cooler but still at a very high luminosity—a red supergiant.

- As each successive stage of nuclear fusion initiates in the core, the star briefly returns back toward the main sequence, but not all the way. It appears as a yellow supergiant. The star swings back and forth in the diagram multiple times as the star proceeds through different stages of nuclear fusion. Finally, once the star has fused all the way to iron in the core, the star goes supernova and disappears from the H–R diagram.

Suggested Reading

DeVorkin, *Henry Norris Russell*.

Las Cumbres Observatory Global Telescope, "Hertzsprung–Russell Diagram Simulator."

Questions to Consider

1. Why do stars spend so much of their lives, about 90%, on the main sequence in the Hertsprung–Russell diagram? What is happening within them that makes them so stable for so long?

2. In what ways is the H–R diagram, and the main sequence in particular, important to astronomy beyond the life cycles of stars?

Stars in Middle Age
Lecture 14—Transcript

Already in this course we've discussed the basic idea that all stars are born, live out their lives, and eventually they die. We've also seen that the mass of a star is its most important physical attribute, determining all of its other physical characteristics, such as its temperature and its luminosity. We've also introduced the Hertzsprung–Russell diagram as a tool for studying the physical processes of stars and how they change at different stages of their lives.

In the last couple of lectures, we've also looked in detail at our own Sun as an example of what stars are like in middle age, the adult stage of their life cycles, during which they stably and comfortably perform nuclear fusion in their cores to generate the heat and pressure with which they are able to withstand the crush of gravity. For a star like our Sun, this middle age of adulthood lasts a good, long time, about 10 billion years.

In this lecture, we will develop this understanding further to trace the evolution of stars through the later stages of their lives in more detail. We will see that the manner in which they evolve is entirely predetermined by their masses and that every star moves through a specific set of characteristic stages from birth to death.

So let's jump in by introducing the Hertzsprung–Russell diagram once again. Recall that the H–R diagram is a graph in which the properties of stars are represented by their temperatures on the horizontal axis and their luminosities on the vertical axis. As we have discussed, one of the most important features of this diagram is what we call the main sequence, a diagonal swath on which 90% of stars are found. All stars begin their lives at some place on this main sequence. Precisely where is determined by their mass, their birth weight. And because stars spend about 90% of their lifetime on the main sequence, this means that the mass of the star determines the temperature and luminosity that the star will have throughout the majority of its life.

In addition, a star's temperature and luminosity together determine its size through a relationship called the Stefan-Boltzmann law. That relationship states that the total luminosity of a star, call it L, equals the amount of light radiated by each square meter of the star's surface, times the total number of square meters of surface area the star has.

The total amount of energy radiated by one square meter of the star's surface is given by the Planck radiation law, which says that E (the amount of light radiated each second by one square meter of radiating surface) is a constant, σ, times the temperature to the fourth power. So that's the energy of one square meter. Now, how many square meters are there in the surface of a star? That's just the surface area of a sphere, A, which you may know is four times pi (π), times the radius squared. So, altogether, the Stefan-Boltzmann law says that the total luminosity of the star is L, which is the product of the luminosity per square meter, E, times the total surface area in square meters, A. And so that equals σ times the temperature to the 4th power times 4π times the radius squared: $E = 4\pi r^2 \sigma T^4$.

Using this formula, we see that the quantities graphed in the H–R diagram—temperature and luminosity—are together connected to the radius, or the size, of the star. If we re-write the Stefan-Boltzmann Law so that we have the star's radius on side and the star's luminosity and temperature on the other side, we can see this directly. For a star of any particular combination of temperature and luminosity, this formula dictates what the star's radius must be. That's because the star's temperature determines how much luminosity it can radiate per square meter, and so it has to have a certain radius in order to have enough surface area to radiate the total luminosity of the star. Finally, and importantly, the star's temperature and luminosity are themselves determined by the star's mass. Remember that mass is a star's DNA. So then altogether we see that the mass of the star is what dictates all three of these basic stellar properties: temperature, luminosity, and size.

More practically speaking, the Stefan-Boltzmann law also means that for a star at any place in the H–R diagram, representing a specific combination of temperature and luminosity, we can directly infer the size of the star. For example, a star at the upper end of the main sequence is very luminous, because it is both very hot and very large. And these stars are very massive, weighing in at well over ten times the mass of our Sun. At the bottom end of the main sequence, the stars are very dim, because they are both cool and small. And these stars have very low masses, as low as about one tenth the mass of our Sun.

The H–R diagram also shows some stars at a couple of places other than the diagonal swath of the main sequence. There are stars at the upper right, meaning that they are cool but extremely luminous. From the Stefan-Boltzmann law we see that these stars must be extremely large so as to compensate for their cool temperatures yet still radiate extremely high luminosities. These are the red giant stars, so called because their cool temperatures cause them to have a red color, but they have enormous sizes, up to 100 times or more the size of our Sun.

At the lower left on the H–R diagram, we see stars that are very hot, yet dim, which from the Stefan-Boltzmann law means that they must have tiny sizes, so that they cannot radiate much light despite having extremely high temperatures. These are the white dwarfs.

What is the relationship between stars in these different parts of the H–R diagram? Fundamentally, these represent stars at different stages of evolution. Stars start on the main sequence, then move up and to the right into the red giants, then finally swoop over the left and down into the white dwarfs. We'll return to the H–R diagram in a few moments, but first let's describe what characteristics define these different stages and what's actually happening to a star when it transitions from one of these stages to another.

We start with the main sequence. This is where stars begin the main, productive, middle years of their lives. After completing the birth process, a star of a given mass will be a main sequence star, having a particular combination of temperature, size, and luminosity. For example, we're looking a G-type star on the main sequence. In other words, a star like our Sun always has a temperature of about 6000 degrees Kelvin, a luminosity of one solar luminosity, and consequently, a radius of one solar radius. The mass of such a star is always one solar mass.

For comparison, a B-type star like Spica is more massive, always with a mass of about 10 solar masses, and so it always has a temperature of about 20,000 degrees Kelvin, a radius of about 7 solar radii, and a luminosity of about 10,000 solar luminosities. And, as an example on the lower mass end, a K-type star, such as Alpha Centauri B, has a mass of about 0.9 solar masses, a temperature of 5200 degrees Kelvin, a radius of 0.85 solar radii, and a total luminosity of about half the Sun's luminosity.

During this long, stable phase of life as a main-sequence star, the star is in a happy state of gravitational equilibrium. Gravity is bearing down, always trying to crush the star, but the star is easily able to withstand the pressure by generating the right amount of heat in its core to push back against gravity. Recall that the way the star does this is through the process of nuclear fusion, in which hydrogen in the core of the star is converted into helium, and energy is released. The energy that is released is as predicted by Einstein's $E = mc^2$, the conversion of a portion of the mass of the hydrogen atoms into pure energy, the newly created helium atoms weighing a bit less than the four hydrogen atoms that went into making them. For a star like the Sun, this process is able to continue steadily for about 10 billion years.

In other words, what defines the long-lived main sequence stage of a star's life is that it has a sufficient amount of hydrogen in its core to be able to fuse helium, thereby generating the energy that it needs to create the heat and pressure required to hold itself up against gravity. That state of gravitational equilibrium is highly stable, and the star remains unchanged throughout this long middle stage of life.

Now, we've already said that a star's mass determines where along the H–R diagram main sequence it will be. In addition, the star's mass determines how long the star will remain a main-sequence star. In other words, how massive a star is determines how quickly it will use up the store of hydrogen in its core. It turns out that how quickly a star will use up the hydrogen in its core—converting it all to helium—depends on the square of the star's mass. To see why this is, let's do just a little bit of math. There are two factors to consider. First, how much energy can a star of a given mass produce through fusion? And second, how quickly does that star use up its fusion energy? The first factor is dependent simply on the mass of the star; the more massive the star, the more massive its core, and the more hydrogen mass is available to convert into helium. So all things being equal, a more massive star has, well, more fuel to burn.

But all things are not equal. A more massive star burns its fuel more fiercely. To be specific, the luminosity of a main-sequence star is proportional to its mass to the third power. So a more massive star has more mass to burn, but it burns through that fuel much, much more quickly. With one factor proportional to the mass, and the other factor proportional to the mass cubed, we end up with a final answer of the lifetime of the star being inversely proportional to the mass squared.

So now we can estimate how long stars of different masses will last as stable, main-sequence stars. The upshot is that, whereas a star like our Sun will last for about 10 billion years, a star 10 times more massive will last 10 squared—or 100 times—shorter, about 100 million years. The most massive stars, with masses of perhaps 30 times our Sun's mass, will last a thousand times shorter, or only about 10 million years. Such stars last as main sequence stars for barely as long as the birth process takes.

In contrast, a star with the minimum stellar mass of about one-tenth the Sun's mass will last 100 times as long as the Sun, or about a trillion years. The entire universe is only about 14 billion years old. So these lowest-mass stars last as main-sequence stars essentially forever. Indeed, every star less massive than our Sun that has ever been made is still a main-sequence star, and will be, for all intents and purposes, forever.

So if earthlings were looking for a more permanent home to migrate to as our Sun approaches its end, they should be looking for a K- or M-type main-sequence star. Could life be viable around such a star? Well, since K- and M-type stars are cooler than the Sun, a planet needs to orbit that star much closer than our Earth orbits the Sun in order that the temperature of the planet's surface be warm enough to sustain liquid water, which of course is essential to life as we know it. So in terms of finding a stellar neighborhood amenable for establishing a truly permanent comfortable home, it's the usual mantra of real estate: location, location, location.

But coming back to the here and now, for stars of the Sun's mass and heavier, their lifetimes are, relative to the age of the universe, finite. There comes a time when these stars use up the store of hydrogen in their core, having converted it all to helium. The most massive stars do this very rapidly; the less massive stars do it more slowly. But for all of these stars they do run out of gas and the end does come. What happens then? The first thing that happens is that the star stops fusing hydrogen into helium. And at that instant, gravity, which has been there all along, bearing down on the star, begins taking advantage of the star's reduced heat output and begins choking the star, shrinking its core toward oblivion.

But this squeezing actually acts to give the star new life, at least for a short time. By compressing the star's core, which is now made of helium, gravity heats up the star's core to an extremely high temperature. At first, this compression and heating causes the core to actually generate even more heat than it did when it was happily fusing hydrogen into helium. And that causes the core to levitate its outer layers so that they expand outward, and the star as a whole swells even as its core is compressing.

As the surface swells, it cools. And so we have a red giant star, cool at the surface, but enormously luminous because of the immense surface area from which it can radiate. A relatively low-mass star like our Sun will become such a red giant. A much more massive star will undergo the same process, but by virtue of its even more swollen size and even more prodigious luminosity, it will become what we call a red supergiant.

But how does this resuscitate the star? What happens at this stage is that gravity has compressed and heated the core of the star to such a degree that a new round of nuclear fusions is able to ignite. Only this time, instead of hydrogen being fused into helium, we have those helium atoms being fused into carbon. Three helium atoms fused together makes one carbon. And as you'll recall from our previous discussion of this process, the resulting carbon atoms weigh a little bit less than the three helium atoms that make them up, leading to a conversion of mass to energy according to $E = mc^2$. The moment at which the fusion of helium into carbon ignites is sudden. And like a defibrillator jolting a heart back into life, the star undergoes what is known as a helium flash. Instantly, the core pulses back to life as it is once again able to generate its own energy and heat with which to hold itself up against gravity. The star is now a stable red giant, or red supergiant, depending on its mass.

But this resuscitated state does not last for long, astronomically speaking. For a star like the Sun, which had managed to support itself through fusion of hydrogen into helium for 10 billion years, this stage lasts only about 10% as long, about one billion years. Why does the fusion of helium to carbon sustain the star for such a much shorter period of time, compared to the hydrogen-to-helium fusion? The basic answer is that fusing helium to carbon is much less efficient at producing energy than is hydrogen to helium. To understand that, we need to consider what physicists call the binding energy curve of the elements.

Here we're looking at a graph of such a binding energy curve of the elements, and what the graph is showing is how much energy is available to be liberated by building up the masses of different elements through the fusion of their nuclei. As we climb the binding energy curve from lighter elements to heavier ones, more energy is released. Hydrogen is off of the graph down at zero. As we climb the curve from hydrogen, we see that the jump up to helium represents the single, biggest jump in the entire curve. Now, if we were to try to go from helium to any of the next elements, say, lithium, or beryllium, or boron, that would be a drop in the curve, and that would actually require energy to be added for fusion to occur. So that kind of fusion occurs very rarely. However, the next jump from helium to carbon, we see does represent a small climb on the curve, and so, energy can be liberated, but much smaller than that from hydrogen to helium. Nitrogen, also produced as minor additional fusion process, can also release a bit of energy. Note that the binding energy curve continues to climb a bit higher, but then it peaks and turns around. We'll refer back to this important feature later.

So what happens when the helium in the star's core has all been converted to carbon? The star once again is forced by gravity to undergo a change. The moment that the helium-to-carbon fusion in the core stops, gravity is there, bearing down, and the core, once again, is compressed. As before, as the core shrinks, it heats up to very high temperatures. And as before, the enormous heat output of the collapsing core drives the surface of the star outward again. The star swells and swells; its surface cools, but radiating with an enormous luminosity thanks to the increased surface area from which it can radiate.

An interesting thing happens at this point. The increasing heat pouring out of the collapsing core drives up the temperature of the surrounding layers, which immediately around the carbon core is made of helium, and above that, is made mostly of hydrogen. The helium layer is now hot and dense enough to start fusing into carbon, and a layer of the hydrogen above that is able to start fusing into helium. Suddenly the star lurches to life again.

But here we must differentiate between stars like the Sun and those that are much more massive. For stars like the Sun, this lurch to life is, in fact, the end of the star, whereas for a much more massive star, the star is able to keep going. The dividing line is at about eight times the mass of the Sun. For a star less massive than that dividing line, such as our own Sun, that sudden lurch to life signals the death of the star. By generating an enormous burst of energy all at once, the deposition of that energy from the core into the surrounding layers of the star levitates those outer layers away from the core. In a very short time, those outer layers are puffed away from the star entirely, leaving behind the hot, exposed core of the star surrounded by an expanding bubble of what was the star's outer layers. That expanding bubble can be seen as what we refer to as a planetary nebula, eerily beautiful objects that we will discuss in detail in our next lecture. The exposed hot core of the star, made of nearly pure carbon, remains at the center of the planetary nebula. It is destined to end its life as a carbon ember, slowly cooling and fading into oblivion, what we call a white dwarf. White dwarfs are fascinating objects that we'll discuss in detail in a later lecture, along with other types of stellar corpses.

Now let's come back to pick up the story for the more massive stars, stars more than eight times the mass of the Sun. When those stars reach the stage where they have completely used up the helium in their core, having fused it into carbon, they resuscitate once again. Only now, the core is fusing carbon together with helium to make oxygen. Surrounding the core, the layer of helium that was fused from hydrogen when the star was a main-sequence star, is now hot enough to fuse into carbon. And surrounding that, a layer of pristine hydrogen is hot enough to fuse into helium, the same way that the core did earlier.

The star will continue to go through a series of fusion in the core, followed by a brief collapse under gravity to a hotter and denser state, followed by a sudden resuscitation as a new wave of fusion is able to begin. And surrounding the core are an increasing number of fusing layers as well. Each of these episodes of resuscitation is briefer and briefer, as the star grows more and more desperate to fuse ever heavier elements to generate the heat necessary to hold itself up against gravity.

Let's revisit the binding energy curve that we looked at earlier. As we follow the binding energy curve up from the helium, we see the next jump to carbon that we've already discussed. The next step up from there is to oxygen, but it is a relatively small upward jump, meaning that the star can extract only a relatively small amount of energy from the fusion of carbon to oxygen. The next steps up from there are sodium, sulfur, calcium, and magnesium. Each step becoming smaller and smaller, each phase of fusion extracting less and less energy and therefore sustaining the star for an ever briefer period of time.

Finally, at the peak of the binding energy curve, we have iron. Once the star has fused successively heavier elements and has a core made of pure iron, it cannot resuscitate itself again, because there is no heavier element into which the star could fuse that iron and produce energy in the process. In fact, if the star were to fuse the iron into a heavier element, say nickel, it would have to put energy in, because now we are dropping back down the binding energy curve. Or to put it another way, fusing to an element heavier than iron yields a negative amount of energy. So that is not going to help the star generate the heat it needs to hold itself up against gravity.

Finally then, a massive star at this stage is collapsed under gravity for good. The core becomes a new bizarre kind of corpse, either a neutron star or a black hole. But in the process, the outer layers of the star are detonated as a supernova explosion, one of the most energetic events in the entire universe. We'll devote another lecture to describing these cataclysmic dying breaths of stars and the exotic corpses they leave behind. But here let's take the process that we've described for the middle life and near-death stages of stars' lives in the context of the Hertzsprung–Russell diagram.

For a star like our Sun with a mass of one solar mass, the star, of course, begins its life on the main sequence in the diagram. It will sit there happily for about 10 billion years. Then, as hydrogen in the star's core is exhausted and completely converted to helium, the star evolves to the right and up in the diagram, as it becomes a red giant star, its surface cooling, but its luminosity increasing dramatically. When the red giant's core compresses enough to become hot enough to start fusing helium to carbon, the star shifts a bit to the left as its surface becomes a bit hotter, and the star stably fuses helium to carbon for a while. Finally, once the helium is fully depleted and converted to carbon in the core, the star again shifts up and to the right, becoming a cool but extremely luminous red giant, and then at last sheds it outer layers as a planetary nebula, revealing the inert, hot carbon core, which becomes a white dwarf at the far lower left in the diagram.

In contrast, for a much more massive star, say 10 times the mass of our Sun, the star, again, starts out on the main sequence. But as it completes each stage of nuclear fusion, it swings far to the right, its surface much cooler, but still at a very high luminosity, a red supergiant. As each successive stage of nuclear fusion initiates in the core, the star briefly returns back toward the main sequence, but not all the way; it appears as a yellow supergiant. The star swings back and forth in the diagram multiple times as the star proceeds through different stages of nuclear fusion. Finally, once the star has fused all the way to iron in its core, the star goes supernova, and disappears from the H–R diagram.

In this lecture, we've looked at the stars during the stable, long-lived portion of their life cycle that we call the main sequence stage. We've seen that during this stage the stars are able, through the energy they generate in fusion reactions, to hold strong against gravity. But as they use up their stores of fusionable material, they undergo a series of shorter-lived resuscitations during which they are briefly able to withstand gravity again for a time before they finally die. Importantly, it is in the resuscitations that stars experience before death that the stars manufacture many of the elements so essential to us, including carbon and oxygen. It is lucky for us that the stars do not, as the poet once said, go gentle into that good night, but rather, they rage against the dying of the light.

Stellar Death
Lecture 15

In this lecture, you will learn about the wondrous, beautiful, and dramatic deaths of stars. Approximately 90% of all stars will end their lives in the silent majesty of a planetary nebula, sprinkling their ashes—including those essential carbon atoms—into the surrounding space. The rest of the stars, the most massive stars, will end their lives dramatically and cataclysmically in supernova explosions. In these explosions, the nuclear products of the stars' lives, including elements as heavy as iron, are sprayed into the surrounding space. And in their fiery deaths, the rest of the elements of the periodic table are synthesized. From these ashes, new life—all life, as we know it—emerges.

Planetary Nebulae

- Planetary nebulae, representing the scattered ashes of dead stars, are not only wondrous to behold, but they are also the mechanism by which stars, in their deaths, supply the cosmos with the legacy of their productive lives, sprinkling the surrounding space with carbon atoms that can become gathered up in the next generation of forming stars to be used, perhaps, for life-forms such as ourselves. Indeed, these graveyards themselves can be the fertilized ground where new enriched stars can begin their lives.

- Planetary nebulae, unlike the name suggests, have nothing to do with planets. The first planetary nebulae to be seen were seen through early telescopes that did not have very good angular resolution. Those early images appeared like small fuzzy blobs of color—not too dissimilar from what some of the distant planets in our solar system looked like to those telescopes, such as the planet Uranus. Hence the name "planetary nebula," and the name stuck.

- Perhaps the most striking features of the planetary nebulae are their amazing colors and shapes. The intensity of light we see from different parts of the nebula is an indication of the amount of emitting material along our line of sight in that part of the nebula. That's because the gas within the nebula overall is highly rarefied—what astronomers call optically thin. So, the light patterns that we see within the nebulae are tracing the underlying pattern of how the gases within the nebulae are distributed spatially.

- But this can lead to some misinterpretation. In some cases, a nebula's ringlike shape suggests a structure like a smoke ring when, in fact, what we're seeing is a bubble. It appears ringlike because the bubble is very thin, very tenuous. So, we see just a small amount of light from the thin part of the bubble directly in front of us—the approaching part of the bubble. In contrast, at the edges, the curvature of the bubble means that we see through more of the bubble in those directions, and with more light-emitting material along those lines of sight, we see more intense light. In other cases, the ringlike appearance really is a ring.

- More generally, what do the shapes and structures of planetary nebulae tell us about the physical mechanisms responsible for their creation? This is an active area of research, and there are still more questions than answers. But some exciting answers are emerging.

The Stellar Death Process for Very Massive Stars
- Remember that up through the point that the stars become red giants fusing helium to carbon in their cores, the story for the highest-mass stars is the same as for the lower-mass Sun-like stars. However, after that point, the stories depart in important and dramatic ways.

- A massive star is able to repeatedly resuscitate itself through a series of rounds of fusion, each involving the products of the previous round, building up to heavier and heavier elements. Once the star is at the point that it is making iron in its core, it is finally at the end. That's because iron is the heaviest element that can be fused from lighter elements and still produce a bit of energy in the process.

- At this point, the interior of the star—now a supergiant—has a so-called onion structure. At the very center is the core, now made of iron. Surrounding that is a layer of silicon fusing from the previous round of fusion. Around that is a layer of magnesium fusing. Around that is neon, then oxygen, then carbon, then helium. Each, like rings in a tree representing a record of the past, is a record of previous episodes of fusion. Finally, out beyond the helium layer, are the original outer layers of the star that never were involved in fusion in the core, remaining primarily hydrogen.

- With the core fused entirely into iron, fusion stops, and instantly, gravity pounces for the kill. The core of the star collapses in an instant. The manner in which gravity destroys the star is itself fascinating. This will end in a violent supernova explosion, but, counterintuitively, this first involves an implosion.

- At the very center of the core, the rapid collapse of the core causes the iron atoms to smash together and break apart into their constituent protons, neutrons, and electrons. This fission of the iron atoms actually saps energy from the core—for elements lighter than iron, the fusion process actually requires energy to be put in—so this only accelerates the star's demise.

- Finally, the newly liberated protons and electrons come together to form neutrons. Besides these new neutrons, there are also all the other neutrons that were liberated from the disintegrated iron atoms. Essentially, the result is that the rapidly collapsing core is turned into pure neutrons. Neutrons, by virtue of being electrically neutral, do not repel one another the way that ordinary matter does, so a ball of pure neutrons can collapse down to become the densest form of matter known: a neutron star.

- This process of rapid collapse of the star's core into a superdense neutron star happens so quickly that it is only after the creation of the neutron star that the outer layers of the star realize that the floor has dropped out from under them. So, the outer layers now begin to fall inward. The layer immediately around the neutron star at the center falls in first. When it reaches the new floor, the ultrahard surface of the neutron star, it bounces back outward toward the next layers, which are still free-falling in.

- This creates a powerful collision, which drives the material rapidly outward only to encounter the next layer free-falling in, and so on—until, finally, the collective wallop of these collisions drives the entire corpus of the star exploding out into space. That is a supernova explosion, and it all starts with an implosion and a bounce.

- A supernova explosion is the most energetically powerful event in the universe after the big bang. At the moment of the supernova, and for weeks thereafter, the explosion shines with more luminosity than all of the rest of the stars in the galaxy shining combined. That powerful explosion blows the body of the star to smithereens, expelling it into space and leaving behind the neutron star at the center of the hot, expanding shell of gas. Years later, the still-expanding shell of gas is what we refer to as a supernova remnant. These remnants have a beauty all their own, similar in some respects to beautiful planetary nebulae.

- Some supernova remnants are very disordered in appearance, whereas others are more symmetric, more orderly. It turns out that the more symmetric ones arise from the detonation of a white dwarf star. The less symmetric ones come from the demise of a massive star through the death process that has been presented in this lecture. Massive stars die violently and somewhat ungracefully.

- The most recent supernova to have been seen in the vicinity of our galaxy, although technically not within our galaxy, was in 1987. That supernova went off in one of the Milky Way's satellite galaxies, and it was near enough that we've been able to study it in exquisite detail since the initial explosion. From its asymmetry, it's pretty clear that it was a massive star that met its end when the fusion process concluded. And in this case, we have the advantage of images of the star before it exploded, confirming that the progenitor was a massive star.

- These massive-star supernova explosions continue to be important even after the explosion itself. In fact, it is only in the aftermath of these supernova explosions that all of the elements of the periodic table heavier than iron are produced. Why does that happen? The basic answer is because it can.

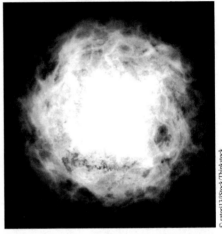

One way a star can die is in a supernova explosion, in which elements are sprayed into the surrounding space.

- Fusion of elements heavier than iron from lighter elements requires an input of energy. During the life of a star, it wouldn't make any sense for the star to perform such heavy-element fusion because the star would be sacrificing the very energy that it is desperately trying to produce in order to keep itself alive against the crush of gravity.

- But, now that the star is dead, and with an overwhelming abundance of energy available in the supernova explosion, these energy-costly fusion reactions can and do occur—for no other reason than that they are not prevented from occurring.

- These advanced-fusion reactions have to wrap up quickly, because the enormous free energy of the supernova is rapidly dissipated. Consequently, these very heavy elements—from cobalt and nickel to uranium—end up occurring in only trace amounts. However, some of these trace elements, such as copper and zinc, are important for life. In the very deaths of these massive stars we find the very zest upon which life as we know it depends.

Suggested Reading

Jastrow, *Red Giants and White Dwarfs*.

Kwok, *Cosmic Butterflies*.

Questions to Consider

1. How does the fact that a low-mass star becomes a red giant prior to its death actually make it easier for the dying star to gently expel its planetary nebula?

2. A planetary nebula virtually always has a white dwarf as it center, but there are examples of supernova remnants that do not have a neutron star at the center. How might that happen?

Stellar Death
Lecture 15—Transcript

Throughout their lives, stars are locked in a fierce battle with gravity. Because a star has mass, and lots of it, gravity is always present, relentlessly bearing down on the star, doing the only thing gravity knows to do, constantly compressing the star. For the long middle phase of their lives, stars are able to withstand the crush of gravity by performing nuclear fusion in their cores, generating sufficient heat with which to generate the outward pressure needed to keep gravity at bay. For a star like the Sun, this tug of war with gravity is a balanced standoff, a ten billion year stalemate. Over the course of that long stalemate, the star leads a productive life, if a pressured one. The nuclear fusion at its heart converts mere hydrogen into helium. Later on, when the energy of that fusion process has been fully tapped, the star gets its second wind through the fusion of that helium into carbon.

You could say that in its retirement and approaching old age, the star, now a red giant still locked in its fight with gravity, manages to lead an even more productive life. That carbon that the star produces as a red giant is by far the most important element that stars produce, at least as far as life on earth is concerned. But the star cannot hold against gravity forever. For a very massive star, more than eight times the mass of our Sun, the star will in the last raging moments of its life sustain itself against gravity for a brief, final period of time as it flails rapidly through fusion of heavier and heavier elements, producing all of the elements up to iron.

But for a star like the Sun, once the power of fusing helium to carbon is fully tapped, the star does not have the ability to resuscitate its fusing heart again. Gravity, that grim reaper, strikes its final blow. The now-dead carbon core of the star is crushed into a bizarre object we call a white dwarf, and the outer layers of the star are quietly sloughed off, producing an eerily beautiful planetary nebula.

As we'll see, these nebulae, representing the scattered ashes of the dead star, are not only wondrous to behold; they are also the mechanism by which these stars, in their deaths, supply the cosmos with the legacy of their productive lives, sprinkling the surrounding space with carbon atoms that can become gathered up in the next generation of forming stars to be used, perhaps, for life-forms such as ourselves. Indeed, as we'll see, these graveyards themselves can be the fertilized ground where new enriched stars can begin their lives.

In my mind there is no better way to start understanding planetary nebulae than to just look at them. They are wondrous things, so mysterious in their luminous, silent beauty, like bioluminescent jellyfish in a dark cosmic sea. Let's enjoy some examples.

This first example is one that maybe you recognize. It's called the Ring Nebula. It has a relatively simple structure, fairly round in appearance, and it has a glowing edge, and it appears to be a filled bubble with different colors glowing within. And at the very center, as a pinpoint of white light, is that white dwarf at the middle. This next example is the Cat's Eye Nebula; you can see, perhaps, why it's called that. It has a more complex structure than what we saw in the previous example, reminiscent, perhaps, of a spirograph pattern, or what you might see from a spinning top warbling around on a piece of paper. And again, at the very center, shining brilliantly, is that white dwarf. And in this third example, we have what is called the Siamese Squid Nebula; you can see why it's called that. This has a very different structure than the other two. This has a more elongated, bipolar structure, sort of like two barrels placed end to end, and sure enough, at the center shining brilliantly, is that white dwarf. So I think you'll agree. These are strikingly beautiful things, these dead suns, sprinkling their ashes into the sky. Arguably, even more beautiful in their deaths than when they lived.

Before we discuss these in detail, first a historical footnote. You may be wondering why these are called planetary nebulae. What do they have to do with planets? The answer is they have nothing at all to do with planets. The first planetary nebulae to be seen, such as the Ring Nebula, were seen through early telescopes that did not have very good angular resolution. Those early images appeared like small, fuzzy blobs of color, not too dissimilar from what some of the distant planets in our solar system looked like to those telescopes, such as the planet Uranus. Hence, the name planetary nebula, and the name stuck.

Perhaps the most striking features of the planetary nebulae are their amazing colors and shapes. What can we learn from this information? What can we learn about the nature and origin of the planetary nebulae when we apply our understanding of light as a tool? The first thing to say is that the intensity of light we see from different parts of the nebula is an indication of the amount of emitting material along our line of sight in that part of the nebula. That's because the gas within the nebula overall is highly rarefied, what astronomers call optically thin. So the light patterns that we see within the nebulae are tracing the underlying pattern of how the gases within the nebulae are distributed spatially.

But this can lead to some misinterpretation. In some cases, a nebula's ring-like shape suggests a structure like a smoke ring, when in fact what we're seeing is a bubble. It appears ring like because the bubble is very thin, very tenuous. So we see just a small amount of light from the thin part of the bubble directly in front of us, the approaching part of the bubble. In contrast, at the edges, the curvature of the bubble means that we see through more of the bubble in those directions, and with more light-emitting material along those lines of sight, we see more intense light. An analogy for this is the shadow cast by a balloon. The shadow appears darkest at the edges because the light has to pass through more rubber as it skims the edges of the balloon, whereas light passing through the center of the balloon only has to go through the thin surface in front and in back. But it's not always that simple, and in some cases, the ring-like appearance really is a ring, as we'll see.

But more generally, what do the shapes and structures of planetary nebulae tell us about the physical mechanisms responsible for their creation? This is an active area of research, and there are still more questions than answers. But some exciting answers are emerging. Let's look at more planetary nebulae, this time, with an eye toward the underlying mechanisms that sculpt them. The Siamese Squid is an example of a so-called butterfly, bow-tie, or bipolar nebulae. It consists of two lobes of fluorescing material conjoined at the central star. Each lobe is expanding outwards, pushed by the pressure of a hot interior heated by extreme winds from the dying star's surface. Two other beautiful examples of bipolar nebulae are the Ant Nebula and the Double Bubble Nebula.

Why do these nebulae develop into two-lobed structures and not simple round bubbles? Apparently, a tiny, dense disk surrounds the dying star. Think of the disk as an equatorial belt or waistband. As the star's winds ram into the disk, material is deflected to the polar axis. One of the simplest ways to model this is to assume that the star has expelled a fast wind, of about 1000 miles per second, into the denser barrel of material that the star ejected previously. The walls of the barrel serve the same function as the nozzle on a jet engine, channeling the gas and compressing it so that it emerges with a rapid speed and in a narrow stream. As the gas escapes along the axis of the barrel, it tries to push ambient material out of its way. The interaction between the escaping gas and its surroundings forms a jet, not unlike the exhaust behind a jet engine. The supersonic shock along the walls of the jet causes the local gas there to radiate, forming the structure shown in the Siamese Squid nebula. Apparently, the jet is following in the path of former ejections seen beyond this image.

This animation from NASA shows how this type of structure might happen if the dying star has a companion object, such as a binary star, in a nearby orbit. A nearby companion star, or perhaps a large planet, becomes engulfed by the expanding, dying star. The star or planet may just find itself inside the star or close to its edge. The gravitational forces exerted by the companion act like an eggbeater, tossing out the loose outer layers of the dying giant star into an equatorial disk and eventually shaping the outflow of the material into the shape of the planetary nebula as we now see it. So as we've seen before, the presence of a sibling star can dramatically affect a star. We saw in a previous lecture how the dynamics of sibling stars can affect the manner in which the stars form. And now we see that this sibling influence continues right on to death.

The Dandelion Nebula and the Spirograph Nebula are examples of so-called elliptical planetary nebulae with very striking and unusual shapes. The Dandelion nebula looks like the puff ball of a dandelion, and gives the impression of an explosion, rather than an organized ejection, as for the other nebulae. The Spirograph Nebula is peculiar, because its outermost shell almost seems carefully inscribed.

The Saturn Nebula and the Blinking Eye Nebula are planetary nebulae that contain so-called FLIERs. These FLIERs are strange red-colored features that resemble flying sparks at the edge of the expanding nebula. The red FLIERs are known to have higher velocities than the gas in which they are embedded. If so, they would leave a wake, as does a speedboat in still water. We expect to find that the narrow part of the wake points outward, in the direction of the speedboat's (FLIERs') motion.

Although we do see structures in the planetary nebulae that might resemble wakes, they point in precisely the reverse direction from that which was expected. This might mean that the bright red knots of material are falling inward. But this seems highly unlikely for stars that are observed to be ejecting gas in high-speed winds. The bottom line is, no satisfactory explanation for the FLIERs has yet emerged from scientific studies.

The Cat's Eye nebula is one of the most extensively investigated planetary nebulae. The bull's eye pattern tells us that up until 1000 years ago the star ejected its mass in isotropic pulses, meaning pulses equally in different directions, as if the mass were coming from a regularly pulsating balloon. Each ring is actually the edge of a spherical bubble seen projected onto the sky; that's why it appears bright along its outer edge. Observations suggest that each bubble contains one-twentieth to one-tenth of the Sun's mass. The Ring Nebula seen in infrared light reveals similar evidence for multiple past ejection bubbles, warming the dust in the surrounding interstellar space and causing that dust to glow in the infrared.

What do the structures of the Cat's Eye and Ring nebulae imply? Since they're larger than the bright cores of the nebulae that they surround, the rings are almost certainly material ejected episodically before the main and bright core of the nebula formed. This means that the star that ejected the nebulae first quivered and shivered and made these concentric rings. Then something big happened, and the density and mechanism for ejecting the mass changed abruptly. This is when the core of the nebula was formed, probably between 1000 and 2000 years ago.

This bubble blowing in the Cat's Eye happened repeatedly, perhaps once every 1500 years or so. Each of these spherical shells contains about 5 to

10% of the Sun's mass. Then the pattern of mass loss suddenly changed, and the Cat's Eye emerged in one large event. The puzzle is, why?

The rhythmic ringing of a dying star is expected as the last of its nuclear fuel is suddenly triggered into ignition by the increasing crush of gravity, much like the juice ejected by squeezing an orange with increasing force. Each expulsion of juice temporarily relieves the internal pressure inside the orange. Similarly, each ejection of mass temporarily stops the combustion of the final dregs of the star's remaining fuel.

But why, some 1000 years ago, did the pattern of ejection change so radically and strongly in the Cat's Eye Nebula? We can only conjecture at this point. It's possible that an orbiting star or giant planet falls onto the dying star. It hits the surface with such force that its atoms ignite in a large conflagration. Somehow, the burst of heat drives the remnants of the dying star into space in fantastic patterns. Clearly, we still have much to learn about these fascinating entities and the processes by which they spread their ashes into the surrounding space.

My own research group was involved in a discovery that speaks directly to these ideas. This is a planetary nebula called SuWt 2. Again, we see a ring-like morphology and a bright hot source at the center. What's interesting in this case is that the bright central source is not a white dwarf. There is a white dwarf in the planetary nebula, but this planetary nebula is relatively old, and so the white dwarf has largely cooled and faded from view. What is that bright central source? Turns out it's an eclipsing binary star system made up of two hot A-type stars. And from our detailed analysis of those eclipsing sibling stars, we discovered something very interesting; the two stars are identical. They have masses that are the same to a fraction of a percent. Like human twins, such identical twins do occur in stars but are extremely rare.

What we've found in this case is that the two sibling stars probably were not originally so identical. In other words, they were not born as identical twins but became identical later on. How could this happen? Our analysis reveals that the third star in the system, the star that eventually expelled the planetary nebula we now see, partially engulfed the other two stars when it swelled to become a red giant shortly before its death. At that point, the two binary stars were literally swimming in the outer layers of the red giant, and in that process, any difference in the masses of the two stars became equalized, so that they now appear as identical twins. In turn, the binary system moving through the outer layers of the red giant acted like the blades of a beater in batter, sculpting and shaping the eventual planetary nebula into the ring shape we see now.

Finally, bringing the story full circle, we now have examples of the conditions for new stellar birth within the planetary nebula stellar graveyard. The Helix Nebula is a beautiful planetary nebula that has been imaged in detail by the Hubble space telescope. Like many of the other examples we've looked at in this lecture, the Helix Nebula shows a round bubble-like structure with a white dwarf at the center. But in the outskirts of the expanding bubble of ash-filled gas from the dying star, we see knots of material that in many ways resemble the protostellar cocoons we saw when discussing the stellar birth process. Each one of these knots is approximately the size of our solar system. And there are some 10,000 of these knots.

To be clear, there is not enough mass in these knots to form full-fledged solar systems. Each one contains a mass comparable to the Earth, or perhaps a little bit more. What will become of each of these planetary mass knots is unknown. Perhaps they will be dissolved by the intense radiation from the white dwarf at the center of the nebula. But what is significant here is that the physical processes of birth by which material becomes compressed and entrained in a way that resembles the stellar birth process are present and intimately connected with the stellar death process.

Let's shift our attention now to the stellar death process for very massive stars, stars at least eight times the mass of the Sun. Remember that up through the point that the stars become red giants fusing helium to carbon in their cores, the story for the highest mass stars is the same as for the lower-mass, Sun-like stars that we've discussed so far. But after that point, the stories depart in important and dramatic ways.

As we've described in a previous lecture, a massive star is able to repeatedly resuscitate itself through a series of rounds of fusion, each involving the products of the previous round, building up to heavier and heavier elements. Once the star is at the point that it is making iron in its core, it is finally at the end. That's because iron, as you'll recall, is the heaviest element that can be fused from lighter elements and still produce a bit of energy in the process.

At this point, the interior of the star, now a supergiant, has a so-called onion structure. At the very center is the core, now made of iron. Surrounding that is a layer of silicon fusing from the previous round of fusion. Around that is a layer of magnesium that's fusing. And around that is neon, then oxygen, then carbon, then helium. Each, like rings in a tree representing a record of the past, is a record of previous episodes of fusion. Finally, out beyond the helium layer, are the original outer layers of the star that never were involved in fusion in the core, remaining primarily hydrogen.

With the core fused entirely into iron, fusion stops, and instantly gravity pounces for the kill. The core of the star collapses in an instant. The manner in which gravity destroys the star is itself fascinating. You already know that this will end in a violent supernova explosion, but counterintuitively this first involves an implosion. At the very center of the core, the rapid collapse of the core causes the iron atoms to smash together and break apart into their constituent protons, neutrons, and electrons. This fission of the iron atoms actually saps energy from the core. Remember that for elements lighter than iron, the fusion process actually requires energy to be put in, so this only accelerates the star's demise.

Finally, the newly liberated protons and electrons come together to form neutrons. And of course, besides these new neutrons, you also have all the other neutrons that were liberated from the disintegrated iron atoms. So what's the result? Well, essentially, the rapidly collapsing core is now turned into pure neutrons. Now, neutrons, by virtue of being electrically neutral, do not repel one another the way that ordinary matter does. And so a ball of pure neutrons can collapse down to become the densest form of matter known, what we call a neutron star. We'll talk about neutron stars in detail in a later lecture.

This process of rapid collapse of the star's core into a super-dense neutron star happens so fast, that it is only after the creation of the neutron star that the outer layers of the star realize that the floor has dropped out from under them. So the outer layers now begin to fall inward. The layer immediately around the neutron star at the center falls in first. When it reaches the new floor, the ultra-hard surface of the neutron star, it bounces back outward toward the next layers, which are still free-falling in. This creates a powerful collision, like an outward swinging bat striking an incoming fastball. And this collision drives that material rapidly outward, only to encounter the next layer free-falling in, and so on and so on. Until finally, the collective wallop of these collisions drives the entire corpus of the star exploding out into space. That is a supernova explosion, and it all starts with an implosion and a bounce.

A supernova explosion is the most energetically powerful event in the universe, after the big bang. At the moment of the supernova, and for weeks thereafter, the explosion shines with more luminosity than all of the rest of the stars in the galaxy shining combined. That powerful explosion blows the body of the star to smithereens, expelling it into space and leaving behind the neutron star at the center of the hot, expanding shell of gas. Years later, the still-expanding shell of gas is what we refer to as a supernova remnant. These remnants have a beauty all their own, similar, in some respects, to the beautiful planetary nebulae we looked at earlier.

Let's look at some examples. The first example we have here is the Crab nebula. This is a supernova that went off in the year 1054 and was actually observed and noted by Chinese astronomers. Note the very chaotic shape of the structure of this supernova remnant, and you'll also notice, shining at the center, the neutron star left behind.

This next example is Kepler's supernova. It went off in the year 1604 and was one of the last supernovae to be seen in our galaxy. To be clear, other supernovae have occurred in our galaxy since then, but most of those other supernovae were obscured by dust in our galaxy and were not seen directly when they occurred. So this was the most recent one to be seen directly by human eyes in our own Milky Way Galaxy. But notice the roughly spherical shape of this one. And again, notice that neutron star at center.

From just these examples, we already see some interesting differences. Some supernova remnants are very disordered in appearance, like the Crab Nebula, whereas others are more symmetric, more orderly, like Kepler's supernova. It turns out that the more symmetric ones arise from the detonation of a white dwarf star through a process that we'll discuss in a next lecture. The less symmetric ones, like the Crab Nebula, come from the demise of a massive star through the death process that we've discussed in this lecture. You could say that massive stars die violently and somewhat ungracefully. The most recent supernova to have been seen in the vicinity of our galaxy, although technically not within it, was in 1987. That supernova went off in one of the Milky Way's satellite galaxies, and it was near enough that we've been able to study it in exquisite detail since the initial explosion.

Here we're seeing a before and after comparison of the supernova called Supernova 1987A. You'll notice in the left image there's an arrow pointing to a star, which was the star right before it blew up. And in the right image you see what that star looked like when it actually exploded as a supernova; quite a difference. Now, the Hubble space telescope has taken detailed images of the supernova remnant in the year since it exploded, and we see it has this very interesting nested-ring structure and, overall, has an asymmetric appearance. Three-dimensional models have been used to reconstruct what the actual structure of the supernova remnant was like, and as we see here, what we find is that that double-ring structure is actually due to a bipolar-type nebula of the kind that we looked at earlier in this lecture.

So what would you guess was the origin of this supernova? From its asymmetry it's pretty clear to us now that this was a massive star that met its end when the fusion process came to an end. And in this case, we have the advantage of images of the star before it exploded, confirming that, indeed, the progenitor was a massive star. In a future lecture, we'll return to a detailed discussion of the shapes of different supernova remnants and how they physically affect the surrounding galactic material to create new generations of stars.

These massive star supernova explosions continue to be important even after the explosion itself. In fact, it is only in the aftermath of these supernova explosions that all of the elements of the periodic table heavier than iron are produced. Why does that happen? The basic answer is, well, because it can. As we've seen, fusion of elements heavier than iron from lighter elements requires an input of energy. During the life of a star, it wouldn't make any sense for the star to perform such heavy-element fusion, because the star would be sacrificing the very energy that it desperately needs in trying to hold itself up against the crush of gravity. But now dead, and with an overwhelming abundance of energy available in the supernova explosion, these energy-costly fusion reactions can occur, and so they do for no other reason than that they are not prevented from occurring.

These advanced fusion reactions have to wrap up quickly, because the enormous free energy of the supernova is rapidly dissipated. Consequently, these very heavy elements, from cobalt and nickel to uranium, end up occurring in only trace amounts. But some of these trace elements are important for life, elements such as copper and zinc, the spice of life, if you will. In the very deaths of these massive stars, we find the very zest upon which life as we know it depends.

In this lecture, we've witnessed the wondrous, beautiful, and dramatic deaths of stars; 90% of all stars will end their lives in the silent majesty of a planetary nebula, sprinkling their ashes, including those essential carbon atoms, into the surrounding space. The rest of the stars, the most massive stars, will end their lives dramatically and cataclysmically in supernova explosions. In these explosions, the nuclear products of the stars' lives, including elements as heavy as iron, are sprayed into the surrounding space. And in their fiery deaths, the rest of the elements of the periodic table are synthesized. From these ashes, new life, all life, as we know it, emerges.

Stellar Corpses—Diamonds in the Sky
Lecture 16

In this lecture, you will learn about each of the three different types of stellar corpses: white dwarfs, neutron stars, and black holes. As you will learn, these stellar corpses continue to battle with gravity, an ongoing struggle that follows the stars into and beyond the grave. These amazing objects reveal an entirely new class of physical laws, from the microscopic physics of quantum mechanics to the fantastic new realm of gravitational wave physics.

White Dwarfs and Neutron Stars

- Our first stop in the stellar graveyard are white dwarfs, which are the exposed cores of stars like the Sun that have since died, puffing out beautiful planetary nebulae. White dwarfs are tightly packed, extremely dense balls of carbon, which is what the red giant star had fused from helium atoms in the final resuscitation of its life. White dwarfs have half of a Sun's mass of carbon packed into a volume the size of the Earth. Truly, these are diamonds in the sky.

- When first exposed as the dead star releases its outer layers into an expanding planetary nebula, the white dwarf is extremely hot, approaching a million degrees. Then, it slowly cools, as its heat is steadily radiated away. That is the fate of the white dwarf—to sparkle with ever less luster, eventually fading from view forever. However, there is a very special circumstance under which a cold, dead white dwarf can shine brilliantly one last time.

- Our next stop in the graveyard is neutron stars. Despite the name, these are not stars at all but, rather, the corpses of massive stars that have since detonated as supernovae. A neutron star is unimaginably dense. Ordinary matter, even matter as dense as lead, is almost entirely empty space. That's because in ordinary matter, the individual atoms are kept very widely separated by the electromagnetic repulsion of the electrons in those atoms.

The distance between the nucleus of an atom and its electrons is comparable in scale to the distance between the Sun and the Earth in our solar system.

- But in a neutron star, there is no repulsion between the particles, because there are only electrically neutral neutrons. So, the neutrons are essentially in direct contact with one another. In the entire universe, there is no form of matter more compact—that is, except for black holes. But unlike a black hole, a neutron star still has an actual physical size.

- A neutron star has anywhere from the mass of the Sun to 3 times that mass—remember that it started out as a very massive star—and has the physical size of about 10 kilometers, comparable to a midsize city. Whereas water has a density of 1 gram per cubic centimeter, and iron has a density of about 5 grams per cubic centimeter, and the core of the Sun has a density of about 100 grams per cubic centimeter, a neutron star has a density of about 100 trillion grams per cubic centimeter.

- Another remarkable feature of neutron stars is that they spin extremely fast. A slow neutron star spins perhaps once per second—which is fast when you consider that a neutron star weighs at least as much as the Sun and is the size of a city. A more typical neutron star spins 30 times per second. The fastest, known as the millisecond pulsars, spin almost 1,000 times per second.

- Why do neutron stars spin so fast? This is a consequence of the law of conservation of angular momentum. That law, similar to the conservation of energy, says in essence that the spin energy of an object is also conserved. The way this works is that a large object spinning slowly is equivalent to a small object spinning rapidly.

- A neutron star represents the highly collapsed, compressed remnant of a star that was a million times larger. That star may have spun relatively slowly, perhaps once per month or so. But now, that initial slow spin has been amplified a million times due to the million-fold decrease in size. As fantastic as these objects are, the explanation for their amazing spin is really just fairly ordinary physics.

- But other aspects of the physical laws describing neutron stars and white dwarfs are anything but ordinary. White dwarfs and neutron stars, despite being dead, are nonetheless able to hold themselves up against the crush of gravity. These stellar corpses are highly compressed, having lost their ability to push back with the heat of fusion, but they still have a nonzero physical size. They hold themselves up through a bizarre form of pressure known as degeneracy pressure.

Degeneracy Pressure
- In ordinary matter, there is a direct relationship between the temperature of an object and the speed with which the atoms in it move. That energy of motion is the pressure of heat, and it is how stars push back against gravity ordinarily. But with degeneracy pressure, the motion of the particles in an object does not depend on the temperature or the heat content of the object. In fact, the object can be completely cold and still exert this type of pressure.

- One of the strange aspects of the physics of quantum mechanics is the Heisenberg uncertainty principle. Quantum mechanics in general is the set of physical laws that describe how matter behaves at the quantum level, at the microscopic level of individual particles. According to the Heisenberg uncertainty principle, it is not possible to simultaneously know the position and the speed of a particle with exactitude. The more exactly we localize a particle in space, the less certain we can be about its speed.

- As a particle's position in space becomes more tightly confined, the larger the spread in the possible speeds it can have. And the larger the spread in speeds that a group of particles has, the stronger the pressure exerted by those particles. The pressure exerted by a gas is determined by the speed of the fastest particles in the gas. Those fast-moving particles carry more energy of motion, and when they hit against something, they impart more energy in collision. That's what pressure is—a collisional pushing of particles against one another due to their energy of motion. Faster-moving particles push harder; they exert stronger pressure.

- In the case of the Heisenberg uncertainty principle, the larger spread of particle motions means that there are more particles with faster motions, so overall the particles in the gas exert a stronger pressure. In other words, merely by packing particles closely together on the scale of individual particles—such as the space between neutrons in a neutron star—the speed of the particles increases, so the pressure they exert increases, increasing their ability to push back against gravity.

- In the case of a white dwarf, it is the collective microscopic action of the electrons in the carbon atoms that exert the degeneracy pressure to hold the white dwarf up. In the case of a neutron star, it is the collective microscopic action of the neutrons that exert the degeneracy pressure. Without this quantum mechanical pressure, these stellar corpses would become black holes.

Black Holes
- A neutron star cannot be any heavier than 3 times the mass of the Sun—any heavier and the degeneracy pressure is broken, and nothing can prevent gravity from finally having its way, crushing the star into nonexistence, creating what is perhaps the most fantastic object of all: a black hole. Black holes represent gravity's ultimate victory over a star, when even degeneracy pressure cannot save the star.

- Black holes don't have a size, but they do have mass. In fact, all of the mass that was there before the stellar corpse finally gave out is there still. Only gravity has crushed that mass down to zero size. Infinite density but finite mass is the incredible mind-bending nature of a black hole.

- There is a common misconception that black holes are lurking, menacing, predators just waiting to suck in anything and everything

The death of a massive star can form a black hole, which is a cosmic body of extremely intense gravity.

around them, like gigantic vacuum cleaners in space. In fact, in the immediate vicinity of a black hole, the laws of physics break down. There is a small region immediately around a black hole where the gravity of the black hole is so intense that not even light can pass by without being pulled forever into the black hole.

- The black hole's sphere of influence is described by the so-called Schwarzschild radius, which, for a black hole weighing the mass of the Sun, is 3 kilometers. In other words, inside of 3 kilometers, nothing—not even light—can escape, and within that space, we do not have the physics to describe the inner workings of the hole. But from a distance larger than the Schwartzschild radius, the black hole just feels like any other object with that same amount of mass.

- The Schwartzschild radius is sometimes also referred to as the photon sphere around the black hole. That's because, at that exact distance from the black hole, a photon of light would have its ordinarily straight trajectory bent all the way around so that it would orbit the black hole.

- How do we know that black holes actually exist? After all, we can't see a black hole directly. It doesn't radiate because not even light can escape from within the Schwartzschild radius. However, we can use the light emitted by objects in the very near vicinity of black holes in order to indirectly weigh the black hole.

- Such systems are known as X-ray binaries. These are systems in which a black hole is orbited by another star and matter from the companion star spills onto the black hole. As it does, the spilling material spirals in faster and faster, like water spiraling down a drain, and in the process, heats up to such a degree that it radiates as X-ray light.

- By measuring the speed of the inspiraling material—using the Doppler effect—we can use the laws of physics to ascertain what the mass of the central object must be. Then, comparing the mass of the unseen central object with its size, as determined by the proximity that the inspiraling material achieves before falling in, we can determine whether the unseen object satisfies the Schwartzschild radius criterion for being a black hole.

- Looking to the future, there are efforts under way to develop an entirely new class of telescope that could someday detect the merger of two black holes. Some black holes may be part of binary star systems. Occasionally, those binary systems will involve "twins" that will both end up as black holes. Those sibling ghost stars would be completely undetectable, as neither can radiate light. However, they can in principle be detected with future gravitational radiation telescopes.

- Already, thanks to advances in very high angular resolution imaging with adaptive optics systems on large telescopes, we can directly study an extremely massive black hole that resides at the very center of our galaxy. That massive black hole probably formed through the merger over time of smaller black holes going back to the formation of our galaxy as a whole.

Suggested Reading

Shapiro, *Black Holes, White Dwarfs and Neutron Stars*.

Tyson, *Death by Black Hole*.

Questions to Consider

1. How do the laws of quantum mechanics—the microscopic physics of individual particles—explain the nature of stellar corpses (white dwarfs and neutrons stars)?

2. Given the relative numbers of stars in the galaxy born with different masses, which of the different types of stellar corpses are the most common in our galaxy?

Stellar Corpses—Diamonds in the Sky
Lecture 16—Transcript

So far in this course we've discussed the life cycle of stars from their birth, through the long-lived productive stages of their lives, as well as their glorious and sometimes violent deaths. And yet the remains of stars can be even more fantastic than the stars themselves. From white dwarfs to neutron stars to black holes, these amazing objects stretch the laws of physics and the imagination. Let's spend some time getting to know the incredible corpses of stars, the ghosts of stars past.

There are essentially three types of stellar corpses left behind by stars in three different mass categories. Low mass stars, like our Sun, leave behind white dwarfs. High mass stars, weighing in at about eight times the mass of our Sun or more, leave behind neutron stars. And the most extremely massive stars, weighing in at perhaps 20 times the mass of our Sun, leave behind black holes.

In this lecture, we'll look at each of these different types of stellar corpses in detail. Think of it as an autopsy. We'll use the evidence of the remains to determine the conditions of each star's demise. We'll also discuss the bizarre laws of physics that govern these dead stars. As we'll see, these stellar corpses continue to do battle with gravity, an ongoing struggle that follows the stars into and beyond the grave. So let's dive in to this amazing aspect of the stellar life cycle and take a wild ride into the stellar afterlife.

Our first stop in the stellar graveyard are the white dwarfs. These are the exposed cores of stars, like the Sun, that have since died, puffing out those beautiful planetary nebulae that we examined in our previous lecture. Look at the center of a planetary nebula, and there, shining with a cold beauty like a pearl, is the white dwarf. Or rather, sparkling like a diamond. Because what are white dwarfs made of after all? White dwarfs are tightly packed, extremely dense balls of carbon, which is what the red giant star had fused from helium atoms in the final resuscitation of its life. Those carbon atoms are extremely densely packed, packed as closely together as the laws of physics allow. In a white dwarf, we have half a Sun's mass of carbon packed into a volume the size of the Earth. Truly, these are diamonds in the sky.

When first exposed as the dead star releases its outer layers into an expanding planetary nebula, the white dwarf is extremely hot, approaching a million degrees. It's like a baked potato just pulled from the oven and placed on the counter. At first it has the temperature of the oven that it was baking in, and is too hot to touch. Then it slowly cools as its heat is steadily radiated away. That is the fate of the white dwarf, to sparkle with ever less luster, eventually fading from view forever. Unless... unless... As we'll see in our next lecture, there is a special circumstance under which a cold, dead white dwarf can shine brilliantly one last time.

But for now, let's move to our next stop in the graveyard, neutron stars. Despite the name, these are not stars at all, but rather the corpses of massive stars that have since detonated as supernovae. To remind you, in the final stage of a massive star's life, the star's core, having fused all the way to iron, finally collapses under gravity, causing the atoms to completely dissociate into their constituent parts: protons, electrons, and neutrons. The protons and electrons come together to form additional neutrons so that pure neutrons are all that remain in the rapidly collapsing core. That ball of pure neutrons collapses to the point that it represents the densest form of matter known in the universe. At that point, the outer layers of the star, also collapsing under gravity, hit and bounce off of the neutron star at the core, exploding as a supernova. The majority of the star's mass is expelled into space in the supernova fireball. But left behind at the center of the expanding supernova remnant, is that ultra-dense ball of pure neutrons, the neutron star.

A neutron star is unimaginably dense. Understand, ordinary matter, even matter as dense as lead, is almost entirely empty space. That's because in ordinary matter, the individual atoms are kept very widely separated by the electromagnetic repulsion of the electrons in those atoms. The distance between the nucleus of an atom and its electrons is comparable in scale to the distance between the Sun and the Earth in our solar system. Another way to visualize this is that if an atom were the size of a football field, the nucleus would be the size of a grape seed at the 50 yard line. So truly, ordinary matter is more of a froth than the solid we imagine it to be.

But in a neutron star, there is no repulsion between the particles, because there are only electrically neutral neutrons. So the neutrons are, essentially, in direct contact with one another. In all the universe, there is no form of matter more compact. That is, except for the truly bizarre black holes that we'll talk about in a bit. But unlike a black hole, a neutron star still has an actual physical size.

A neutron star has anywhere from the mass of the Sun to three times that mass, remember it started out as a very massive star, and has the physical size of about 10 kilometers, comparable to a midsize city. What does that mean for the density of a neutron star? Whereas water has a density of one gram per cubic centimeter, and iron has a density of about 5 grams per cubic centimeter, and the core of the Sun has a density of about 100 grams per cubic centimeter, a neutron star has a density of about 100 trillion grams per cubic centimeter. One sugar-cube-size chunk of a neutron star would weigh more than all the people who have ever lived. It's quite a thought, all of humanity stuffed into a sugar cube. But such is the amazing nature of matter compressed to the limits of the laws of physics.

Another remarkable feature of neutron stars is that they spin extremely fast. A slow neutron star spins perhaps once per second. That might not seem so impressive, but I dare you to try it. Try counting and spinning so that on each beat you're back facing forward. Not so easy! Now, consider that a neutron star weighs at least as much as the Sun and is the size of a city. Spinning once per second! And that's a slow one. A more typical neutron star, like the one in the center of the crab nebula supernova remnant, spins thirty times per second. The fastest, known as the millisecond pulsars, spin almost a thousand times per second. That's about the speed of the blades in your kitchen blender! We'll talk about pulsars in detail in another lecture.

But why do neutron stars spin so fast? This is a consequence of the law of conservation of angular momentum. That law, similar to the conservation of energy, says in essence that the spin energy of an object is also conserved. The way this works is that a large object spinning slowly is equivalent to a small object spinning rapidly. This is the very law of physics that explains how spinning ice skaters can wind up spinning so fast by pulling in their arms. When going into the spin, the skaters extend their arms and legs, and so start out big and slow. Then, by pulling their arms and legs in close, they become smaller, and the law of conservation of angular momentum then forces their spin to speed up.

A neutron star represents the highly collapsed, compressed remnant of a star that was a million times larger. That star may have spun relatively slowly, perhaps once per month or so. But now that initial spin, that initial slow spin, has been amplified a million fold due to the million-fold decrease in size. As fantastic as these objects are, the explanation for their amazing spin is really just fairly ordinary physics. But other aspects of the physical laws describing neutron stars and white dwarfs are anything but ordinary. White dwarfs and neutron stars, despite being dead, are nonetheless able to hold themselves up against the crush of gravity.

Think about it. Through the entire life story of the stars, we've continually returned to this essential point. Stars are constantly having to hold themselves up against gravity, and during most of their lives, they do this by generating heat and pressure through fusion to push back against gravity. The stellar corpses that we're discussing now are highly compressed, having lost their ability to push back with the heat of fusion, but they do still have a non-zero physical size. How are they holding themselves up now? They do it through a bizarre form of pressure known as degeneracy pressure. To understand degeneracy pressure, we'll need to get into the weird physics of quantum mechanics. I think you're going to enjoy this.

In ordinary matter, there is a direct relationship between the temperature of an object and the speed with which the atoms within it move. That energy of motion is the pressure of heat, and that is how stars push back against gravity ordinarily. But with degeneracy pressure, the motion of the particles in an object does not depend on the temperature or the heat content of the object. In fact, the object can be completely cold and still exert this type of pressure.

To understand why, let's introduce one of the strange aspects of the physics of quantum mechanics, what we refer to as the Heisenberg Uncertainty Principle. Quantum mechanics in general is the set of physical laws that describe how matter behaves at the quantum level, at the microscopic level of individual particles. According to the Heisenberg Uncertainty Principle, it is not possible to simultaneously know the position and the speed of a particle with exactitude. The more exactly we localize a particle in space, the less certain we can be about its speed.

Another way of expressing this, is to say that as a particle's position in space becomes more tightly confined, the larger the spread in possible speeds it can have. And the larger the spread in speeds that a group of particles has, the stronger is the pressure exerted by those particles. That might not be intuitive, so let me explain. The pressure exerted by a gas is determined by the speed of the fastest particles in the gas. Those-fast moving particles carry more energy of motion, and when they hit against something, they impart more energy in collision. That's what pressure is, a collisional pushing of particles against one another due to their energy of motion. Faster moving particles push harder, they exert stronger pressure.

In the case of the Heisenberg Uncertainty Principle, the larger spread of particle motions means that there are more particles with faster motions, and so, overall, the particles in the gas exert a stronger pressure. So in other words, merely by packing particles closely together on the scale of individual particles, such as the space between neutrons in a neutron star, the speed of the particles increases, and so the pressure they exert increases, increasing their ability to push back against gravity.

If this sounds strange, well, it is! Quantum mechanics is an aspect of physics that most physicists learn to come to grips with despite its strangeness. But it is also a strangely elegant thing that the exotic nature of some of the most extreme objects in the universe is explainable only through the weird physics of the very small, microscopic world of individual subatomic particles. In the case of a white dwarf, it is the collective microscopic action of the electrons in the carbon atoms that exert the degeneracy pressure to hold the entire white dwarf up. In the case of a neutron star, it is the collective microscopic action of the neutrons that exert the degeneracy pressure. Without this quantum mechanical pressure, these stellar corpses would become, well, black holes! How would that happen? Well, degeneracy pressure is bizarre and amazing, but this form of pressure does have a limit, and that limit can be broken. To see how this works, let's think about what happens to a neutron star as we slowly add to its mass.

In the case of an ordinary star, the more massive the star, the larger is its size. That's a consequence of the ordinary physics of ordinary pressure produced by ordinary heat. But with degeneracy pressure, remember that the source of the pressure is not heat, but rather, the quantum mechanical confinement of the particles in close proximity to one another. So, if we add mass to a neutron star, the crush of gravity becomes stronger, and the neutron star has to generate more pressure. To do that, it has to confine the neutrons even more tightly, and so the neutron star shrinks to crank up the density and thereby force the particles to move faster.

But there is a limit to this; that limit is the speed of light. A neutron star can shrink and shrink to increasingly drive up the density and the motions of the neutrons but only to the point that the neutrons are moving at the speed of light. At that point, there is nothing more that the neutron star can do to generate additional pressure. This is the fundamental limit to how massive a neutron star can be, and it turns out to be about three times the mass of the Sun.

For a white dwarf, whose degeneracy pressure is driven by the electrons in the carbon atoms, the limit is 1.4 times the mass of the Sun. This limiting mass is known as the Chandrasekhar limit, named after the physicist who first showed that quantum mechanics provides the explanation for the physical structure of these stellar corpses. What happens when a white dwarf exceeds that limit? Well, we'll return to that in our next lecture. For now, let's focus on neutron stars.

As I noted a moment ago, a neutron star cannot be any heavier than three times the mass of the Sun; not one iota. Any heavier, and the degeneracy pressure is broken, and nothing, nothing can prevent gravity from finally having its way, crushing the star into non-existence. Creating what is perhaps the most fantastic object of all, a black hole. Black holes represent gravity's ultimate victory over a star, when even degeneracy pressure cannot save the star. Black holes are the stuff of imagination and of science fiction. And yet they are very, very real.

In fact, I shouldn't have referred a moment ago to black holes as non-existent. What black holes don't have is a size, but they do have mass. In fact, all of the mass that was there before the stellar corpse finally gave out is still there, only gravity has crushed that mass down to zero size. Infinite density but finite mass, that is the incredible mind-bending nature of a black hole.

There is a common misconception that black holes are lurking, menacing, predators just waiting to suck in anything and everything around them, like gigantic vacuum cleaners in space. In fact, in the immediate vicinity of a black hole, the laws of physics break down. There is a small region immediately around a black hole where the gravity of the black hole is so intense that not even light can pass by without being pulled forever into the black hole. But just how big is the black hole's sphere of influence, really?

That sphere of influence is described by the so-called Schwarzschild radius. The Schwarzschild radius for a black hole weighing the mass of the Sun is three kilometers. In other words, inside of three kilometers, nothing, not even light, can escape. And within that space, we do not have the physics to describe the inner workings of the black hole. But beyond the Schwarzschild radius, the black hole is actually not very strange at all. From a distance larger than the Schwarzschild radius, the black hole just feels like any other object with that same amount of mass. If our Sun were to suddenly become a black hole, right now, the Earth's orbit would be completely unchanged. As far as the Sun's gravitational influence on the Earth is concerned, nothing would be different. There would still just be the same amount of mass forcing the Earth to orbit with its exact same motion. Of course, the light and heat from the Sun would be gone, but never mind that.

But coming back to the Schwarzschild radius and within, things are very, very strange. The Schwarzschild radius is sometimes also referred to as the photon sphere around the black hole. That's because, at that exact distance from the black hole, a photon of light would have its ordinarily straight trajectory bent all the way around so that it would actually orbit the black hole. This is an incredible thing to think about. Imagine orbiting a black hole in the photon sphere, and imagine shining a flashlight directly ahead. The photons from the flashlight would orbit around the black hole and illuminate the back of your head. Then, that light reflecting off the back of your head would orbit back around to your eyes, so you shine the flashlight ahead of you and see the back of your own head!

Einstein famously described gravity as being "nothing more than geometry." In Einstein's worldview, the traditional description of gravity by Newton as a "force" was more ad hoc than necessary. Rather than appealing to force fields like Star Wars, Einstein reimagined gravity in terms of the geometry of space. Imagine space as a stretchy fabric. Place a massive object on that fabric, and the fabric stretches to produce a deep well or pucker. Why do the planets orbit the Sun on circular paths? According to Einsteinian gravity, they do so simply because that is what a straight trajectory on a puckered surface looks like. So in the Einsteinian description, a black hole really is a hole of sorts. A hole in the fabric of space. A place where the stretchy fabric is so intensely curved that it is infinitely deep, and so an object traveling on an otherwise straight trajectory that comes too close is forced to fall in, straight down, no climbing back out.

Now, how do we know that these black holes actually exist? After all, we can't see a black hole directly. It doesn't radiate, because not even light can escape from within the Schwarzschild radius. However, we can use the light emitted by objects in the very near vicinity of black holes in order to indirectly weigh the black hole. Such systems are known as X-ray binaries. These are systems in which a black hole is orbited by another star, and matter from the companion star spills onto the black hole. As it does, the spilling material spirals in faster and faster, like water spiraling down a drain, and in the process, it heats up to such a degree that it radiates as X-ray light.

By measuring the speed of the inspiraling material using the Doppler Effect, we can use the laws of physics to ascertain what the mass of the central object must be. Then, comparing the mass of the unseen central object with its size, as determined by the proximity that the inspiraling material achieves before falling in, we can determine whether the unseen object satisfies the Schwarzschild radius criterion for being a black hole. A number of such X-ray binaries have been discovered with X-ray telescopes such as the Chandra X-ray telescope, named after the same physicist who first established the maximum mass for white dwarfs and neutron stars.

How do X-ray binaries come to be? Well, start by recalling that many, if not most, stars are born with siblings. But recall also that only rarely are the siblings of identical mass. More commonly, the siblings are not identical; one will be more massive than the other. Consequently, the more massive sibling will proceed through its life cycle first and dying first. Should that first sibling have been massive enough to end up as a black hole, its sibling will then for a time find itself orbiting a black hole. Then, when the sibling begins to end its own life, it will swell up as a red giant, and its now-swollen surface approaches the black hole; matter from the surface of the sibling star can now spill onto the black hole. And this then produces the X-ray binary phenomenon that we observe.

Looking to the future, there are efforts underway to develop an entirely new class of telescope that could someday detect the merger of two black holes. As we've discussed, some black holes may be part of binary star systems. Occasionally, those binary systems will involve twins that will both end up as black holes. Those sibling ghost stars would be completely undetectable, as neither can radiate light. Except, they can, in principle, be detected with future gravitational radiation telescopes. These are telescopes that detect not light, but the passing ripple of a disturbance in the Einsteinian fabric of space. When two black holes merge, they produce a short-lived, but intense lurch in the fabric of space that then propagates outward as a wave, like that produced by a rock splashing into a pond. These future gravitational wave telescopes will be searching for those rippling space signals, providing the first non-electromagnetic means of studying the stars.

Already, thanks to advances in very high, angular-resolution imaging with adaptive optics systems on large telescopes, we can directly study an extremely massive black hole that resides at the very center of our galaxy. That massive black hole probably formed through the merger over time of smaller black holes going back to the formation of our galaxy as a whole. Let's look at how astronomers can study this amazing black hole at our galaxy's center.

In this animation, we are seeing individual stars, represented as small rainbow blobs, as they orbit around a massive but unseen object at the position of the five-pointed star in the image. Each frame in the animation is a real snapshot of the stars taken at a particular time, indicated in years at the upper left. So think of this animation like a flip book, the individual pictures showing the changing relative positions of the stars over time. The entire sequence of images spans approximately 16 years in time, so while for some stars we end up tracing only a portion of their orbits, for a few, we are able to see their entire orbits about the massive unseen object at the center.

In particular, notice the star labeled S0-2, which goes all the way around its 15-year orbit. Because we trace the entire orbit, we can apply Newton's laws of orbits, just as we discussed in a previous lecture, for measuring the masses of binary stars. We have here a measurement of the star's orbital period and the size of its orbit, so we can directly calculate the mass of the unseen body at the center, and that comes out to be 3 million times the mass of the Sun. So whatever the unseen object at the center is, it's clearly very massive.

Next, look at the star labeled S0-16. It doesn't quite make it all the way around its orbit in the 16 years that it's been watched, but importantly, it comes extremely close to the central, unseen body, similar to the distance of Pluto from our Sun. So now we know that the unseen body at the center is extremely massive, and, it must be extremely compact, small enough to fit within a space comparable to our solar system. So together, the extremely large mass and the extremely compact size, tell us that the unseen body must be a supermassive black hole. The orbits of the stars about this supermassive black hole at the center of our galaxy can also be represented three dimensionally. As adaptive optics imaging techniques with large telescopes continue to improve, our understanding of the bizarre stellar corpses left behind at the end of the stellar life cycle will become even more detailed.

In this lecture, we've taken a pretty big step away from the familiar. We've visited the stellar nether world, the corpses of stars, white dwarfs, neutron stars, and black holes. These amazing objects reveal an entirely new class of physical laws, from the microscopic physics of quantum mechanics to the fantastic new realm of gravitational wave physics. As fantastic as the stars may be while living, they become even more fascinating to us after death.

Dying Breaths—Cepheids and Supernovae
Lecture 17

Because stars very often begin their lives with siblings, they also often end their lives still bound to their siblings by gravity. The death throes of the longer-lived sibling can breathe new life into its long-dead sibling, for a brief but powerfully spectacular encore. In addition, the final moments of any star's life often involve hiccups and gasps, pulsations and mini-explosions, that provide some of our most valuable methods for measuring the distances to other galaxies and, therefore, have an importance to astronomy and to our understanding of the universe far beyond the stellar life cycle itself.

The Cepheid Instability Strip and Cepheid Variable Stars

- Less massive stars like our Sun end their lives as white dwarfs, the inert carbon remains of the core of the once-living star, now exposed at the center of a planetary nebula. White dwarfs are glittering diamonds that are destined to fade into obscurity as their ashes spread gently into space. More massive stars die dramatically in violent supernova explosions, the most energetic events in the universe, leaving behind a compact ball of pure neutrons—a neutron star—as the rest of the star is blown to smithereens, its material flung far and wide.

- In the lead-up to its death, before it becomes a white dwarf, a low-mass star goes through a stage in its life cycle when it is a red giant. At this stage, the surface of the star is swollen to an enormous size as helium begins to fuse into carbon in the core. For stars of a certain mass, about 2 or 3 times the mass of our Sun, they will pass through a special region of the Hertzsprung–Russell diagram as they evolve away from the main sequence toward becoming a red giant. This region is known as the Cepheid instability strip.

- The instability strip simply represents a particular combination of stellar properties—temperature and radius and luminosity—that cause stars for a brief time to be pulsationally unstable. Whereas stars in their main sequence phase of life are stable, stars that have the physical properties to place them in the instability strip are unstable and are prone to oscillate strongly. These oscillations turn out to have an important predictable character that we can use.

- Stars like the Sun that are still in the main sequence phase of life are highly resilient and are able to maintain their composure in the face of adversity. If the star should experience any kind of disturbance, the overall properties of the star remain remarkably stable. That's because any minor change in the internal structure of the star is immediately compensated and corrected so that the star very quickly returns to its normal configuration.

- A star in the instability strip pulsates—it swells to become larger, and then shrinks to become smaller, then larger, then smaller, and so on. Accompanying this swelling and shrinking is a change in the star's luminosity. The temperature of the surface of the star doesn't change much during these oscillations, so the change in size directly leads to a change in brightness. When the star is larger, it is brighter; when it is smaller, it is fainter. These oscillating, pulsating stars are called Cepheid variable stars and RR Lyrae stars, named after the first stars to have been recognized as undergoing this phenomenon.

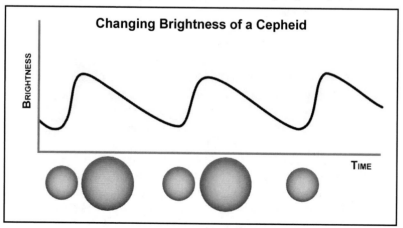

Changing Brightness of a Cepheid

BRIGHTNESS

TIME

- An important consequence of these pulsations in dying stars is that they occur with a regularity that can be related to their average intrinsic luminosity. The regularity of their pulsations is called the pulsation period. By looking at a graph of the changing brightness of a Cepheid or RR Lyrae star, we can identify the pulsation period as the time between successive peaks in their brightness.

- Cepheid and RR Lyrae stars that pulsate with longer cadences are intrinsically more luminous than the more rapidly oscillating ones. This so-called period-luminosity relationship is incredibly useful and important.

- Because these pulsating stars are giants, they are bright, which means that we can see them over great distances. And because they pulsate in a steady fashion, it is easy to pick them out. And because their pulsation periods are in the range of 1 to 50 days, it is relatively easy to measure the period of pulsation for any given Cepheid or RR Lyrae star.

- Finally, by combining the measured period of pulsation with the known relationship between period and luminosity, their distances can be measured quite accurately. In fact, the first measurement of the distance to a galaxy other than our own was made possible through this technique.

- Soon after Henrietta Leavitt discovered and published the relationship between the Cepheid stars' pulsations and their luminosities, Edwin Hubble identified a Cepheid variable star in our neighboring galaxy Andromeda and was able to measure its distance. He did this by measuring the regular, periodic pattern of brightening and dimming in the stars over the course of several weeks.

- Hubble's discovery was the one of the first definitive pieces of evidence that galaxies like Andromeda were in fact island universes of billions of stars all their own, totally separate and vastly distant from our own Milky Way Galaxy. So, these dying breaths of stars like the Sun have been an invaluable tool to our broader understanding of the universe.

- But if the stars provide crucially important service in their dying gasps, they can provide an even more important service after their deaths—that is, as long as they have a sibling companion star to bring them back to life for one brief, fiery, resplendent moment.

White Dwarf Supernovae

- When stars are born, they frequently—perhaps even most often—are born with at least one sibling. The two twins orbit one another, bound to one another by gravity. Beginning their lives as main sequence stars, the twins are relatively small, so they fit comfortably into the confines of their mutual orbit.

- Later on, the more massive of the twins ends its life, swelling to become a red giant, then sloughing off its outer layers as a planetary nebula, leaving behind the white dwarf corpse. The swelling of the senior sibling as a red giant may allow some of its material to spill onto the junior sibling, but this bit of sloshing of material does not fundamentally alter the course of the junior sibling's life. Perhaps it finds itself having inherited a bit more mass from its senior sibling, but the impact on the junior sibling's life cycle is minor.

- However, when it becomes the junior sibling's turn to face death, and it swells to become a red giant, the situation is very different. Here, material from the junior sibling spirals onto the white dwarf corpse of the first sibling, slowly adding mass to the white dwarf. At first, the addition of mass to the white dwarf simply crushes the already-compact white dwarf a bit more, as the white dwarf squeezes its carbon atoms evermore tightly in order to force the quantum mechanical degeneracy pressure to push back on the increased gravity more strongly.

- But once a sufficient layer of fresh material has accumulated on the surface of the white dwarf, a runaway burst of nuclear fusion can occur, turning the freshly accreted hydrogen from the donor star rapidly into helium and carbon and causing the white dwarf to briefly shine brilliantly for perhaps a month or two. The amount of mass involved in these mini-eruptions is tiny, but it is enough to make the white dwarf's brightness increase by 100 times or more. The short brightening of the white dwarf can lead to the brief appearance in the sky of a new star, or *nova* in the original Latin.

- These novae are often recurring. As each episode of flash fusion completes, a next layer of fusionable material is deposited onto the now slightly more massive white dwarf, bursting again and again. These nova episodes continue until one of two things happen. Either the donor star ends its life to become a white dwarf as well, in which case the novae stop and the two dead siblings spend the rest of eternity in a ghostly dance. Or the senior sibling's white dwarf gets pushed over the brink, beyond the maximum mass that a white dwarf can sustain with degeneracy pressure. And if that happens, a very special and important event—a white dwarf supernova—occurs.

- A white dwarf supernova is a special type of supernova explosion that only occurs when a white dwarf stellar corpse is pushed over the degeneracy pressure limit by a binary companion star. These supernova explosions are important because they allow us to measure distances to the farthest reaches of the universe.

- The laws of physics allow a white dwarf to hold itself up against gravity by utilizing a strange form of pressure known as degeneracy pressure. Degeneracy pressure arises from the weird rules of quantum mechanics, which include a rule known as the Heisenberg uncertainty principle. That rule states that the more that particles become localized in space, the more uncertain their speeds become.

- Therefore, a white dwarf can force its constituent particles to move faster and faster by compressing the particles to greater and greater densities. This generates ever-greater pressure with which to push back against gravity. There is no heat involved in this—

just compression and the rules of quantum mechanics driving up the pressure.

- However, this process has a fundamental limit, which is the ultimate speed limit in the universe: the speed of light. This speed limit for the particles in a white dwarf corresponds to a certain maximum mass—namely, 1.4 times the mass of the Sun. That is the so-called Chandrasekhar limit for a white dwarf.

- Whereas a massive star's supernova can involve the detonation of a range of different amounts of material depending on the mass of the star, a white dwarf supernova always involves precisely the same amount of material—namely, 1.4 times the mass of the Sun. As a result, a white dwarf supernova always releases the same amount of explosive energy, because the same amount of explosives is involved every time.

- This property of a white dwarf supernova means that, if we have a way of reliably recognizing a white dwarf supernova as opposed to a massive-star supernova, then whenever we see one, we know precisely how luminous it is, and we can use that in comparison to how bright it appears to accurately determine the distance to it.

- The distance of objects in the sky can be calculated precisely using the inverse square law of light. This is, in fact, one of the most important tools by which astronomers measure accurate distances to some of the farthest galaxies. This works because supernovae are intrinsically the most luminous events in the universe, so they can be seen across the vast stretches of the universe. Thankfully, nature has provided a type of supernova that is always and everywhere the same—a benchmark.

- There are two ways to determine that one distant flash of light is the result of a white dwarf detonating as opposed to a massive star ending its life. The first way is to look at the light spectrum from the supernova flash. The vast majority of the body of the star involved in the explosion is the outer layers of the star, which were never processed through the star's stages of fusion. That material is mostly hydrogen. Therefore, look at the light spectrum of a massive star's supernova, and you'll see very clearly the light features of hydrogen.

- The other way is to use the shape of the explosion. Long after the supernova goes off, the exploded material can be seen as an expanding shell of material that we call a supernova remnant—some of which are highly spherically symmetrical while others have much more complex morphologies. Research has shown that the white dwarf supernovae are measurably more symmetric than those produced by massive star deaths.

Suggested Reading

Bartusiak, *The Day We Found the Universe*.

Mann, *Shadow of a Star*.

Questions to Consider

1. What might be some reasons that more slowly pulsating Cepheid stars are also more luminous?

2. What might be some reasons that white dwarf supernovae produce more symmetric and spherical supernova remnants than massive star supernovae?

Dying Breaths—Cepheids and Supernovae
Lecture 17—Transcript

We've seen that the life cycle of stars ends in a death that, depending on the mass of the star, can be silent and beautiful or powerful, dramatic, and violent. The corpses that the stars leave behind, whether white dwarf or neutron star, are fantastic and strange, governed by a strange realm of physics befitting these netherworld objects. But as we'll see in this lecture, these corpses are not necessarily the last hurrah for the star.

Because stars very often begin their lives with siblings, they also often end their lives still bound to their siblings by gravity. As we'll see, the death throes of the longer-lived sibling can breathe new life into its long-dead sibling, for a brief but powerfully spectacular encore. In addition, the final moments of any star's life often involve hiccups and gasps, pulsations, and mini-explosions. As we'll see, these pulsations and explosions provide some of our most valuable methods for measuring the distance to other galaxies, and so have an importance to astronomy and to our understanding of the universe far beyond the stellar life cycle itself.

To get us started, let's briefly review the stellar corpses left behind by different types of stars. Less massive stars, like our Sun, end their lives as white dwarfs, the inert carbon remains of the core of the once-living star now exposed at the center of a planetary nebula. White dwarfs are glittering diamonds that are destined to fade into obscurity as their ashes spread gently into space. More massive stars die dramatically in violent supernova explosions, the most energetic events in the universe, leaving behind a bizarre, compact ball of pure neutrons—a neutron star—as the rest of the star is blown to smithereens, its material flung far and wide.

You'll also recall that in the lead up to its death, before it becomes a white dwarf, a low-mass star goes through a stage in its life cycle when it is a red giant. At this stage, the surface of the star is swollen to an enormous size as helium begins to fuse into carbon in the core. Now, for stars of a certain mass, about two or three times the mass of our Sun, they will pass through a special region of the Hertzsprung-Russell diagram as they evolve away from the main sequence toward becoming a red giant. This region is known as the Cepheid instability strip.

The instability strip simply represents a particular combination of stellar properties—temperature and radius and luminosity—that cause stars, for a brief time, to be what we call pulsationally unstable. Whereas stars in their main sequence phase of life are stable, stars that have the physical properties to place them in the instability strip are unstable and are prone to oscillate strongly. These oscillations turn out to have an important predictable character that we can use.

Stars like the Sun that are still in the main sequence phase of life are highly resilient and are able to maintain their composure in the face of adversity. If the star should experience any kind of disturbance, perhaps a gravitational disturbance by a passing star or a temporary internal disturbance due to some glitch in the fusion process, call it indigestion, the overall properties of the star remain remarkably stable. That's because any minor change in the internal structure of the star is immediately compensated and corrected so that the star very quickly returns to its normal configuration.

For example, imagine if the energy generation due to fusion in the core of the Sun were to suddenly increase by a bit for some reason. That would raise the temperature in the core, driving up the temperature in the outer layers, and puffing out the surface slightly. But this would cause the surface to cool a bit and the internal pressure to drop slightly, quickly bringing the fusion rate back down to what it was. So there is, essentially, an internal regulation valve that keeps the star very stable in terms of its physical properties at all times. Another way to think of it is as a car with good shock absorbers; if the car goes over a bump, it'll bounce once and then quickly right itself. But when that same star finds itself in the instability strip, it's as if the shock absorbers are shot. Instead of smoothing out the ride, the car bounces wildly over any little bump. And if it goes over a series of bumps with just the right spacing, the resulting oscillations of the car can be overwhelming.

For a star in the instability strip, this analogy translates into actual oscillations of the surface of the star. The star literally pulsates; it swells to become larger, then shrinks to become smaller, then larger, then smaller, and so on. Accompanying this swelling and shrinking is a change in the star's luminosity. The temperature of the surface of the star doesn't change much at all during these oscillations, and so the change in size directly leads to a change in brightness. When the star is larger, it's brighter, and when it's smaller, it's fainter. These oscillating, pulsating stars are called Cepheid variable stars and RR Lyrae stars, named after the first stars to have been recognized as undergoing this phenomenon.

An important consequence of these pulsations in dying stars is that they occur with a regularity that can be related to their average intrinsic luminosity. The regularity of their pulsations is called the pulsation period. Looking at a graph of the changing brightness of a Cepheid or RR Lyrae star, we can identify the pulsation period as the time between successive peaks in their brightness. It turns out that the Cepheid and RR Lyrae stars that pulsate with longer cadences are intrinsically more luminous than the more rapidly oscillating ones. This so-called period-luminosity relationship is incredibly useful and important.

Because these pulsating stars are giants, they are bright, which means we can see them over great distances. And because they pulsate in a steady fashion, it's easy to pick them out. And because their pulsation periods are in the range of about 1 to 50 days, it is relatively easy to measure the period of pulsation for any given Cepheid or RR Lyrae star. Finally, by combining the measured period of pulsation with the known relationship between period and luminosity, their distances can be measured quite accurately.

In fact, the first measurement of the distance to a galaxy other than our own was made possible through this technique. Soon after Henrietta Leavitt discovered and published the relationship between the Cepheid stars' pulsations and their luminosities, Edwin Hubble identified a Cepheid variable star in our neighboring galaxy, Andromeda, and was able to measure its distance. He did this by measuring the regular, periodic pattern of brightening and dimming in the stars over the course of several weeks, as you see in these, his actual data graphs.

For example, with the star labeled number 30, one peak in the star's brightness occurs at what Hubble called day 0, and then that peak brightness occurs again about 45 days later. That's a relatively slow Cepheid variable, and according to Leavitt's relation, would indicate a relatively luminous Cepheid star. In contrast, the star that Hubble labeled number 34, repeats its peak brightness level after just 22 days or so, indicating an intrinsically less luminous Cepheid star. You can see Hubble's own excitement at his discovery of the first of these from the exclamatory note that he wrote directly on the photographic plate that he took of the galaxy. That's the "VAR!" written at the upper right, Hubble's shorthand for a Cepheid variable star. He knew he was on to something very important.

You see, one of the biggest questions of the day had to do with the very nature of galaxies. Astronomers simply didn't know how far away they were, so they didn't know whether they represented entire systems of stars entirely outside our Milky Way galaxy, or if they represented a kind of nebula within our galaxy. This was, in fact, such a hotly debated topic, that there is a famous series of research articles representing what astronomers now refer to as the Great Debate. The debate was between two astronomers—Harlow Shapley and Heber Curtis. Shapley argued that Andromeda and other galaxies were simply nebulae within our galaxy. In other words, the universe, according to Shapley, was coextensive with the Milky Way. Heber Curtis countered that the galaxies were distinct entities far outside our galaxy.

There were good arguments on both sides of this debate. What was missing to clinch the proof one way or the other was a clear, accurate measurement of the distance to at least one of those galaxies. All that was known was that the galaxies, whatever they were, were far enough away that we couldn't perform the technique of triangulation—or parallax—on them. But that by itself wasn't saying much. In the end, Hubble's discovery of the Cepheid variable star in Andromeda would help to prove Curtis right. Those galaxies were not nebulae within the Milky Way, but entirely distinct systems of stars far, far beyond our Milky Way.

So that's why Hubble was so excited to find a Cepheid variable star in the galaxy Andromeda. He knew that the resulting precise distance measurement to that galaxy would at last help answer this raging debate about the very nature of galaxies. Hubble's discovery, therefore, was one of the first definitive pieces of evidence that galaxies, like Andromeda, were, in fact, island universes of billions of stars all their own, totally separate and vastly distant from our own Milky Way galaxy. So you see, these dying breaths of stars like the Sun have been an invaluable tool to our broader understanding of the universe.

But if the stars provide crucially important service in their dying gasps, they can provide an even more important service after their deaths, that is, as long as they have a sibling companion star to bring them back to life for one brief, fiery, resplendent moment. Let's look at what happens to a white dwarf that has a sibling. First, recall that when stars are born, they frequently, and perhaps even most often, are born with at least one sibling star. The two twins orbit one another, bound to one another by gravity. Beginning their lives as main-sequence stars, the twins are relatively small, as stars go, and so they fit comfortably into the confines of their mutual orbit.

Later on, the more massive of the twins ends its life, swelling to become a red giant, then sloughing off its outer layers as a planetary nebula, leaving behind the white dwarf corpse. The swelling of the senior sibling as a red giant may allow some of its material to spill onto the junior sibling, but this bit of sloshing of material does not fundamentally alter the course of the junior sibling's life. Perhaps it finds itself having inherited a bit more mass from its senior sibling, but the impact on the junior sibling's life cycle is minor.

However, when it becomes the junior sibling's turn to face death and it swells to become a red giant, the situation is very different. Here, material from the junior sibling spirals onto the white dwarf corpse of the first sibling, slowly adding mass to the white dwarf. At first, the addition of mass to the white dwarf simply crushes the already-compact white dwarf a bit more, as the white dwarf squeezes its carbon atoms ever more tightly in order to force the quantum mechanical degeneracy pressure to push back on the increased gravity more strongly.

But once a sufficient layer of fresh material has accumulated on the surface of the white dwarf, a runaway burst of nuclear fusion can occur, turning the freshly accreted hydrogen from the donor star rapidly into helium and carbon and causing the white dwarf to briefly shine brilliantly for perhaps a month or two. The amount of mass involved in these mini eruptions is tiny, perhaps only one ten-thousandth of a solar mass, but it's enough to make the white dwarf's brightness increase by 100 times or more. The brief brightening of the white dwarf can lead to the brief appearance in the sky of a new star, or nova, in the original Latin.

These novae are often recurring. As each episode of flash fusion completes, a next layer of fusionable material is deposited onto the now slightly more massive white dwarf, bursting again and again. These nova episodes continue until one of two things happen. Either the donor star ends its life to become a white dwarf as well, in which case the novae stop and the two dead siblings spend the rest of eternity in a ghostly dance. Or else, the senior sibling's white dwarf gets pushed over the brink, beyond the maximum mass that a white dwarf can sustain with degeneracy pressure. And if that happens, we get a very special and important event, a white dwarf supernova. A white dwarf supernova is a special type of supernova explosion that only occurs when a white dwarf stellar corpse is pushed over the degeneracy pressure limit by a binary companion star. These supernova explosions are important, because they allow us to measure distances to the farthest reaches of the universe.

Recall that the laws of physics allow a white dwarf to hold itself up against gravity by utilizing a strange form of pressure known as degeneracy pressure. Degeneracy pressure arises from the weird rules of quantum mechanics, which include a rule known as the Heisenberg Uncertainty Principle. That rule states that the more that particles become localized in space, the more uncertain their speeds become. And so a white dwarf can force its constituent particles to move faster and faster by compressing the particles to greater and greater densities. This generates an ever greater pressure with which to push back against gravity. There is no heat involved in this, just compression and the rules of quantum mechanics driving up the pressure.

But as we learned when we first discussed degeneracy pressure, this process has a fundamental limit, which is the ultimate speed limit in the universe, the speed of light. And it turns out that this speed limit for the particles in a white dwarf corresponds to a certain maximum mass for the white dwarf, namely, 1.4 times the mass of the Sun. That is the so-called Chandrasekhar limit for a white dwarf.

So, picture a white dwarf that has a red giant for a sibling, and picture material from that sibling giant spilling onto the white dwarf. The mass of the white dwarf builds and builds right up to the moment when its total mass exceeds the limit of 1.4 times the mass of the Sun. At this moment, the white dwarf is crushed by gravity once and for all. And in the process, it goes through the same collapse and bounce process that we saw when we looked at the deaths of massive stars. The result? A white dwarf supernova explosion.

However, and this is the important part, whereas a massive star's supernova can involve the detonation of a range of different amounts of material depending on the mass of the star, a white dwarf supernova always involves precisely the same amount of material, namely 1.4 times the mass of the Sun. As a result, a white dwarf supernova always releases the same amount of explosive energy, because the same amount of explosives is involved every time.

This property of a white dwarf supernova means that, if we have a way of reliably recognizing a white dwarf supernova, as opposed to a massive star supernova, then, whenever we see one, we know precisely how luminous it is. And we can use that, in comparison to how bright it appears, to accurately determine the distance to it. Remember that how bright an object appears to us at Earth is due to the combination of its true luminosity and its distance from us. A star of a given intrinsic luminosity can appear very bright if it's very close or very dim if it's very far away. In the case of a white dwarf supernova, we know the true intrinsic luminosity precisely, and so the only thing that can explain how bright or dim it appears to us, is its distance. And as you know, the distance of objects in the sky can be calculated precisely using the inverse square law of light.

This is, in fact, one of the most important tools by which astronomers measure accurate distances to some of the farthest-away galaxies. This works because supernovae are intrinsically the most luminous events in the universe, and so they can be seen across the vast stretches of the universe. Thankfully, nature has provided a type of supernova that is always and everywhere the same, a benchmark, a beacon with which we may plumb the depths of space.

But how do we tell a white dwarf supernova from the other kind? How can we determine that one distant flash of light is the result of a white dwarf detonating, as opposed to a massive star ending its life? There are two ways, one intuitive, the other less so. The first and most obvious way to tell that you have a white dwarf supernova on your hands is to look at the light spectrum from the supernova flash. Think about the nature of the material involved in a massive star's supernova, and you'll quickly realize that the vast majority of the body of the star involved in the explosion is the outer layers of the star, which were never processed through the star's stages of fusion. That material is, therefore, mostly hydrogen. And so, look at the light spectrum of a massive star's supernova, and you'll see very clearly the light features of hydrogen.

But when a white dwarf supernova goes off, what is it that was exploded? It is the material that made up the white dwarf, almost entirely carbon, the now-inert core of a star that had once fused hydrogen to helium to carbon. And so, look at the light spectrum of a white dwarf supernova, and you'll see the light features of all kinds of elements, from carbon, of course, to many heavier elements fused in the supernova explosion itself, but the one you won't see in abundance is hydrogen. So if you see no hydrogen in the light spectrum of a supernova, it's a very good bet you've got a white dwarf supernova. That makes perfect sense, right?

Well, it turns out there's also another way to tell, using the shape of the explosion. Remember that long after the supernova goes off, the exploded material can be seen as an expanding shell of material that we call a supernova remnant. Some supernova remnants are highly spherically symmetrical; others have much more complex morphologies. Research from my own group has shown that the white dwarf supernovae, also referred to by astronomers as Type Ia supernovae, are measurably more symmetric than those produced by massive star deaths.

The reasons for this big difference in the symmetries of white dwarf supernovae versus massive star supernovae are not yet fully understood. But it appears that for some reason, a massive star's supernova involves another axis of symmetry beyond the simply spherical. This could be because of the massive star's rotation prior to its collapse and explosion. The rotation of the star, in essence, defines a preferred direction in space, an axis, that could in some way help to direct the supernova explosion into a bipolar shape.

Ultimately, if we can understand the reasons behind the structures of supernova explosions, that will be important information for filling in the details behind how supernova explosions happen at all. But getting at this information has been a real challenge. Indeed, sophisticated computer simulations of the supernova explosion mechanism have found it difficult not only to reproduce the observed features of supernovae, but even to be able to make a star explode at all. It turns out it's not so easy to blow up a star. Obviously nature knows perfectly well how to do it. And our most basic conceptual understanding, involving the bounce of the outer layers of the star off of the hard neutron star formed from the collapsed core, must be correct at some basic level. But the devil is in the details, and in many attempts to simulate the detonation of a star, the explosion fizzles out before it even really gets going.

Here's an example of the type of simulation that flummoxed researchers for over two decades. The simulation involves a dead star, and you can see the supernova fireball building deep within its core. But despite all of the bubbling and churning, the explosion is simply unable to break out through the enormous heft of the overlying layers of the star. Just two tenths of a second after the bounce off of the neutron star at the center, the outward propagating shock stalls, traveling out no more than a mere 200 kilometers or so. The star in this type of simulation rumbles mightily deep within, but overall, it simply collapses, no supernova.

But more recent simulations are at last able to get the thing to actually blow up. Let's look at one of these simulations together. Here we're focusing in on the central core of a star about 11 times the mass of the Sun. And as we zoom in to the central core of the star, we see depicted at the very center that hard neutron star corpse that was produced from the initial collapse of the core. And we see the outer layers collapsing in, seen streaming in as shades of color in this simulation. Now, initially here, the explosion has a really hard time getting going. It's, essentially, stalled, and you see the rumbling and churning and bubbling around the neutron star at the center, but the explosion doesn't really seem to want to go. But little by little an asymmetry develops. You see a kind of bipolar shape to the outward-expanding material, and as that asymmetry develops, the explosion finally is able to squirt out towards the top and towards the bottom, and this squirting out of the collapsing and now exploding material from the center will literally tear the star apart. Now, impressively, what you've seen in this entire simulation actually represents just one second of real time.

An important ingredient in these successful simulations has been to better tap the energy of the neutrinos that are generated during the collapse of the star's core into a neutron star. Recall that the collapse of the core leads to the creation of the neutron star at the center, off of which the rest of the star bounces and explodes. That neutron star is the result of protons and electrons in the star's core being smashed together to produce neutrons. Well, one of the byproducts of the proton plus electron equals neutron reaction is the creation of neutrinos.

You remember neutrinos as a byproduct of the fusion process that we discussed in detail in an earlier lecture. In fact, the neutrinos that are produced in the initial collapse that triggers the supernova explosion represent 99% of the total energy of the supernova. The supernova that we see in the form of light represents the most energetic event in the universe. But that visible supernova explosion involves just 1% of the total energy actually released in the supernova. The overwhelming majority of the explosion's energy is released in the form of neutrinos.

The trouble is, those neutrinos do not interact strongly with matter very much at all. So, many early simulations ignored the small amount of those neutrinos whose energy might be tapped to help power the explosion. But when 99% of the total available energy is in those neutrinos, tapping even 1% of their energy represents a doubling—a 100% increase—in the energy that the simulation can use to propel the explosion. So at last, harnessing the sophisticated capabilities of powerful supercomputers running for years, we can successfully do what the stars do in a fraction of a second, blow up!

In this lecture, we've spent some time with the stars during their final gasps, their dying breaths, if you will. In the pulsations that some stars undergo in their final moments—the Cepheid variable stars—we have a tool for measuring the distances to nearby galaxies, as Hubble first did to show that Andromeda is a distinct galaxy apart from our Milky Way. We've also explored the explosive detonations from white dwarfs whose backs are broken by the final straw of material from a donor sibling star. And we've seen that these explosions can serve as a tool to accurately plumb the farthest reaches of the universe. The stars, you might say, are gifts that even in death, keep on giving.

Supernova Remnants and Galactic Geysers
Lecture 18

T he same cataclysmic explosions that signal the dramatic end of a massive star also serve to spread the chemically enriched remains of the star throughout the galaxy and help to trigger new generations of stars and solar systems. In this lecture, you will learn how the remnants of supernova explosions sculpt and compress the gas and dust between the stars. You will also learn that large numbers of exploding stars produce galactic geysers of chemically enriched material, which then rains back down onto the galaxy and provides the substance of new life.

Supernova Remnants

- There are two ways in which a star can end its life as a supernova. One way is that a white dwarf, the stellar corpse of a low-mass star like the Sun, may accumulate material from a sibling star until it reaches a critical mass and undergoes a thermonuclear explosion.

- The other way is when a massive star uses up its nuclear fuel, having alchemized in its core—starting from hydrogen—all of the elements up to iron. At that point, the star is unable to generate more fusion energy with which to hold itself up against gravity, collapses inward under its own gravity, and then explodes through a bounce mechanism. This type of supernova produces remnant expanding bubbles of hot gas that we can still see throughout our galaxy.

- When a star explodes as a supernova, it very energetically expels nearly all of the material that once made up the star's body. The expelled material pushes outward at an extremely high speed, as much as 30,000 kilometers per second, or about 10% of the speed of light. Because of that very energetic speed, the ejected material plows into the surrounding interstellar space supersonically.

- As a result, a strong shockwave forms ahead of the expanding bubble of ejected material, and this shockwave heats the surrounding galactic medium up to temperatures well above millions of degrees. That is so hot that the shock front emits strongly in X-ray light. The shock continuously slows down over time as it sweeps up the ambient medium, but it can expand over hundreds of thousands of years and over tens of light-years in size before its speed finally slows and comes to a halt.

- All supernova remnants go through a series of 5 stages as they expand into the surrounding interstellar space and then ultimately becoming incorporated into it. The first stage is the so-called free expansion stage, during which the ejected material sweeps out through its surroundings essentially unimpeded. This phase continues until the out-moving material sweeps up an amount of ambient material equal to its own mass. This can last up to a few hundred years, depending on the density of the surrounding gas.

- Second is the sweeping up of a shell of shocked interstellar gas. The supernova remnant bubble continues pushing outward, just somewhat more slowly than before. The strong shockwaves and hot shocked gas at the edges of the supernova remnant emit strong X-ray light.

- The third stage is called the snowplow phase, during which the shell of the supernova fireball begins to cool, forming a thin, dense shell surrounding the hot gas in the interior of the expanding bubble. The expanding bubble is slowing down during this third stage, but it continues to be propelled, like a snowplow, by the

© Stocktrek Images/Thinkstock.

The Crab Nebula, estimated to be 10 light-years in diameter, is an expanding remnant of a star's supernova explosion.

pressure of the extremely hot gas within. The gas and dust in the surrounding galactic medium is pushed together and gathered up, like snow and debris collected by a plow.

- In the fourth stage, the material in the supernova remnant's interior cools down, the bubble having expanded to a very large size. The shell around the bubble continues to expand from its own momentum, but slowly now, and the shell begins to break up, the overall supernova remnant becoming less intact.

- Finally, in the fifth stage, the material in the supernova remnant becomes gradually incorporated into the surrounding interstellar medium. At this stage, the supernova remnant has slowed down to a near stop, comparable to the slow speed of the random velocities of gas and dust particles in the surrounding medium. In total, about 30,000 years have elapsed since the initial supernova explosion.

- Because the supernova remnant interacts with the surrounding galactic medium through its plowing and compressing action, the supernova remnant can and does act as a trigger in some cases to initiate new stellar birth. That's because stellar birth takes place in clouds of gas and dust that have become sufficiently dense for gravity to take over and begin collapsing the material down to make new stars.

- With the supernova remnant plowing into its surrounding material, we have a perfect means by which otherwise diffuse galactic material is gathered up and compressed to the types of densities required for gravity to start the stellar birth process.

- The nursery that is giving birth to these stars now has much more in the way of elements such as silicon and iron to make rocks and oxygen to make breathable atmospheres in those newly formed worlds. In all likelihood, this is the type of scenario in which our own Sun and solar system—our Earth—came to possess the elements that make our rocky planet, its oceans, and our oxygen atmosphere possible.

- In the context of understanding how supernovae help to create and spread the heavy elements that are required for creating rocky worlds and for any life upon them, astronomers often need to be able know how a given supernova remnant was produced.

- There are two basic mechanisms for producing supernova explosions: One type of supernova results from the implosion then explosion of a massive star's death; the other type results when a white dwarf is pushed over its maximum possible mass by a sibling star spilling mass onto it. The two types of supernovae will both produce supernova remnants, but the amounts and proportions of heavy elements that they produce may differ.

- When we use computer analysis to quantify the degree of roundness and symmetry of each supernova remnant, we find that there is a stark, measurable difference between the two groups: The massive-star supernovae remnants are measurably more complex and less symmetric than the white dwarf remnants.

Super-Bubbles

- When multiple supernova explosions go off in proximity to one another, the result is an enormous type of supernova remnant that we call a super-bubble, which can dramatically influence the stellar life cycle across the entire galaxy. There are examples of such super-bubbles in our Milky Way Galaxy and in our nearest satellite galaxies. These structures can span enormous distances—up to 1000 light-years or more.

- From measurements of the vast dimensions of these structures and from measurements of the speeds with which they continue to expand in size, astronomers can estimate that these structures may represent the combined supernova remnant action of perhaps 10 or more supernova explosions all going off at nearly the same place and at nearly the same time.

- The stellar birth process might cause such an apparent orchestration of supernova explosions. When the stellar birth process in a stellar nursery produces a cluster of stars—a stellar family—it tends to produce a distinct pattern of birth weights for the stars: lots of little runts and just a few behemoths.

- In most stellar nurseries, which birth perhaps 1000 stars or so, this translates into perhaps 1 or 2 stars massive enough to go supernova within the nursery. However, in some cases, a stellar nursery can birth perhaps 10,000 stars. In such cases, there may be a dozen very massive stars that will all die in supernova explosions. Because those massive stars were born at nearly the same time and from the same stellar nursery, they will die at nearly the same time and in that place—thus, driving a super-bubble.

- Super-bubbles push out into the surrounding space. They carve out large cavities in the surrounding galactic disk of gas and dust and push their hot, exploded stellar material up and out into the surrounding halo of the galaxy. From there, that material can then rain back down to the farthest reaches of the galaxy, where it can be gathered up again by gravity to form new generations of stars and planets.

- The effects of these super-bubbles on a galactic scale are most easily seen by looking at more distant galaxies, where we can take a panoramic view of an entire galaxy at once. The best case studies are galaxies seen edge-on from our perspective, because then we can use the glow of the billions of stars within that galaxy to illuminate from behind the large-scale structures caused by the super-bubbles.

Galactic Geysers
- Just as the geysers on Earth spew hot steam created from water heated beneath the surface up into the surrounding air, which condenses as mineral-rich water raining back down to the surface, helping vegetation to grow, galactic geysers spray hot gas from supernovae up into the galactic halo, which then condenses and rains back down as chemically enriched material to the disk of the galaxy, where it can fertilize new generations of stars.

- We can see the influence of such geysers in our own galaxy by observing the hot gas that fills our galaxy's halo and glows in X-ray light. Images taken with the Chandra X-ray telescope reveal not only that our galaxy's halo is filled to the brim with hot gas originating in supernova super-bubbles, but also that this hot gas extends far, far out from our galaxy. In fact, it completely envelops our galaxy, and beyond that, it completely envelops the Milky Way's satellite galaxies.

- This is important to our picture of how the life cycles of future generations of stars come to be enriched with more abundant heavy elements as produced in supernovae. The supernovae don't merely sprinkle their ashes hither and yon. More generally, the collective action of the many supernova explosions in the galaxy is to bathe, to engulf, the entire galaxy in those ashes, suffusing the galaxy with them.

- No wonder that a solar system such as our own, forming in the suburbs of the galaxy, has been so enriched with heavy elements that it was able to form a rocky, watery planet with enough organic material to permit abundant life.

- Maybe there are suburbs of other galaxies where the conditions for stars sufficiently enriched in heavy elements can form solar systems including rocky planets covered in liquid oceans and surrounded by breathable oxygen-rich atmospheres capable of supporting life as we know it. It is tempting to contemplate what a civilization on such a world at the outskirts of our galaxy or indeed in another galaxy—looking at our own suburb of the Milky Way, supernova remnants all around—might wonder about the potential for life around that otherwise ordinary yellow star we call the Sun.

Suggested Reading

Arnett, *Supernovae and Nucleosynthesis.*

Marschall, *The Supernova Story.*

1. In what ways do supernova explosions influence the life cycles of other stars, including those not yet born?

2. Astronomers sometimes describe galaxies as giant recycling plants. With respect to the life cycle of stars, in what ways is this description accurate, and in what ways is it not?

Supernova Remnants and Galactic Geysers
Lecture 18—Transcript

The poet T. S. Eliot famously wrote that the world ends not with a bang, but a whimper. For human cultures and civilizations, perhaps this is true, but not so for a massive star. As we've seen, these stars end their lives in dramatic, fiery, violent deaths that we call supernova explosions. These are cataclysms of unimaginable proportions, the most energetic events that we know in the universe today. And as the dying star is blown to smithereens, a supernova marks the end of a life lived fast and furious. And yet, while massive stars do go out with a bang, their deaths do not represent the end of their story, but rather, a new beginning.

You see, the same cataclysmic explosions that signal the dramatic end of a massive star also serve to spread the chemically enriched remains of the star throughout the galaxy and help to trigger new generations of stars and solar systems. In this lecture, we'll explore amazing images of the remnants of supernova explosions, and we'll see how they sculpt and compress the gas and dust between the stars. We'll see images of new stars whose birth has been triggered by the supernova deaths of nearby stars. We'll see examples of regions in our Galaxy where large numbers of exploding stars produce galactic geysers of chemically enriched material that then rains back down onto the galaxy, a kind of manna from heaven that literally provides the substance of new life. And we'll see how the explosive energy of the supernova explosions serves to stir and mix these new life-giving elements throughout the galaxy, blending in the new material with the old and creating new potential for enriched chemical lives of new generations of stars and solar systems yet to be.

In our previous lectures, we've seen that there are two ways in which a star can end its life as a supernova. One way is that a white dwarf, the stellar corpse of a low-mass star like the Sun, may accumulate material from a sibling star until it reaches a critical mass and undergoes a thermonuclear explosion. We've seen that these white dwarf supernovae are extremely useful for accurately measuring distances to faraway galaxies, because these supernovae always release the same amount of energy, the same intrinsic luminosity. We'll set these white dwarf supernovae aside for the time being.

But the other way that a star can go supernova is when a massive star uses up its nuclear fuel, having alchemized in its core, starting from hydrogen, all of the elements up to iron. At that point, the star is unable to generate more fusion energy with which to hold itself up against gravity, and so collapses inward under its own gravity, and then explodes through a bounce mechanism that we discussed in detail in a previous lecture. This is the type of supernova that we'll mainly focus on today, specifically the remnant expanding bubbles of hot gas that they produce and that we can still see throughout our galaxy.

Let's look at an example of a supernova remnant, just to identify some of the key features, and then we'll look at several other examples to draw out some of the differences among different supernova remnants. Here we're looking at a supernova called the Tycho supernova, so named because it was observed by Tycho Brahe himself. This is a supernova that went off in the year 1572. Tycho Brahe, who was studying the motions of planets, nonetheless observed the supernova and measured its brightness.

At the center, we see the stellar corpse left behind by that supernova explosion, and surrounding it, an expanding bubble of hot gas. This was the material that was part of the star's body when it blew up. And we see, within the supernova remnant, turbulent motions that are manifested in this scalloped appearance within the gas inside of the bubble. And at the outer edge we see in extremely hot, glowing blue, the shockwave pushing out into the ambient galactic medium.

When a star explodes as a supernova, it very energetically expels nearly all of the material that once made up the star's body. The expelled material pushes outward at an extremely high speed, as much as 30,000 km/s, or about 10% of the speed of light. Because of that very energetic speed, the ejected material plows into the surrounding interstellar space supersonically. Physicists describe the degree to which an object moves through a surrounding gas supersonically by the Mach number; Mach 1 corresponds to breaking the speed of sound. By comparison, a supernova blast wave moves through the surrounding gas of the interstellar medium at Mach 1000 or higher.

As a result, a strong shock wave forms ahead of the expanding bubble of ejected material, and this shock wave heats the surrounding galactic medium up to temperatures well above millions of degrees. That is so hot that the shock front emits strongly in X-ray light. The shock continuously slows down over time as it sweeps up the ambient medium, but it can expand over hundreds of thousands of years and over tens of light-years in size before its speed finally slows and comes to a halt.

Now let's look at a few more examples of supernova remnants to take in their amazing appearances and some of the variety among them. The first example that we're looking at here is the Kepler supernova remnant. This is a supernova that went off in the year 1604. Notice it has a roughly spherical shape, and it's filled with very hot gas. And what we're seeing is the light signatures of different elements that were produced in the star and in the supernova glowing their signature colors within the expanding bubble. And we see a shell that is the shock wave around the perimeter glowing hot, but it's beginning to break up into bead-like structures.

This next example is Supernova 1006; it's a supernova remnant of a supernova that went off in the year 1006 and was described in writings by observers in China, and Japan, Iraq, Egypt, and Europe, and possibly even in North American petroglyphs. Again, as in the last example, this supernova remnant is quite symmetric and spherical in shape. And we see blue knots all through the remnant that are breaking up the edges of the shockwave shell seen in front and behind.

This next example is the Crab supernova remnant. This is a supernova that went off in the year 1054 and was particularly noted by Chinese astronomers at the time. This one has a very different appearance than the others; it's much less symmetric, and its edges are highly broken up as this remnant now begins to disintegrate and incorporate itself into the surrounding galactic medium.

This next example is the Vela supernova remnant. This is a supernova that went off about 12,000 years ago and now stretches about 100 light-years across. This one has now been almost completely incorporated into the surrounding medium.

How do these remnants of supernova explosions actually interact with their surroundings? All supernova remnants go through a series of five stages as they expand out into the surrounding interstellar space and then ultimately become incorporated in it. Let's go through these five stages now in detail. The first is the so-called free-expansion stage, during which the ejected material sweeps out through its surroundings essentially unimpeded, like a rocket through air. This phase continues until the out-moving material sweeps up an amount of ambient material about equal to its own mass. This can last up to a few hundred years, depending on the density of the surrounding gas.

The second is the sweeping up of a shell of shocked interstellar gas. The supernova remnant bubble continues pushing outward, just somewhat more slowly than before. The strong shock waves and the hot shocked gas at the edge of the supernova remnant emit strong X-ray light. The third stage is called the snowplow phase. During the snowplow phase, the shell of the supernova fireball begins to cool, forming a thin, dense shell surrounding the hot gas in the interior of the expanding bubble. Bear in mind, we're talking interstellar scales here; when I say a thin shell, I mean it's approximately one light-year thick. And the hot gas inside the bubble? Well, we're talking a few million degrees Kelvin! In any event, the expanding bubble is slowing down during this third stage, but it continues to be propelled, like a snowplow, by the pressure of the extremely hot gas within. The gas and dust in the surrounding galactic medium is pushed together and gathered up, like snow and debris collected by a plow.

In the fourth stage, the material in the supernova remnant's interior cools down, the bubble having expanded now to a very large size. The shell around the bubble continues to expand from its own momentum, but slowly now, and the shell begins to break up, the overall supernova remnant becoming less intact. Finally, in the fifth stage, the material in the supernova remnant becomes gradually incorporated into the surrounding interstellar medium. At this stage, the supernova remnant has slowed down to a near stop, comparable to the slow speed of the random velocities of gas and dust particles in the surrounding medium. In total, about 30,000 years have elapsed since the initial supernova explosion.

Because the supernova remnant interacts with the surrounding galactic medium through its plowing and compressing action, the supernova remnant can and does act as a trigger in some cases to initiate new stellar birth. That's because, as we've discussed in previous lectures, stellar birth takes place in clouds of gas and dust that have become sufficiently dense for gravity to take over and begin collapsing the material down to make new stars. With the supernova remnant plowing into its surrounding material, we have a perfect means by which otherwise diffuse galactic material is gathered up and compressed to the types of densities required for gravity to start the stellar birth process.

A nice example of this is in the Lambda Orionis supernova remnant. This remnant appears as a large ring of gas and dust surrounding the bright star Lambda Orionis, the head of Orion. To be clear, that bright star was not the progenitor of the supernova that created that large remnant. Rather, astronomers believe that this star previously had a massive stellar sibling that was the first to form in this region, and when it went supernova, it left behind its Lambda Orionis companion and produced the large remnant that now rings it.

Now, when we look in detail at the clumps of gas and dust that protrude from the surrounding supernova remnant ring and jut in toward Lambda Orionis, we find entire litters of stars now being born. For example, in this Spitzer Space Telescope infrared image of the pillar Barnard 30, we see a large number of baby stars beginning to light up deep within and still enshrouded by that large pillar. In other words, the Lambda Orionis supernova remnant is at once a stellar graveyard and a stellar nursery.

The material of the nursery itself has been gathered together by the compression power of the supernova remnant, and its elemental material sprinkled with the chemically enriched stellar ashes of the star that exploded. Because the long-dead, exploded star fused so many heavier elements during its life and death, the new stars that form from that material will presumably have an even greater chance of making rocky worlds such as our own Earth. In other words, the nursery that is giving birth to these stars now has much more in the way of elements, such as silicon and iron, to make rocks and oxygen to make breathable atmospheres on those newly formed worlds.

This, in all likelihood, is the type of scenario in which our own Sun and solar system—and our Earth—came to possess the elements that make our rocky planet, its oceans, and our oxygen atmosphere, possible.

In the context of understanding how supernovae help to create and spread the heavy elements that are required for creating rocky worlds and for any life upon them, astronomers often need to be able know how a given supernova remnant was actually produced. Recall that there are two basic mechanisms for producing supernova explosions. One type of supernova results from the implosion, then explosion, of a massive star's death; the other type results when a white dwarf is pushed over its maximum possible mass by a sibling star spilling mass onto it. The two types of supernovae will both produce supernova remnants, but the amounts and proportions of heavy elements that they produce may differ. So how might we determine which type of supernova explosion left behind a particular supernova remnant?

The ideal way would be to simply measure the light spectrum of the supernova soon after it explodes and check whether there is a strong signature of hydrogen in that light spectrum. As you'll recall from our previous lectures, massive star supernovae generally exhibit strong hydrogen signatures in their light spectra, because the massive star that blew up still had mostly unfused hydrogen in its outer layers. In contrast, white dwarf supernovae generally show little hydrogen in their spectra, because the white dwarf that detonated was made of almost pure carbon.

The problem is that in our own galaxy we can expect a supernova only once every century or so. So while astronomers routinely observe lots of supernovae in surveys of millions of other galaxies, we just don't have the luxury of waiting around for the next one to go off in the Milky Way. In contrast, the supernova remnants of long-ago supernova explosions can be observed in our galaxy for thousands of years, so we'd really like a way of determining what type of supernova went off that produced these remnants long ago.

My own research group has been involved in developing a new technique to determine which type of supernova produced a given supernova remnant, even when the supernova remnant is thousands of years old. Think of it as a mummy autopsy. And the beauty of this technique is that it requires only an examination of the shape of the supernova remnant. Let me show you how this works. In this set of images, we are looking at 15 different supernova remnants from different parts of our galaxy. Importantly, for each one, we've been able to independently determine whether the supernova progenitor of that remnant was a massive star or a white dwarf by using sensitive measurements of the light spectra. The ones whose numbers are shown with a red color are the ones established to have been white dwarf supernovae, while those whose numbers are shown with a blue color are the ones established to have been massive-star supernovae. Do you notice any differences between the two groups?

Your eyes are probably telling you that the first set, the white dwarf supernova remnants, appear generally rounder and overall simpler, whereas the second set, the massive star supernova remnants, appear generally more complex and asymmetric. Indeed, when we use computer analysis to quantify the degree of roundness and symmetry of each supernova remnant, we find that there is a stark, measurable difference between the two groups in just that sense; the massive star supernovae remnants are measurably more complex and less symmetric than the white dwarf remnants.

This research is important because it's allowing astronomers to determine the type of supernova—a white dwarf explosion or a massive star explosion—that produced each of the hundreds of supernova remnants that we can currently see throughout the galaxy, and indeed, supernova remnants in other galaxies. Since these different types of supernovae produce a different mix of elements, understanding how common are the different types of supernovae will help astronomers to more precisely trace the origin of the mix of elements present in the Earth and in the hundreds of other solar systems that are now being discovered around other stars.

Now that we've seen how we can discern the different types of supernova remnants and have looked at what happens when an individual supernova remnant plows into the surrounding galactic material, let's broaden our scope to consider what happens when multiple supernova explosions go off in proximity to one another. The result is an enormous type of supernova remnant that we call a super bubble, and these super bubbles can dramatically influence the stellar life cycle across the entire galaxy.

We can see examples of such super bubbles here in our Milky Way galaxy and in our nearest satellite galaxies. These structures can span enormous distances, up to 1000 light-years or more! From measurements of the vast dimensions of these structures and from measurements of the speeds with which they continue to expand in size, astronomers can estimate that these structures may represent the combined supernova remnant action of perhaps 10 or more supernova explosions all going off at nearly the same place and at nearly the same time.

What might cause such an apparent orchestration of supernova explosions? Well, it's really just the stellar birth process. Recall from our previous lectures that when the stellar birth process in a stellar nursery produces a cluster of stars—a stellar family, if you will—it tends to produce a distinct pattern of birth weights for those stars, lots of little runts and just a few behemoths. In most stellar nurseries, which birth, perhaps, 1000 stars or so, this translates into, perhaps, one or two stars massive enough to go supernova within the nursery. But in some cases, a stellar nursery can birth stars with enthusiasm, birthing, perhaps, 10,000 stars. In such cases, there may be a dozen very massive stars that will all die in supernova explosions. Because those massive stars were born at nearly the same time and from the same stellar nursery, they will die at nearly the same time and in that same place, thus, driving a super bubble.

But what are the effects of these super bubbles? Well, essentially, they push out into the surrounding space; they carve out large cavities in the surrounding galactic disk of gas and dust; and they push their hot, exploded stellar material up and out into the surrounding halo of the galaxy. From there, that material can then rain back down to the farthest reaches of the galaxy, where it can be gathered up again by gravity to form new generations of stars and planets.

The effects of these super bubbles on a galactic scale are most easily seen by looking at more distant galaxies, where we can take a panoramic view of an entire galaxy at once. And the best case studies are galaxies seen edge on from our perspective, because then, we can use the glow of the billions of stars within that galaxy to illuminate from behind the large scale structures caused by the super bubbles.

In the galaxy NGC 891, a galaxy much like our own Milky Way, we have a splendid example. In this image we see NGC 891 edge on. All across the disk of the galaxy we see clouds of gas and dust—appearing in dark silhouette—stretched vertically above and below the galaxy's disk. Those vertical structures, sometimes referred to as galactic chimneys, extend to heights of hundreds of light-years above and below the disk of the galaxy, evidence of material within the galaxy that has been levitated to great heights.

These chimneys are the effects of supernova super bubbles. Super bubbles within the galaxy have pushed out vertical walls of gas and dust and have blown them open. And the open chimneys themselves are the channels that enable hot ejected material from supernova explosions to pour up and out into the larger halo of the galaxy. These outpouring bursts of hot gas into the galactic halo are what astronomers refer to as galactic geysers.

Just as the geysers here on earth spew hot steam created from water heated beneath the surface up into the surrounding air, then condensing as mineral-rich water raining back down to the surface helping vegetation to grow, so do these galactic geysers spray hot gas from supernovae up into the galactic halo, then condensing and raining back down as chemically enriched material down to the disk of the galaxy where it can fertilize new generations of stars. At the galactic level, we can visualize what this might look like with multiple such geysers going at once, as depicted in this illustration.

We can actually see the influence of such geysers in our own galaxy by observing the hot gas that fills our galaxy's halo and glows in X-ray light. Images taken with the Chandra X-ray Space Telescope reveal not only that our galaxy's halo is filled to the brim with hot gas originating in supernova super bubbles, but that this hot gas extends far, far out from our galaxy. In fact, it completely envelops our galaxy, and beyond that, it completely envelops the Milky Way's satellite galaxies.

This is important to our picture of how the life cycles of future generations of stars come to be enriched with more abundant heavy elements as produced in supernovae. The supernovae don't merely sprinkle their ashes hither and yon. More generally, the collective action of the many supernova explosions in the galaxy is to literally bathe, to engulf, the entire galaxy in those ashes, suffusing the galaxy with them. No wonder that a solar system such as our own, forming in the suburbs of the galaxy, has been so enriched with heavy elements that it was able to form a rocky, watery planet with enough organic material to permit abundant life.

And who knows? In a galaxy bathed in carbon and silicon and oxygen and iron maybe there are suburbs of other galaxies where the conditions for stars sufficiently enriched in heavy elements can form solar systems, including rocky planets covered in liquid oceans and surrounded by breathable oxygen-rich atmospheres capable of supporting life as we know it. It is tempting to contemplate what a civilization on such a world at the outskirts of our galaxy, or indeed in another galaxy, might wonder looking at our own suburb of the Milky Way. Supernova remnants all around might wonder about the potential for life around that otherwise ordinary yellow star we call the Sun.

In this lecture, we've taken our understanding of the intimate connection between stellar death and new stellar birth to a much broader level. Supernova explosions, whether through the deaths of massive stars or the detonation of white dwarfs, produce supernova remnants that plow into and dramatically alter the surrounding galactic medium. They can directly trigger the onset of new stellar nurseries nearby. They can carve out large chimneys in the galaxy through which they can inject their hot ashes far above and all around the galaxy. Those ashes, raining back down into the galaxy, drenching the galaxy with the enriched elements produced by stars long dead, provide the elemental basis for new stars, new worlds like our own, and the very substance of life.

Stillborn Stars
Lecture 19

In this lecture, you will get to know the stars with the least potential, those that never make it as stars at all. Brown dwarfs for decades eluded discovery, but now we know that they do exist and that they appear to represent the failure of the stellar birth process. At the same time, these failed stars have taught us much about the stellar birth process and about the properties of stars more generally. It's little consolation to these stillborn stars, destined to fade totally into obscurity and darkness, but we've learned much from them about how and why nature sometimes succeeds, and sometimes fails, to birth a star.

Brown Dwarfs

- Not all stellar embryos ultimately make it as full-fledged stars. These recently discovered "failed stars," known as brown dwarfs, represent a strange type of netherworld, neither star nor planet. Put simply, a brown dwarf is an object that does not have enough mass to initiate nuclear fusion in its core. That mass turns out to be about 8% of the mass of our Sun.

- An object with a mass of just over 8% of the Sun's mass will turn on as a star, fusing hydrogen to helium in its core and going through the regular sequence of life stages for low-mass stars. But if an object emerges from the birth process with just a little bit less than that mass, it will never achieve the necessary temperature in its core to initiate fusion.

- Brown dwarfs are in many respects like Jupiter—relatively massive objects that, while not shining brilliantly as stars, do glow with the dim remnant warmth of their formation. So why don't we just call brown dwarfs planets?

- Fundamentally, a planet is an object that forms from the disk of gas and dust encircling a star and comes to orbit that star. A brown dwarf, on the other hand, starts out going through the birth process of a star but, for some reason, doesn't emerge from the stellar womb with a high enough birth weight to make it as a star.

- At the same time, we can also separate brown dwarfs from planets based on their mass. Where Jupiter has a mass of 0.10% of the Sun's mass, brown dwarfs are generally regarded as having masses of at least 10 times Jupiter's mass. In other words, brown dwarfs are something betwixt and between stars and planets.

- The reason that there is a minimum mass that a brown dwarf can have is that, for brown dwarfs more massive than 10 times Jupiter's mass, they are able to very briefly fuse deuterium into helium. Deuterium is sometimes called heavy hydrogen; it is a form of hydrogen in which the nucleus already includes 1 neutron in addition to the solitary proton of ordinary hydrogen.

- By adding 1 proton, the deuterium becomes a "light helium," having 2 protons and 1 neutron instead of the usual 2 plus 2. This "light fusion" process does provide a bit of energy to the brown dwarf. It's only for a brief time of perhaps 100 million years or so, but it does serve as another characteristic that distinguishes brown dwarfs from planets.

- In fact, the fusion process—or the lack thereof—serves as one of our most direct methods for confirming the existence of brown dwarfs. This is important, because at first blush, many brown dwarfs can appear so similar to very low-mass stars that for decades the hunt for brown dwarfs was stymied by the uncanny resemblance. After all, the lowest-mass stars, weighing in at a mere 8% of the mass of the Sun, are red and faint, just like brown dwarfs.

- Despite the name, brown dwarfs are not brown. To the eye and to a telescope's camera, they appear deep red or magenta in color—the same as the lowest-mass stars, which we call red dwarfs.

The Physical Nature of Brown Dwarfs

- How can we distinguish a brown dwarf (a failed star) from a red dwarf (a small but legitimate star) if they both appear more or less the same in their actual colors? This is where the power of dissecting the light spectrum comes to the rescue. In particular, the presence of the element lithium in an object's atmosphere is a very strong confirmation that it is too low in mass to be a star. Most stars are born with a certain amount of lithium, inherited from the clouds of gas and dust from which they are formed.

- Lithium turns out to be a very fragile element, at least as far as nuclear reactions are concerned. Even at relatively low temperatures, and well before most stars ignite full-on hydrogen fusion, they are able to fuse lithium into beryllium. Consequently, true stars, as they are completing their birth process and certainly by the time they start their middle lives as main sequence stars, have already burned up all of their lithium.

- It was realized early in the search for brown dwarfs that an object with a mass lower than that of a true star will not become warm enough to fuse lithium. So, if we see the spectral signature of lithium in a star, that tells us that it is not a star but, rather, an object that never became hot enough in its interior to fuse one of the easiest elements of all. Therefore, it must be a brown dwarf.

- In ordinary stars, the lower in mass, the dimmer the luminosity and the cooler the temperature, so the redder the color. In other words, with ordinary stars, the dimmest stars are the reddest ones. However, the opposite is true for brown dwarfs.

- The reason for the strange and counterintuitive colors of brown dwarfs has to do with the fact that these objects are so cold that instead of having atoms in their atmospheres, their atmospheres are dominated by cold molecules, such as methane, ammonia, and even water vapor. These molecules absorb so much of the reddest wavelengths of light that the brown dwarf ends up appearing blue. In fact, the coolest known brown dwarfs are among the bluest.

- Amazingly, these very cool brown dwarfs have temperatures comparable to our own bodies at room temperature—about 300 degrees Kelvin. And because they are so cool and dim, most of the known brown dwarfs are very close by. Indeed, some of the closest known objects to us are brown dwarfs, the nearest of which is only 6 light-years away.

- An important clue to the physical nature of brown dwarfs comes from examining their physical sizes in comparison to their masses. There is a direct relationship between the mass of a star and its size—more massive stars are bigger—and this relationship continues all the way down to even planets like Jupiter. However, brown dwarfs, despite having masses that place them above planets like Jupiter, are actually smaller than Jupiter.

- Just like white dwarfs and neutron stars, brown dwarfs are dead stars that hold themselves up against gravity using degeneracy pressure, which uses the strange properties of quantum mechanics to generate pressure by compressing the particles ever more tightly.

- But how do we know the sizes of brown dwarfs? How do we know the relationships between their basic properties? Stars that are parts of eclipsing binary systems are among our most important informants of the properties of stars. Eclipsing binary systems are pairs of stars that periodically pass directly in front of one another, eclipsing one another, and from those mutual eclipses, we can measure things like the sizes of the stars.

- Unfortunately, there is at this time only one known example of an eclipsing binary system in which both siblings are brown dwarfs. So, that one system serves as a kind of Rosetta stone for understanding the properties of brown dwarfs more generally. From that system, in the stellar nursery in Orion, we have been able to directly measure the properties of the two brown dwarfs in the system. Moreover, because the system is in a stellar nursery, it gives us a glimpse into the process by which stars are stillborn.

The Formation of Brown Dwarfs

- Brown dwarfs are something between star and planet but are neither. Their masses are in between those of planets and stars, but what does their birth process—or stillbirth process—look like? Do they form like stars, collapsing under gravity and attempting but failing to ignite fusion, or do they form like planets, in the disks of gas and dust in orbit about stars?

- This was a basic question that persisted for some time. However, the evidence now appears to be quite incontrovertibly in favor of the starlike formation scenario. There are several lines of evidence for this interpretation.

- First, we now have exemplars of brown dwarfs in stellar nurseries that are ringed by disks of gas and dust all their own. Such cases demonstrate that brown dwarfs undergo a gravitational collapse very much like that which stars experience as they form.

- A second piece of evidence is that we now have exemplars of brown dwarfs with actual planets orbiting them. This makes sense given that we've seen brown dwarfs ringed by disks of gas and dust. But the discovery of planets around brown dwarfs lends further support to the idea that brown dwarfs form like stars. It further indicates that failed stars in some cases also represent failed solar systems.

- Why does nature sometimes decide to abort the stellar birth process and produce a stillborn brown dwarf instead? It appears that the dynamics of the star-formation process itself may be at fault. Interactions between forming sibling stars can lead to ejections of some stars, kicking them out not only from the sibling family, but in some cases from the stellar nursery altogether.

- A currently favored explanation for why some stars may fail to fully form is that this dynamical ejection process cuts some protostars off from the reservoir of material that they would otherwise have fed from. Because they can't feed from this reservoir, they are unable to build up enough mass to become full-fledged stars. In this sense, brown dwarfs really represent aborted stars as opposed to stillborn stars, having been forcibly removed from the womb that would have presumably otherwise nurtured them to full term and to status as a full-fledged star.

- Regardless of the cause of a brown dwarf's existence, what is undeniable is its ultimate fate. A brown dwarf, unable to generate its own light and heat through fusion, will fade over time, becoming ever cooler and dimmer, eventually fading entirely from view.

- But in the meantime, as long as we can see and study them, these stillborn stars provide us our most important and direct way of understanding the stellar birth process. By studying brown dwarfs, we gain insight into the circumstances under which some stars ignite while others fail to form at all.

Suggested Reading

Boss, *International Astronomical Union Working Group on Extrasolar Planets Definition of a 'Planet.'*

Burgasser, "Brown Dwarfs."

Questions to Consider

1. In light of how they come to be formed, are brown dwarfs best thought of as stillborn stars or as aborted stars?

2. What role does gravity, which normally drives the stellar birth process, play in the failure of brown dwarfs to become full-fledged stars?

Stillborn Stars

Lecture 19—Transcript

We've seen some pretty amazing things in this course, stellar nurseries, stellar explosions, stellar corpses. But you haven't seen anything yet. There are still quite some thrills to be had in this carnival that is the life cycles of stars—strange titanic stars that may have constituted the very first stars in the universe; failed stillborn stars that we call brown dwarfs; exotic magnetism that seems to pervade many stars, including our own Sun. Like the freak show tent at a carnival of years past, the life cycle of stars includes some freak sideshows of its own.

So far in this course, we've examined the main highways through the stellar life cycle. We've witnessed the normal process of stellar birth, driven by the gravitational collapse of gigantic clouds of gas and dust. We've seen how most stars are these days born into the open star clusters of hundreds to thousands of stars that slowly disperse over time, and that in previous eras stars were born into tight-knit families that we still see today as the globular star clusters orbiting the center of our galaxy.

We've also seen the normal course of a star's most productive stage of life. For a star like our Sun, that course begins with the long span of its middle life as a main sequence star, when it is fusing hydrogen to helium in its core. Then it proceeds to the period when the star, having become a red giant, fuses helium into carbon, until it finally dies quietly as a beautiful planetary nebula, spreading its ashes into space, and leaving behind a glittering diamond in the sky that we call a white dwarf.

For more massive stars, the fusion process can continue all the way to the element iron, at which point the star explodes in a violent supernova, leaving behind a bizarre neutron star at the center of an expanding supernova remnant, and fusing the rest of the elements of the periodic table in the process. We've even seen that in some extreme cases, the end of a massive star's life can be an ultra-strange black hole, the ultimate capitulation to gravity.

Importantly, we've learned that the particular path that a given star will travel through life depends principally on one thing, the mass with which the star is born. And so it may not surprise you to learn that there are stars whose birth masses are, in fact, so small that they fail to ever light up as stars at all. These stillborn stars will be our first stop on a tour that we'll be taking over the next few lectures, a tour in which we'll be looking at a series of unusual phenomena in the lives of stars. Along the way, we'll discover that these various phenomena actually define the frontiers of stellar research.

Not all stellar embryos ultimately make it as full-fledged stars. These recently discovered failed stars, known as brown dwarfs, represent a strange type of netherworld, neither star nor planet. To fully understand how stars are born, we also need to understand when, why, and how often the star-formation process fails. So let's talk about brown dwarfs.

First we need to define what a brown dwarf is. Put simply, a brown dwarf is an object that does not have enough mass to initiate nuclear fusion in its core. That mass turns out to be about 8% of the mass of our Sun. An object with a mass of just over 8% of the Sun's mass will turn on as a star, fusing hydrogen to helium in its core, and going through the regular sequence of life stages for low-mass stars that we've discussed. But if an object emerges from the birth process with just a little bit less than that mass, it will never achieve the necessary temperature in its core to initiate fusion.

Brown dwarfs are, in many respects, like Jupiter, relatively massive objects that, while not shining brilliantly as stars, do glow with the dim remnant warmth of their formation. So why don't we just call brown dwarfs planets? The answer has to do with the process of their formation. Fundamentally, a planet is an object that forms from the disk of gas and dust encircling a star, and so comes to orbit that star. A brown dwarf, on the other hand, starts out going through the birth process of a star, but for some reason does not emerge from the stellar womb with a high enough birth weight to make it as a star.

At the same time, we can also separate brown dwarfs from planets based on their mass. Where Jupiter has a mass of one-tenth of a percent of the Sun's mass, brown dwarfs are generally regarded as having masses of at least ten times Jupiter's mass. In other words, brown dwarfs are something betwixt and between stars and planets. The reason that there is a minimum mass that a brown dwarf can have is that, for brown dwarfs more massive than ten times Jupiter's mass, they are able to very briefly fuse deuterium into helium. Deuterium is sometimes called heavy hydrogen; it is a form of hydrogen in which the nucleus already includes one neutron in addition to the solitary proton of ordinary hydrogen. By adding one proton, the deuterium becomes a light helium, having two protons and one neutron instead of the usual two plus two. This light fusion process does provide a bit of energy to the brown dwarf. Yes, it's only for a brief time of perhaps 100 million years or so. But it does serve as another characteristic that distinguishes brown dwarfs from planets.

In fact, the fusion process, or the lack thereof, serves as one of our most direct methods for confirming the existence of brown dwarfs. This is important, because at first blush, many brown dwarfs can appear so similar to very-low-mass stars, that for decades, the hunt for brown dwarfs was stymied by this uncanny resemblance. After all, the lowest mass stars, weighing in at a mere 8% of the mass of the Sun, are red and faint, just like brown dwarfs. Despite the name, brown dwarfs are not brown. To the eye and to a telescope's camera they appear deep red or magenta in color, the same as the lowest mass stars, which we call red dwarfs.

So as an aside, you may be curious to know how brown dwarfs got their name. Brown dwarfs had been first theorized to exist in the 1960s but were initially termed black dwarfs because of the idea that they would become so cool and dim over time that they would not be visible. However, the term black dwarf ended up being used to describe white dwarfs that eventually cool and fade from view. So brown dwarf was adopted as an alternative to describe the faint, stillborn stars, and the name has stuck ever since.

In any event, how can we distinguish a brown dwarf, a failed star, from a red dwarf, a small but legitimate star, if they both appear more or less the same in their actual colors? This is where the power of dissecting the light spectrum comes to the rescue. In particular, the presence of the element lithium in an object's atmosphere is a very strong confirmation that it is too low in mass to be a star. Most stars are born with a certain amount of lithium, inherited from the clouds of gas and dust from which they are formed. Lithium turns out to be a very fragile element, at least as far as nuclear reactions are concerned. Even at relatively low temperatures, and well before most stars ignite full on hydrogen fusion, they are able to fuse lithium into beryllium. Consequently, true stars, as they are completing their birth process, and certainly by the time they start their middle lives as main-sequence stars, have already burned up all of their lithium.

It was realized early in the search for brown dwarfs that an object with a mass lower than that of a true star will not become warm enough to fuse lithium. And so, if we see the spectral signature of lithium in a star, that tells us that it is not a star at all, but rather an object that never became hot enough in its interior to fuse one of the easiest elements of all, so it must be a brown dwarf.

One of the first objects to be definitively confirmed as a brown dwarf using the presence of lithium in its spectrum was Gliese 229B. The name itself reveals much about this remarkable object. The Gliese designation refers to the name of a catalog of very cool, red M-type stars nearby to the Sun. This object happens to be number 229 in that catalog. The B in the name is an indication that this is, in fact, a binary system. The brown dwarf is a sibling companion of the originally catalogued Gliese 229, now called Gliese 229A. And just to emphasize how incredibly faint brown dwarfs are, in this image, the faint speck is the brown dwarf. And that incredibly bright looking star to the left? That is actually itself a very faint red dwarf star, the original Gliese 229 star. As the Gliese 229 system suggests, brown dwarfs come with siblings, just like full-fledged stars do. And the siblings are not always living stars as the M-type sibling in Gliese 229. In many cases, the siblings are themselves stillborn stars, sometimes twins, sometimes even triples.

Here is an example of one of the many binary brown dwarf systems now known. This one was discovered using the infrared adaptive optics camera on the Canada-France-Hawaii Telescope. At first blush, this might look like any one of countless binary stars in the sky, one sibling brighter than the other, and the two siblings having different colors reflecting their different temperatures. But if you think about this image for a moment, you might start to scratch your head at something that seems amiss. In ordinary stars, the lower in mass you go, the dimmer the luminosity, and the cooler the temperature, and so the redder the color. In other words, with ordinary stars, the dimmest stars are the reddest ones.

But in this binary brown-dwarf system, the dimmer sibling is bluer. Does that mean it is actually hotter than its brighter sibling? Actually, no. The reason for the strange and counter-intuitive colors of brown dwarfs has to do with the fact that these objects are so cold, that instead of having atoms in their atmospheres, their atmospheres are dominated by cold molecules, molecules such as methane, ammonia, even water vapor. These molecules absorb so much of the reddest wavelengths of light that the brown dwarf ends up appearing blue. In fact, the coolest known brown dwarfs are among the bluest.

Amazingly, these very cool brown dwarfs have temperatures comparable to our own bodies at room temperature, about 300 degrees Kelvin. These are objects that you could walk right up to and touch and not even notice. And because they are so cool and dim, most of the known brown dwarfs are very close by. Indeed, some of the closest known objects to us are brown dwarfs, the nearest of which is only six light-years away.

An important clue to the physical nature of brown dwarfs comes from examining their physical sizes in comparison to their masses. As we've previously learned, there is a direct relation between the mass of a star and its size—more massive stars are bigger. And this relation continues all the way down to even planets, like Jupiter. However, brown dwarfs, despite having masses that place them above planets like Jupiter, are actually smaller than Jupiter.

Can you recall the physical effect we've discussed in which objects become smaller as they become more massive in order to hold themselves up? If you said degeneracy pressure, you'd be correct. Indeed, just like white dwarfs and neutron stars, brown dwarfs are dead stars that hold themselves up against gravity using degeneracy pressure. Do you remember how degeneracy pressure works? Well, basically, it uses the strange properties of quantum mechanics to generate pressure by compressing the particles ever more tightly.

But how do we know the sizes of brown dwarfs? How do we know the relationships between their basic properties? As you may recall from our previous lectures, stars that are parts of eclipsing binary systems are among our most important informants of the properties of stars. Eclipsing binary systems are pairs of stars that periodically pass directly in front of one another, eclipsing one another, and from those mutual eclipses we can measure things like the sizes of the stars.

Unfortunately, there is at this time only one known example of an eclipsing binary system in which both siblings are brown dwarfs. So that one system is pretty important, and it serves as a kind of Rosetta Stone for understanding the properties of brown dwarfs more generally. From that system, in the stellar nursery in Orion, we have been able to directly measure the properties of the two brown dwarfs in the system. Moreover, because the system is in a stellar nursery, it gives us a glimpse into the process by which stars are stillborn. I was fortunate to lead the team that made this important discovery. Let's take a look at this unique system, an amazing pair of twin failed stars, and let's find out what this system teaches us about the nature of these strange objects.

The story begins with me as a young Ph.D. student at the University of Wisconsin. My graduate adviser and I began a project to search for eclipsing binary star systems in the Orion Nebula stellar nursery. We decided to start this project because, up to that time, there had been barely a handful of eclipsing binary star systems discovered within stellar nurseries. Recall that eclipsing binary star systems are one of the only tools that astronomers have for directly measuring the masses and the diameters of stars. With only a few such systems having been discovered in stellar nurseries up to that time, this meant that there was very limited knowledge of what the true sizes of very young stars actually are.

So we set about trying to find some newborn eclipsing binary stars. We did this by monitoring the brightnesses of thousands of stars in the Orion Nebula stellar nursery by which we could identify any eclipsing star systems through the detection of a repeated dimming of a star. We needed to collect lots of measurements of many stars' brightnesses over a long period of time, because eclipsing binary systems are rare, and because when they do eclipse, their eclipses are usually very brief. So by monitoring lots of stars, we increased our chances of finding at least one of these rare eclipsers; and by monitoring over a long period of time, we increased our chances of actually catching any short-lived eclipse events that could be easy to miss. So we used a telescope at the Kitt Peak National Observatory in Arizona, as well as a telescope in the Chilean Andes, increasing our ability to continually collect data.

So, fast forward to about 10 years later. Now a professor at Vanderbilt with students of my own, my research began to systematically analyze the millions of brightness measurements that I had amassed over those ten years for a few thousand stars. We quickly found one relatively bright eclipsing binary system and published an article about it. At the same time, there was a very faint object in our data that appeared to also be undergoing periodic eclipses.

At first we were pleased to have discovered what we thought to be a second eclipsing binary star system in this important stellar nursery, but then were confused by the fact that, if this were a system of two stars in the nursery, it should not appear so faint. So we worried that perhaps it was a bad luck case of spotting a much more distant eclipsing binary that just happened to be along the line of sight to the Orion stellar nursery.

However, we decided to make Doppler shift measurements of the system using the infrared spectrograph on the Gemini telescope in the Chilean Andes. After measuring the masses of the two stars from the Doppler shifts of their light spectra, we were shocked and amazed to find that both siblings in this eclipsing system have masses well below the minimum allowed mass for a star.

In fact, the two brown dwarfs have masses of 60 and 30 times the mass of Jupiter. That's just 6% and 3% the mass of the Sun, placing both well below that 8% threshold that I mentioned earlier for a star to be a legitimate star. So we were able to show definitively that these are both bona fide brown dwarfs, unable to initiate fusion in their cores, a pair of stillborn stellar siblings. Never before had the birth weight of any brown dwarf been measured directly, and here we had been able to measure the birth weights of two at once!

However, the joy at our realization that we had two brown dwarfs on our hands was accompanied by bafflement at the other properties of the system. Most perplexingly, the temperatures of the two brown dwarfs turn out to be reversed relative to the masses. That is, the more massive brown dwarf sibling is cooler than its lower-mass companion. Such a reversal of temperatures with mass had not been expected nor predicted. What was going on?

Our follow-up analysis of the system found that the brown dwarfs are magnetic. And more to the point, the heavier brown dwarf is much, much more magnetic than the lower-mass brown dwarf. Somehow, the magnetism of the heavier brown dwarf was making it much cooler than it should be. Since then, we have used this system to determine that magnetic brown dwarfs have their radii increased and their temperatures suppressed by that magnetism. In a nutshell, the reason for this appears to be that the strong magnetic field threading through the interior of the brown dwarf causes its surface to have dark, cool regions, perhaps similar in nature to sunspots. By causing the surface of the brown dwarf to be a bit cooler than it would be otherwise, these spots then force the brown dwarf to readjust its size so as to have enough surface area from which to radiate the total energy welling up from within as the brown dwarf steadily cools over time.

Significantly, in the course of solving the mystery of this unique eclipsing brown dwarf system, we were able to develop an explanation for a more general phenomenon. The effect of magnetism on the temperatures of the brown dwarfs we discovered also explains why some objects that appear to be brown dwarfs might actually be low-mass stars whose temperatures have been suppressed by their magnetism.

That's important, because a question we would like to be able to ultimately answer is: How common are brown dwarfs relative to full-fledged stars? In other words, how often does the stellar birth process fail? It's a bit like using the infant mortality rate of a country to assess something basic about how well that country is taking care of its own. So, what's the current report card for how well stellar nurseries do at successfully birthing their young? What's the infant mortality rate for stellar nurseries?

As you'll recall from our previous lectures, stellar nurseries form stars with a distinct pattern of masses that we call a power law, in which for every star birthed with the mass of our Sun, there are 100 stars birthed with just one tenth the mass of our Sun. In other words, the stellar birth process very strongly favors the lightest birth weights. Does this pattern continue down into the brown dwarfs? Do stillborn stars outnumber legitimate stars?

For a time, there was some evidence that the answer might be yes, because census studies of the number of newborn stars with different temperatures showed that there were indeed a very large number of stars whose temperatures were so cool that they must be brown dwarfs. However, now that we understand that magnetism in stars can cause their surfaces to be cooler than they would be otherwise, we understand that many of those putative brown dwarfs are, in fact, legitimate stars that just happen to be highly magnetized. To be clear, there are real brown dwarfs out there, as our discovery of the eclipsing brown dwarf pair helped to prove. But overall, the infant mortality rate of stellar nurseries appears to be at most 10% or so.

Now that we've described brown dwarfs generally and discussed in detail what we've been able to learn about them from that one special eclipsing brown dwarf pair, there is a final question for us to consider. We've described brown dwarfs as representing something between star and planet, but neither. So, do brown dwarfs fundamentally form like planets or like stars? We've already seen that their masses are in between those of planets and stars. But what does their birth process, or still-birth process, if you will, actually look like? Do they form like stars do, collapsing under gravity and attempting but failing to ignite fusion? Or do they form like planets do, in the disks of gas and dust in orbit about stars?

This was a basic question that persisted for some time. However, the evidence now appears to be quite incontrovertibly in favor of the star-like formation scenario. There are several lines of evidence for this interpretation. First, we now have exemplars of brown dwarfs in stellar nurseries that are ringed by disks of gas and dust all their own. Such cases demonstrate that brown dwarfs undergo a gravitational collapse very much like that which stars experience as they form.

A second piece of evidence is that we now have exemplars of brown dwarfs with actual planets orbiting them. This makes sense, given that we've seen brown dwarfs ringed by disks of gas and dust. But the discovery of planets around brown dwarfs lends further support to the idea that brown dwarfs form like stars. It further indicates that failed stars in some cases also represent failed solar systems.

So why does nature sometimes decide to abort the stellar birth process and produce a stillborn brown dwarf instead? It appears that the dynamics of the star-formation process itself may be at fault. Recall that in a previous lecture we looked at computer simulations of forming stars in a stellar nursery, and we saw how interactions between forming sibling stars can lead to ejections of some stars, kicking them out not only from the sibling family, but in some cases, from the stellar nursery altogether.

A currently favored explanation for why some stars may fail to fully form is that this dynamical ejection process cuts some protostars off from the reservoir of material that they would otherwise have fed from. Since they can't feed from this reservoir, they are unable to build up enough mass to become full-fledged stars. In this sense, brown dwarfs really represent aborted stars as opposed to stillborn stars, having been forcibly removed from the womb that would have presumably otherwise nurtured them to full term and to status as a full-fledged star.

Regardless of the cause of a brown dwarf's existence, what is undeniable is its ultimate fate. A brown dwarf, unable to generate its own light and heat through fusion, will fade over time, becoming ever cooler and dimmer, eventually fading entirely from view. So brown dwarfs might be best thought of as eventual black dwarfs—dark, unchanging, forever dead. But in the meantime, so long as we can see and study them, these stillborn stars provide us our most important and direct way of understanding the stellar birth process. By studying brown dwarfs, we gain insight into the circumstances under which some stars ignite while others fail to form at all.

In this lecture, we've gotten to know the stars with the least potential, the true runts of the litter, those that never make it as stars at all. Brown dwarfs for decades eluded discovery, but now we know that they do exist, and we know that they appear to represent the failure of the stellar birth process. At the same time, these failed stars have taught us much about the stellar birth process and about the properties of stars more generally, including the crucial role of magnetism in setting the basic properties of stars. It's little consolation to these stillborn stars, destined to fade totally into obscurity and darkness, but we've learned much from them about how and why nature sometimes succeeds and sometimes fails to birth a star.

The Dark Mystery of the First Stars
Lecture 20

I n this lecture, you will learn about the biggest, most massive stars that might have populated the universe's very beginnings and that would have initiated the stellar life cycle for the rest of time. There are multiple good lines of evidence that such a strange population of stellar beasts roamed the early universe. To understand those first ancestral stars, we might ultimately have to penetrate the hidden nature of the strange dark matter particles that still fill the universe, whose gravity we can sense but which we have not even seen with our own eyes.

The Oldest Stars in Our Galaxy

- For decades, astronomers have worked in an ongoing effort to find the truly oldest stars in our galaxy because we can use them to directly measure what the chemical composition of the universe was like at the time that our galaxy formed 12 to 14 billion years ago, soon after the big bang.

- But searching for the oldest relic stars in our galaxy is challenging because there is not much about a star that, on the face of it, gives us a hint that it is much older than any of the other billions of stars swarming around the galaxy.

- To find the oldest stars, we first have to restrict our attention to relatively low-mass stars—because the lowest-mass stars live the longest. If a star like our Sun formed when the galaxy was very young, it would already have ended its life, because stars like our Sun live to be about 10 billion years old, whereas the galaxy is more like 14 billion years old. In contrast, virtually every star ever born weighing less than about 80% of the Sun's mass is still around, because such low-mass stars live longer than the galaxy is old.

- The problem is that such low-mass stars constitute about 75% of all stars in the galaxy, and the galaxy has been birthing such stars continuously since it began. So, we're talking about roughly 150 billion of these low-mass stars. In addition, those 150 billion low-mass stars all look pretty much the same, because they are all still main sequence stars. They have similar temperatures and colors and luminosities. The answer is to look for a star that's jumping out at you and then use the light spectrum of that star to verify that it has little to no heavy elements in its makeup.

- The oldest stars in our galaxy were born at a time in the galaxy's history when the galaxy as a whole had a much less orderly shape. Before the flattened disk of the Milky Way formed through gravitational collapse, our galaxy had a more nearly spherical shape. So, the stars that were born then moved through the galaxy in a more disorderly spherical swarm about the center of the galaxy, in contrast to the more orderly circular rotation of stars like our Sun in the disk of the Milky Way today.

- Those early stars that are still around today continue to orbit the galaxy in that spherically distributed, more haphazard fashion, making up what we call the halo of our galaxy, extending to large distances all around it. And because they orbit the galaxy in that way, they periodically plunge down through the disk of the galaxy to the other side and then eventually plunge back up through the disk, over and over through the eons.

- As a result of this, some of the nearest stars to the Sun right now happen to be these very old halo stars that are nearby—not because they are true neighbors of our solar system residing in the disk of the galaxy, but because they are interlopers plunging rapidly through the disk on their way to the other side.

- These stars turn out to be relatively easy to spot because we can see them tearing by us. And they tear by in a very peculiar direction—straight up or down instead of along the direction of motion of our Sun and of other stars in our neighborhood.

- The star that currently holds the record as the oldest star in our galaxy is called HD 140283, or the Methuselah star. It's a fairly unremarkable star, but it was recognized about 100 years ago because of its remarkably fast and odd direction of motion. It is moving at 800,000 miles per hour, which is so fast that the Hubble Space Telescope could measure its motion in the sky over the course of just a few hours. It is also very bright, because it just so happens to be plunging through the disk of our galaxy very close to the Sun's position in the disk.

- The latest measurements using the Hubble Space Telescope have allowed astronomers to precisely pinpoint the star's distance—190.1 light-years—which in turn has permitted its luminosity to be accurately measured from its apparent brightness. That luminosity, together with its color or temperature, lets us use the Hertzsprung–Russell diagram to measure exactly where the star lies in relation to the main sequence.

- This star is just entering the final stages of its life, leaving the main sequence stage and turning into a red giant. Therefore, by placing it accurately in the H–R Diagram, it is possible to determine its age: 14.5 billion years, plus or minus 0.8 billion years. With the age of our universe being 13.8 billion years, the measurement of Methuselah's age puts it right at the maximum possible age for a star in our universe, meaning that this star must have formed just as the galaxy itself was forming and very soon after the big bang.

- From careful measurements of the elemental signatures in its light spectrum, scientists have determined that Methuselah contains just a smidgen of heavy elements, a mere $1/250^{th}$ of the amount contained in the Sun—but that's more than zero. And this tells us definitively that there must have been a generation of stars that came even before Methuselah and its cohort to create that smidgen of heavy elements for Methuselah to inherit.

- Any such prior generation of stars would have had to exist sometime after the big bang; they could not have existed prior to the start of the universe as a whole. So, with Methuselah's age having been pinned to be at least 13.7 billion years old, and the universe having begun 13.8 billion years ago, there is a tiny window of time—about 0.1 billion, or 100 million years—during which time a generation of pre-Methuselah stars could have existed.

- In addition, the measurements of Methuselah's detailed elemental abundances tell us that there is something peculiar about this star's chemical makeup: It has much more oxygen relative to its hydrogen than what we currently see in the universe's mix of chemical elements.

Dark Stars

- Because none of the first stars that formed in the universe are around anymore, it is not possible to directly observe them. So, the only way to understand what they must have been like is through detailed calculations and computer simulations. And until recently, such simulations couldn't be done because of the sheer computing power required.

- You might think that the stellar birth process at the beginning of the universe would be simpler than it is today because the universe soon after the big bang was a lot less complex than it is now. Almost all of the matter in the universe was in the simplest form possible—simple hydrogen and helium gas. There were no heavier elements, and there were no dust particles. And there were no existing stars with their magnetic fields and stellar winds to complicate things.

- The complication is that in the early universe, the ordinary matter with which we are familiar—stuff like hydrogen gas—was not the form of matter that controlled things, gravitationally speaking. The dominant form of matter was what we call dark matter, a mysterious form of matter that exerts the same gravity as ordinary matter but whose physical properties we are only beginning to understand.

- Current estimates are that dark matter actually comprises 90% of the total matter content of the universe and would have also dominated the matter content of the universe back when the very first stars were forming.

- So, at the universe's beginning, from the standpoint of gravity and its influence in the formation of the first stars, imagine the universe as a place principally filled with dark matter. Regions where the dark matter was denser were the places where the ordinary matter out of which stars form—hydrogen and helium in the young universe—collected. In addition to the nature of dark matter still being largely unknown, these regions of denser dark matter would have been enormous, further exacerbating the challenge of simulating these regions computationally.

- In recent years, however, advances in computational power have made it possible for scientists to simulate the formation of the earliest stars, and the simulations tell us that the masses of stars formed in the dense dark matter regions of the early universe would have been about 100 to 300 times as massive as the Sun—an extraordinary amount of mass for a star.

- This idea remains controversial. In fact, the most recent simulations indicate that the first stars might have been less hefty, around 40 times the mass of the Sun. The consensus is that the very first stars in the early universe must have been much more massive than the stars that we observe forming today. There is also an emerging consensus among astronomers that there must have been at least some first stars that were more massive than 140 times the mass of the Sun.

- A number of astronomers are becoming convinced that some of the first gargantuan stars may have been made not of ordinary matter, but mainly out of dark matter. If correct, then the first stars to form in the early universe would have been what some are calling dark stars, fueled by an altogether different engine than stars as we know them today.

- Ordinary stars like the Sun today shine because they are fueled by nuclear fusion in their core that converts hydrogen to helium. But these theoretical dark stars would have run on dark matter particles colliding and annihilating each other.

- Dark matter is a substance that astronomers have not yet observed directly—that's why we call it "dark"—but we infer its existence because we can measure its gravitational effects on visible matter. We don't yet know exactly what dark matter is, but it must be some type of elementary particle that we have not yet observed in particle colliders.

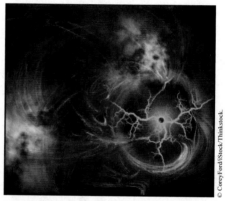

© Corey Ford/iStock/Thinkstock.

Dark matter, a mysterious substance, makes up 26.5% of the matter-energy composition of the universe.

- The findings of this area of research so far suggest a dramatic altering of the theoretical framework for the formation of the first stars. Dark stars are an entirely new kind of star from anything we've known or conceived of before.

- According to dark stars theory, the first stars are thought to have formed inside clouds of dark matter, when hydrogen and helium gases cooled to a temperature at which nuclear fusion could begin. In more conventional theories of the first stars, dark matter doesn't affect this process except to provide the gravity to bring the gases together.

- However, in dark stars theory, the dark matter concentrations are high enough for the particles in the dark matter clouds to collide with and annihilate each other, destroying themselves and, more importantly, providing a source of internal heat to support the very massive star without relying on nuclear fusion alone.

- As a result, the star can be enormous—perhaps as large 2000 times the size of the Earth's orbit about the Sun. And when these enormous stars finally end their lives as supernovae, perhaps as quickly as 1000 years or even less after they formed, they leave behind the heavier elements such as carbon, nitrogen, and oxygen that became part of the next generation of stars, such as in the Methuselah stars still roaming our galaxy.

Suggested Reading

Larson and Bromm, "The First Stars in the Universe."

Nicolson, *Dark Side of the Universe.*

Questions to Consider

1. In what ways do the very first generation of stars differ from the "ordinary" stars that populate the universe today?

2. Why doesn't dark matter play as much of a crucial role in the lives and deaths of stars now as it did in the very early universe?

The Dark Mystery of the First Stars
Lecture 20—Transcript

In our previous lecture, we looked at the extremes of the stellar life cycle from the standpoint of the smallest, the so-called brown dwarfs, which are such runts that they actually fail to ever turn on as stars at all, doomed forever as something neither star nor planet. In this lecture we'll be looking at the opposite extreme, the very largest stars. As we'll see, this exploration will lead us to examine theories of the very first stars ever to form in the universe, and the role of mysterious dark matter in the makeup of these first behemoth stars.

Indeed, astronomers are engaged in an intense manhunt for the universe's first stars. Those first stars presumably formed from pure hydrogen and helium. Now, think about it, why would that be? Well, basically it's because the big bang would have produced hydrogen and helium, but there had not yet been any stars to forge any other elements through the process of fusion. These first stars would have been responsible for initiating the cycle of stellar birth and death through which the universe became sufficiently chemically enriched to permit later stars like our Sun to form. As well, the enriched chemical products from the stellar ashes of these stars would have made possible the formation of solar systems such as our own and so many others.

What were these first stars like? As we'll see, they were probably unlike anything we've seen since. And the latest theories suggest that the strange form of matter that fills the universe and that would have dominated the universe early on—dark matter—may have been an essential part of these first stars' internal engines. Moreover, the evidence from examining the chemical content of the oldest relic stars we still see in our galaxy tells us that those bizarre, massive, so-called dark stars must have lived out their lives more quickly than anything we've since witnessed. In the process, they would have kick started the stellar life cycle for the rest of the universe.

Let's begin by discussing what we know about the oldest stars in our galaxy and what their elemental makeup tells us about their predecessors. For decades, astronomers have worked in an ongoing effort to find the truly oldest stars in our galaxy, because we can use them to directly measure what the chemical composition of the universe was like at the time that our galaxy formed some 12 to 14 billion years ago, soon after the big bang. But searching for the oldest relic stars in our galaxy is challenging, because there is not much about a star that, on the face of it, gives us a hint that it is much older than any of the other billions of stars swarming around the galaxy.

Think of it like an archaeologist's challenge. You unearth some bones or some relic objects of a bygone civilization, but just looking at those unearthed materials doesn't give you a very accurate idea of exactly how long ago those materials came to be. And going through an archaeologist's laboratory stacked with many bones and many relics from many dig sites, they all kind of look alike, dusty and old. But how old? Which are older than which? And which is the oldest?

To find the oldest stars, we first have to restrict our attention to relatively low-mass stars. That's because, as you'll remember from our previous lectures, the lowest mass stars live the longest. Even a star like our Sun, if such a star formed when the galaxy was very young, it will already have ended its life. Why? Because stars like our Sun live to be about 10 billion years old, whereas the galaxy is more like 14 billion years old. In contrast, virtually every star ever born weighing less than about 80% of the Sun's mass is still around, because such low-mass stars live longer than the galaxy is old.

The problem is, such low-mass stars constitute about 75% of all stars in the galaxy, and the galaxy has been birthing such stars continuously since it began. So we're talking about, say, 150 billion of these low-mass stars. In addition, those 150 billion low-mass stars all look pretty much the same, because they are all still main-sequence stars. They have similar temperatures and colors and luminosities. So the oldest star in the galaxy is buried somewhere among 150 billion other very low-mass stars now swirling about the Milky Way that look pretty much just like it. Talk about a needle in a haystack!

So what's a stellar archaeologist to do? Ironically, the answer is to look for a star that's literally jumping out at you, and then, like the archaeologist performing radio-carbon dating, use the light spectrum of that star to verify that it has little to no heavy elements in its makeup.

You see, the oldest stars in our galaxy were born at a time in the galaxy's history when the galaxy as a whole had a much less orderly shape. Before the flattened disk of the Milky Way formed through gravitational collapse, our galaxy had a more nearly spherical shape. And so the stars that were born then moved through the galaxy in a more disorderly spherical swarm about the center of the galaxy, in contrast to the more orderly circular rotation of stars like our Sun in the disk of the Milky Way today.

Those early stars that are still around today continue to orbit the galaxy in that spherically distributed, more haphazard fashion, making up what we call the halo of our galaxy, extending to large distances all around it. And because they orbit the galaxy in that way, they periodically plunge down through the disk of the galaxy to the other side, then eventually plunge back up through the disk, over and over through the eons. As a result of this, some of the nearest stars to the Sun right now happen to be these very old halo stars that are nearby, not because they are true neighbors of our solar system residing in the disk of the galaxy, but because they are interlopers, plunging rapidly through the disk on their way to the other side.

These stars turn out to be relatively easy to spot, because we can literally see them tearing by us. And not only that, they tear by in a very peculiar direction—straight up or down, instead of along the direction of motion of our Sun and of other stars in our neighborhood. Imagine driving down a highway with lots of other traffic, and suddenly a car comes racing from below and shoots by straight up toward the sky. Even if that car looked like all of the others on the road, you would notice it.

Well, the star that currently holds the record as the oldest star in our galaxy is called HD 140283, a.k.a. the Methuselah star. It's a fairly unremarkable star, all things considered. But it was recognized about 100 years ago because of its remarkably fast and odd direction of motion. It is moving at 800,000 miles per hour, which is so fast that the Hubble Space Telescope could measure its motion in the sky over the course of just a few hours. It is also very bright, because it just so happens to be plunging through the disk of our galaxy very close to the Sun's position in the disk. You can see it yourself in the constellation Libra with a pair of binoculars.

The latest measurements using the Hubble space telescope have allowed astronomers to precisely pinpoint the star's distance—190.1 light-years—which in turn has permitted its luminosity to be accurately measured from its apparent brightness. That luminosity, together with its color or temperature, lets us use the Hertzsprung-Russell diagram to measure exactly where the star lies in relation to the main sequence. This star is just entering the final stages of its life, leaving the main sequence stage, and turning into a red giant. Therefore, by placing it accurately in the H–R diagram, it is possible to accurately determine its age, using the H–R diagram age-dating technique that we discussed in a previous lecture.

How old is Methuselah? 14.5 billion years, plus or minus 0.8 billion years. With the age of our universe being 13.8 billion years, that would seem to make Methuselah impossibly older than the universe itself. But that's why that uncertainty of 0.8 billion years is so important. It's more accurate to say that the age of the Methuselah star can be anywhere in the range of 13.7 to 15.3 billion years, so the accepted 13.8 billion year age of the universe is within that margin of error. In other words, the measurement of Methuselah's age puts it right at the maximum possible age for a star in our universe, meaning that this star must have formed just as the galaxy itself was forming and very soon after the big bang.

And now the 64,000 dollar question. What is Methuselah made of? Is it pure hydrogen and helium supplied by the big bang itself? Or does it possess even a trace of heavier elements? The answer, from careful measurements of the elemental signatures in its light spectrum, is that it contains just a smidgen of heavy elements, a mere one 250th of the amount contained in the Sun. But that's more than zero. And it tells us definitively that there must have been a generation of stars that came even before Methuselah and its cohort to create that smidgen of heavy elements for even poor old Methuselah to inherit.

To be clear, any such prior generation of stars would have had to exist sometime after the big bang. They could not have existed prior to the start of the universe as a whole. So with Methuselah's age having been pinned to be at least 13.7 billion years old, and the universe having begun 13.8 billion years ago, there is a tiny window of time—about 0.1 billion, or 100 million years—during which time a generation of pre-Methuselah stars could have existed.

In addition, the measurements of Methuselah's detailed elemental abundances tell us that there is something peculiar about this star's chemical makeup. It has much more oxygen relative to its hydrogen than what we currently see in the universe's mix of chemical elements. That's an important clue. So let's talk about those mysterious first stars that must have come before HD 140283, the Methuselah star.

Since none of the first stars that formed in the universe are around anymore, it is not possible to directly observe them. So the only way to understand what they must have been like is through detailed calculations and computer simulations. And until recently, such simulations couldn't be done because of the sheer computing power required. It's really quite difficult to perform detailed simulations of how stars in the very early universe would have formed. Early simulations were able to follow the process to the point where a clump of gas about 1% as massive as the Sun had formed. But as you know, that's just tip of the iceberg, not even a stellar embryo, really.

You might think that the stellar birth process at the beginning of the universe would be simpler than it is today, because the universe soon after the big bang was a lot less complex than it is now. Almost all of the matter in the universe was in the simplest form possible—simple hydrogen and helium gas. There were no heavier elements and there were no dust particles. And there were no existing stars with their magnetic fields and stellar winds to complicate things.

The complication is that in the early universe, the ordinary matter with which we are familiar—stuff like hydrogen gas—was not the form of matter that controlled things, gravitationally speaking. The dominant form of matter was what we call dark matter, a mysterious form of matter that exerts the same gravity as ordinary matter but whose physical properties we are only beginning to understand. We'll say a bit more about dark matter in a moment, but for now, current estimates are that dark matter actually comprises 90% of the total matter content of the universe, and would have also dominated the matter content of the universe back when the very first stars were forming.

So, at the universe's beginning, from the standpoint of gravity and its influence in the formation of the first stars, imagine the universe as a place principally filled with dark matter. Regions where the dark matter was denser were the places where the ordinary matter out of which stars form— hydrogen and helium in the young universe—collected. In addition to the nature of dark matter still being largely unknown, these regions of denser dark matter would have been enormous, further exacerbating the challenge of simulating these regions computationally.

In recent years, however, advances in computational power have made it possible for scientists to simulate the formation of the earliest stars. And what the simulations tell us is that the masses of stars formed in the dense dark-matter regions of the early universe would have been about 100 to 300 times as massive as the Sun. That is an extraordinary amount of mass for a star! A very few stars today might approach such a mass, but only very, very rarely. As we've discussed in previous lectures, these days in the universe, far less massive stars like our own Sun are the rule.

So why were the very first stars so massive? The main reason is that the matter out of which they formed was essentially pure hydrogen and helium. Without heavy elements in the mix, it was much harder for the gas to cool off as it collapsed under gravity. That's because the presence of heavy elements in a gas gives that gas more energy transitions for the electrons in the atoms to be able to jump between. Heavy elements, such as iron, by virtue of having many more electrons than hydrogen, have many more possible energy levels that the electrons can occupy. And remember, when an electron jumps from a higher energy level to a lower one, it emits energy in the form of photons. Consequently, there are lots of ways for an iron atom to radiate away energy through emission of photons of lots of different wavelengths. Hydrogen and helium can only radiate at a few specific wavelengths. So in a gas of pure hydrogen and helium, the heat from gravitational collapse builds and builds.

In these early universe stars, the pressure from heat in the gas could become much higher than when stars form today, and so more mass was required for gravity to bear down and balance that very high pressure pushing out from within. Remember what we've discussed several times throughout our course; gravity continually bears down on a star, trying to crush it down to nonexistence, and it is only by generating sufficient pressure through heat in its interior that a star can push back on gravity to hold itself up. With these first stars being capable of generating much higher internal pressures, they could withstand gravity piling on even more mass and bearing down even harder. Thanks to their heat, these enormous early stars held strong.

To be sure, the idea that the very first generation of stars in the universe had a mass around 100 times the mass of the Sun remains controversial. The most recent simulations, in fact, indicate that the first stars may have been slightly less hefty, around 40 times the mass of the Sun. That's because the temperatures in the forming stars rise so high that intense ultraviolet radiation starts to be emitted by the forming star very quickly. This radiation has a feedback effect on the surrounding gas, causing it to be expelled from the forming star instead of collapsing inward to add to the star's heft. Even so, the consensus is that the very first stars in the early universe must have been much more massive than the stars that we observe forming today.

There is also an emerging consensus among astronomers that there must have been at least some first stars that were more massive than at least 140 times the mass of the Sun. This number—140 times the mass of the Sun—is important, because stars more massive than that will end their lives in an unusual type of supernova event called a pair-instability supernova. This is a very different type of supernova from the type we discussed when we were addressing the death of massive stars in our universe today.

Importantly, a pair-instability supernova produces different proportions of heavy elements. Remember that peculiar ratio of oxygen to hydrogen in the Methuselah star that I mentioned earlier? Well, it could well be that its unusual mix of elements, even though there is only a smidgen of them, could have been the result of Methuselah forming from material that had been enriched, not by the type of supernovae that we see today, but by these bizarre pair-instability supernovae.

This is an ongoing and cutting-edge of research, and so the jury is still out on many of the details. In addition, as supercomputers become even more powerful, it will be possible to perform full three-dimensional simulations that should further clarify how these first stars formed. In any case, the galactic archaeology studies of Methuselah stars in our own galaxy have yet to find any stars that formed from pure hydrogen and helium. So whatever the process, it's clear that the first generation of stars must have lived fast and furious and rapidly generated enough heavy elements to create the smidgen traces of things like oxygen and iron that we do see in the Methuselah stars.

It's a bit like learning about dinosaurs. The idea of a gargantuan beast like a *Tyrannosaurus rex* walking the Earth seems incredible. Yet the bones do not lie. If the idea of *T. rex* stars in the early universe seems incredible, just wait, there's more. A number of astronomers are becoming convinced that some of the first gargantuan stars may have been stranger still, made not of ordinary matter, but mainly out of dark matter. If correct, then the first stars to form in the early universe would have been what some are calling dark stars, fueled by an altogether different engine than stars as we know them today. As we've learned in previous lectures, ordinary stars, like the sun today, shine because they are fueled by nuclear fusion in their cores that converts hydrogen to helium. But these theoretical dark stars would have run on dark matter particles colliding and annihilating each other.

I referred to dark matter earlier. But what is it? Dark matter is a substance that astronomers have not yet observed directly, that's why we call it dark. But we infer its existence because we can measure its gravitational effects on visible matter. For example, the orbital motions of stars in our galaxy cannot be explained if the only source of gravity in the galaxy is the stars and gas and dust that we can directly see. The motions in our galaxy require that there must be another component of matter in our galaxy that makes up some 90% of its total matter content.

We don't yet know exactly what dark matter is, but it must be some type of elementary particle that we have not yet observed in particle colliders. I'm not referring here to the Higgs Boson that you've probably read about in the news. That recently discovered particle was believed to exist for a long time and is part of ordinary matter. Hopefully, in the next few years of research at the Large Hadron Collider near Geneva, Switzerland, we will hear that physicists have at last found the mysterious particles that make up the dark matter. But in any case, the fact that this dark matter exerts some gravity tells us that at some basic level it must operate much like any other type of matter. And that means that whatever the dark-matter particles are, they almost certainly have anti-particle counterparts.

The elementary particles with which most of us are already familiar, such as protons and electrons, have anti-matter counterparts. A proton has an anti-matter counterpart called an anti-proton. An electron has an anti-matter counterpart called a positron. These particles share all of the same properties as their ordinary counterparts, except that when a particle and its anti-particle counterpart come together, they convert into pure energy. This process is called particle-anti-particle annihilation. How much energy is produced in a particle-anti-particle annihilation event? Remember $E = mc^2$. The two particles are converted totally into energy, so it's the sum of their two masses, times the speed of light squared. And that's a lot of energy. So highly energetic gamma ray photons are produced in the process, carrying away the original mass energy of the two particles.

With so much dark matter in the universe, astronomers have been re-thinking the formation of the first generation of stars in the early universe through the lens of dark-matter theory. In essence we are asking, what about this enormous reservoir of dark matter? What does it do? And what does it mean for the formation of the very first stars that would have been swimming in the stuff? The findings of this area of research so far suggest a dramatic altering of the theoretical framework for the formation of the first stars. Dark stars are an entirely new kind of star from anything we've known or conceived of before.

According to dark-stars theory, the first stars are thought to have formed inside clouds of dark matter when hydrogen and helium gases cooled to a temperature at which nuclear fusion could begin. In more conventional theories of the first stars that we discussed earlier, dark matter doesn't affect this process, except to provide the gravity to bring the gases together. But in dark-stars theory, the dark matter concentrations are high enough for the particles in the dark matter clouds to collide with and annihilate each other, destroying themselves, and more importantly, providing a source of internal heat to support the very massive star without relying on nuclear fusion alone.

As a result, the star can be enormous, perhaps as large 2000 times the size of the Earth's orbit about the Sun. That's a big star! And when these enormous stars finally end their lives as supernovae, perhaps as quickly as a thousand years or even less after they formed, they leave behind the heavier elements, such as carbon, nitrogen, and oxygen that became part of the next generation of stars, such as in the Methuselah stars still roaming our galaxy. This idea is so intriguing that astronomers are already looking for ways to detect these dark stars, or at least their footprints. The products from the dark matter particles annihilating each other are energetic gamma-rays that might be detectable with current space telescopes.

In this lecture, we've swung the pendulum from brown dwarfs to the far other extreme of the stellar world—the biggest, most massive stars that may have populated the universe's very beginnings and that would have initiated the stellar life cycle for the rest of time. As we've seen, there are multiple good lines of evidence that such a strange population of stellar beasts roamed the early universe, including stellar archaeology studies of the oldest Methuselah stars in our own galaxy. We've also discovered that to understand those first ancestral stars, we may ultimately have to penetrate the hidden nature of the strange dark matter particles that still fill the universe, whose gravity we can sense, but which we have not even seen with our own eyes.

Stars as Magnets
Lecture 21

In this lecture, you will continue to explore the cutting edge of current research into the nature of stars. In particular, you will explore the phenomenon of magnetism in stars. Many of the extreme behaviors of stars can be understood through their magnetic nature. As you will learn, through their spins, stars can act as powerful magnets, affecting not only themselves but also the space around them. In a sense, it is through its magnetism that a star is able to extend its influence far beyond its surface.

Magnetism

- One of the basic properties of magnetism in nature is that magnets never occur with single poles, or monopoles. A basic magnet, whatever it's made of or however its magnetism is generated, has two poles—a north pole and a south pole. It is dipolar.

- Stars like the Sun also have dipolar magnetic fields. There is a point typically near the north rotational pole of the star that is the north magnetic pole and a point typically near the south rotational pole that is the south magnetic pole.

- How do stars, which are gaseous fluids, generate and sustain their magnetic fields? The basic answer is electricity, and the intimate relationship between electricity and magnetism.

- Up until the 19th century, electricity and magnetism were thought to be independent forces, but ever since James Clerk Maxwell devised the laws of electromagnetism, we understand that they are in fact manifestations of a unified force. One of the consequences of that fact is that magnetism and electricity can induce one another.

- For example, in an electromagnet, a strong electrical current is used to create a magnetic field. The way this works is based on one of Maxwell's laws, which states that electrically charged particles moving around in a current produce a magnetic field. The magnetic field produced by an electrical current is oriented perpendicular to the motion of the current. That's precisely what happens within a star—a star is an electromagnet.

- The gas within a star is hot and so is partially or fully ionized. And all stars rotate, or spin. The charged particles within a star, carried along by the star's rotation, constitute an electrical current, and as such, they generate a magnetic field.

- As the star's rotation carries the current around parallel to the star's equator, the magnetic field it generates is oriented perpendicular to the equator, so the poles of the magnetic field are oriented such that the north and south magnetic poles correspond closely to the rotational axis of the star.

- A magnetic field is really just a way of representing the ability of a magnet to act at a distance. Another important aspect of magnetic field lines is that only charged particles feel them; electrically neutral particles don't notice magnetic field lines. A charged particle in the presence of a magnetic field is forced to move along that field line—up or down along it—and cannot move across it. When charged particles, such as electrons, encounter a magnetic field, they slide effortlessly along the magnetic field lines, like beads on a string.

- The Sun is a magnet. The Sun rotates—its period of rotation is about 26 days—and that equatorial rotation represents an electric current that in turn generates a magnetic field emanating from the Sun's poles. The Sun's global magnetic field lines can be visualized as a mesh of tendrils, like hair, that grow out of the north magnetic pole and wrap down around the Sun, then reconnect with the Sun's surface down at the south magnetic pole.

- However, there are two important ways in which the detailed topology of the Sun's magnetic field differs from this simple representation. The Sun's rotation—the same rotation that causes the Sun's magnetic field in the first place—also shapes and distorts the magnetic field in two important ways.

- The first effect is that the magnetic field gets wrapped around the equator and balloons out at the poles. This makes the magnetic field at the equator tend to be tightly confined near the Sun's surface, whereas at the poles it extends far, far away from the Sun's surface. In other words, near the equator, the magnetic field is tightly wound and even kinked up, whereas near the poles it is straight and stiff.

- The second way in which the Sun's rotation alters the Sun's magnetic field has to do with that tight wrapping near the equator. As the magnetic field near the equator becomes more and more wrapped, then it becomes kinked and pokes up out of the Sun's surface. The places where those kinked loops poke up and back into the surface are the places where pairs of sunspots form.

- For those kinked loops protruding from the Sun's surface near the equator, hot ionized gas lifts up along those field lines and is held aloft by the magnetic field, creating amazing solar prominences. These prominences can erupt as energetic flares, driving so-called space weather. As for the straight magnetic field lines emanating from the Sun's poles, charged particles flow along those lines way out from the Sun. This is the solar wind.

- The solar wind has an important influence on the Sun. As particles from the Sun's rotating surface stream along the magnetic field, they end up flung out into the outer reaches of our solar system. These solar wind particles take a tiny, insignificant bit of the Sun's mass with them. But these particles also take with them a bit of the Sun's spin energy—a bit of its angular momentum.

- The resulting steady decrease of the Sun's angular momentum leads to a steady slowing down of the Sun's spin. In fact, 90% of the Sun's original spin has already been lost, and over the remaining course of the Sun's life, it will continue to wind down, probably spinning at only half of its current speed by the time it's done. And as it winds down, its magnetism will steadily decrease, meaning less severe solar storms at Earth.

The Internal Structures of Stars

- Interestingly, stars more massive than about 3 times the Sun's mass generally possess extremely weak magnetic fields. This is surprising because more massive stars tend to be more luminous, more voluminous, and hotter. But there is an important way in which massive stars are deficient in comparison to lower-mass stars, and this difference provides an insight into the real mechanism for the generation of magnetic fields in stars. In short, it has to do with the internal structure of stars.

- One of the key features of the Sun's internal structure is that the outer layers of the Sun form the so-called convective zone, in which the energy from the Sun's core is transported upward to the surface through the roiling circulatory motion, similar to that of a boiling pot of water. Stars less massive than the Sun have an even more substantial outer convective zone relative to their overall size. In contrast, stars more massive than the Sun have very thin outer convective zones, and the most massive stars have no outer convective zone.

- Therefore, the presence of an outer convective zone, and its depth relative to the overall size of the star, is an important ingredient in the generation of a magnetic field. Fundamentally, the convection is responsible for the generation of the electrical currents that are necessary, together with the star's rotation, to create the star's magnetic field. That process of combining convective currents with rotation to create a magnetic field is called a dynamo.

- Have you ever used a device that powers a light or a radio by turning a crank? That's a dynamo, although working in reverse. In this type of dynamo, a magnet is used to generate an electrical current by spinning a loop of wire around that magnet. In a star, it's the same idea—only the electrical current formed by the convection zone is spun around by the star's rotation, and this creates a magnet.

- This relationship between a star's internal structure—in particular, the depth of its outer convection zone—and the strength of its magnetic field also explains the pattern of flaring of X-ray emission that we observe in other stars.

- Perhaps counterintuitively, the stars with the strongest magnetic flares and the most intense bursts of X-ray emission are the lowest-mass stars. But this makes perfect sense, because the lowest-mass stars are also the ones with the deepest convective zones and, therefore, have the strongest dynamos with which to generate the strongest magnetic fields.

- As a consequence of this, massive stars tend to be rapidly spinning—and tend to remain so throughout their lives. Without a strong magnetic field to drive a wind to remove angular momentum, they cannot slow down. On the other hand, less massive stars do slow down with time, just like the Sun. This effect has led to the development of a promising technique, called gyrochronology, for inferring the ages of stars from their spins.

Pulsars were initially discovered because they emanate regular pulses of radio waves.

Pulsars

- The role of magnetism in the birth of stars remains one of the frontiers of research. But magnetic fields matter at the ends of stars' lives, too. Indeed, some of the most extreme phenomena we have observed from stars are from the highly magnetized corpses of dead stars. The best example of this is what we refer to as pulsars.

- Pulsars are neutron stars. But they are neutron stars that we happen to view from Earth at just the right angle so that the intense light they emit from their extremely strong magnetic poles aims right at us as the neutron star spins around. This occurs because the magnetic poles are offset somewhat relative to the spin axis of the neutron star, so the magnetic poles swing around as the neutron star spins.

- One of the first pulsars to be discovered was the neutron star at the center of the Crab Nebula supernova remnant. That pulsar spins about 30 times per second, and in fact, that pulse can even be seen in visible light. But most pulsars radiate their pulse beams primarily in radio light.

- The first pulsar ever discovered came about 30 years after the invention of radio telescopes. Since then, many pulsars have been discovered, and they all spin very fast. The slowest among them spins about once per second, and the fastest—known as the millisecond pulsars—spin about 1000 times per second.

- The regularity of a pulsar's pulse is so perfect that it can be used as one of the most exact timekeeping devices known. As a result of this exquisite timekeeping ability, pulsars have been used to prove Einstein's prediction of gravitational radiation, in which a pair of pulsars orbit one another so fast that they radiate away gravitational energy and are spiraling into one another.

- The pulsar regularity also led to the discovery of the first exoplanet. The first planets discovered outside of our own solar system were discovered not around another living star, but around a pulsar.

Suggested Reading

Choi, "Alien Planets Circling Pulsing Stars May Leave Electric Trails."

McNamara, *Clocks in the Sky.*

1. What does the existence of pulsars imply for how we might modify our strategy to detect true signals from another civilization?

2. What would a civilization on a planet orbiting a pulsar have experienced in the lead-up to the star having become a neutron star? Could such a civilization have survived and still live on the pulsar planet?

Stars as Magnets
Lecture 21—Transcript

In this lecture we'll be continuing our exploration of the cutting edges of current research into the nature of stars. In particular, we'll explore the phenomenon of magnetism in stars. Many of the extreme behaviors of stars can be understood through their magnetic nature. As we'll see, through their spins, stars can act as powerful magnets, affecting not only themselves, but also the space around them. In a sense, it is through its magnetism that a star is able to extend its influence far beyond its surface.

So let's start with a quick reminder about magnets in general and then relate these properties to stars. The goal here is to see that stars don't just behave like magnets, they are magnets literally. The best way to begin visualizing the stars as magnets is to think about a bar magnet. You've probably played with a bar magnet at some point. It is just a piece of metal that has been magnetized through exposure to a strong magnetic field. The bar has two magnetic poles—a north pole and a south pole. This is a generic feature of all magnets. In fact, one of the basic properties of magnetism in nature is that magnets never occur with single poles, or monopoles. A basic magnet, whatever it's made of or however its magnetism is generated, will have two poles; it will be dipolar.

Stars like the Sun have dipolar magnetic fields also. There is a point, typically near the north rotational pole of the star, that is the north magnetic pole, and a point typically near the south rotational pole that is the south magnetic pole. It is as if stars, like the Sun, have bar magnets beneath their surfaces. But of course, stars are not solid, so the analogy of the metal bar magnet only applies so far. How do stars, which as we've seen are gaseous fluids, generate and sustain their magnetic fields? The basic answer is electricity, and the intimate relationship between electricity and magnetism.

Up until the 19th century, electricity and magnetism were thought to be independent forces. To this day, it remains convenient in many applications to describe them separately. But ever since James Clerk Maxwell devised the laws of electromagnetism, we understand that they are, in fact, manifestations of a unified force. One of the consequences of that fact is that magnetism and electricity can induce one another.

For example, in an electromagnet, a strong electrical current is used to create a magnetic field. The way this works is based on one of Maxwell's laws, which states that electrically charged particles moving around in a current produce a magnetic field. The magnetic field produced by an electric current is oriented perpendicular to the motion of the current. That's precisely what happens within a star; a star is an electromagnet.

What do I mean? Well, remember that the gas within a star is hot, and so is partially or fully ionized. And remember that all stars rotate; they spin. What is a current, but a set charged particles moving around in a path? The charged particles within a star, carried along by the star's rotation, constitute an electrical current, and as such, they generate a magnetic field. As the star's rotation carries the current around parallel to the star's equator, the magnetic field it generates is oriented perpendicular to the equator, and so the poles of the magnetic field are oriented such that the north and south magnetic poles correspond closely to the rotational axis of the star.

Now, I keep referring to a magnetic field. What do I mean by that, and what is the best way to visualize it? The first thing to realize is that a magnetic field is really just a way of representing the ability of a magnet to act at a distance. But it can be helpful to visualize it as though it were a set of ropes or tendrils radiating out from one magnetic pole and coming back together at the other pole. Think about the way in which lines of longitude on a globe fan out from one pole and then come back together to a point at the other pole. Magnetic field lines are like that.

Another important aspect of magnetic field lines is that, only charged particles can feel them. Electrically neutral particles don't notice magnetic field lines at all. A charged particle in the presence of a magnetic field is forced to move along that field line, up or down along it, and cannot move across it. When charged particles, such as electrons, encounter a magnetic field, they slide effortlessly along the magnetic field lines, like beads on a string. So remember that image; charged particles moving along a magnetic field slide along like beads on a string. With this understanding, hopefully it will make some sense when I say that the Sun is a magnet. The Sun rotates; its period of rotation is about 26 days; and that equatorial rotation represents an electric current that, in turn, generates a magnetic field emanating from the Sun's poles.

The Sun's global magnetic field lines can be visualized as a mesh of tendrils, like hair, that grow out of the north magnetic pole and wrap down around the Sun, then reconnect with the Sun's surface down at the south magnetic pole. And for many purposes, this simplistic view of the Sun's magnetic topology is sufficient. But in reality, there are two important ways in which the detailed topology of the Sun's magnetic field differs from that of a simple dipolar bar magnet. The Sun's rotation, the same rotation that causes the Sun's magnetic field in the first place, also shapes and distorts the magnetic field in two important ways. The first effect is that the magnetic field gets wrapped around the equator and balloons out at the poles. This makes the magnetic field at the equator tend to be tightly confined near the Sun's surface, whereas at the poles it extends far, far away from the Sun's surface. In other words, near the equator the magnetic field is tightly wound and even kinked up, whereas near the poles it is straight and stiff, like electrified hair.

The second way in which the Sun's rotation alters the Sun's magnetic field has to do with that tight wrapping near the equator that I just mentioned. As the magnetic field near the equator becomes more and more wrapped, then, like a twisted rubber band, it becomes kinked and pokes up out of the Sun's surface. The places where those kinked loops poke up and back in to the surface are the places where pairs of sunspots form.

Now, it's important to remember the point I made earlier, that charged particles can slide effortlessly along a magnetic field line, like beads on a string. For those kinked loops protruding from the Sun's surface near the equator, that means that hot ionized gas lifts up along those field lines, and is held aloft by the magnetic field, creating amazing solar prominences. These prominences can erupt as energetic flares, driving the so-called space weather that we'll discuss in detail in our next lecture. As for the straight magnetic field lines emanating from the Sun's poles, charged particles flow along those lines out, way out from the Sun. This is the solar wind.

The solar wind has an important influence on the Sun. As particles from the Sun's rotating surface stream along the magnetic field, they end up flung out into the outer reaches of our solar system. These solar wind particles take a tiny bit of the Sun's mass with them, about one ten-trillionth of the Sun's mass per year. That's not significant at all; after 10 billion years that doesn't even amount to 1% of the Sun's total mass. But what is significant is that these particles also take with them a bit of the Sun's spin energy, a bit of its angular momentum.

The resulting steady decrease of the Sun's angular momentum leads to a steady slowing down of the Sun's spin. In fact, the Sun now spins about one-tenth as fast as it probably spun shortly after its birth. In other words, 90% of the Sun's original spin has already been lost. And over the remaining course of the Sun's life, it will continue to wind down, probably spinning at only half of its current speed by the time it's done. And as it winds down, its magnetism will steadily decrease as well, meaning less severe solar storms at Earth.

Why is such a tiny loss of mass so effective at winding down the Sun? The reason is that the Sun's long magnetic field lines essentially represent a very long lever arm with which the outflowing particles exert a torque back on the Sun, slowing it down as they go. As Archimedes once famously said, "give me a lever long enough … and I shall move the world." Well, in the Sun's magnetic field, the particles that make up the solar wind have a lever the length of the solar system, and with that lever, those outflowing particles move—or spin down—the Sun.

So what about other stars? With the Sun, we have the great advantage of being able to study the surface in exquisite detail, and we can make movies of the dynamic ways in which the magnetic field changes on the Sun. We can even directly sense the particles that the Sun's magnetic field flings at us. But we can't directly see the surface of any other star. So in order to study the magnetism of stars more distant than the Sun, we have to, once again, use the power of light and the richness of physical information encoded in it.

To measure the strength of a star's magnetic field, we may use an effect known as the Zeeman effect. Remember that the different elements that make up the Sun's atmosphere cause certain wavelengths of light to be absorbed from the Sun's spectrum. The Zeeman effect causes those wavelengths of absorption to become split. And that splitting of the light causes half of the absorbed light to be shifted to slightly bluer wavelengths, and the other half to be shifted to slightly redder wavelengths. In other words, the two halves of the absorbed light get shifted oppositely. So by measuring the light spectrum of a star and measuring the blueshift versus redshift of the two halves of the absorption, we have a measure of the strength of the star's magnetic field.

One of the interesting results from this type of study is that stars of different masses tend to have very different magnetic field topologies. Whereas stars with masses similar to the Sun have magnetic fields more or less like the Sun's, the lowest mass stars tend to have more well-ordered fields, with very simple geometries. Stars somewhat more massive than the Sun have less well-ordered fields, with even more complex geometry than the Sun's.

Interestingly, stars more massive than about three times the Sun's mass generally possess extremely weak magnetic fields. Does that seem surprising? After all, in most other respects, more massive stars tend to be more—more luminous, more voluminous, hotter. But there is an important way in which massive stars are deficient in comparison to lower mass stars, and this difference provides an insight into the real mechanism for the generation of magnetic fields in stars. In short, it has to do with the internal structure of stars.

Think back to our discussion of the internal structure of the Sun. One of the key features of the Sun's internal structure is that the outer layers of the Sun form the so-called convective zone, in which the energy from the Sun's core is transported upward to the surface through the roiling circulatory motion, similar to that of a boiling pot of water. Stars less massive than the Sun have an even more substantial outer convective zone relative to their overall size. In contrast, stars more massive than the Sun have very thin outer convective zones, and the most massive stars have no outer convective zone at all. Clearly then, the presence of an outer convective zone and its depth relative to the overall size of the star is an important ingredient in the generation of a magnetic field. Fundamentally, the convection is responsible for the generation of the electrical currents that are necessary, together with the star's rotation, to create the star's magnetic field. That process of combining convective currents with rotation to create a magnetic field is called a dynamo.

You may have seen or even used a dynamo. Ever use a device that powers a light or a radio by turning a crank? That's a dynamo, although, working in reverse. In this type of dynamo, a magnet is used to generate an electrical current by spinning a loop of wire around that magnet. In a star, it's the same idea, only the electrical current formed by the convection zone is spun around by the star's rotation, and this creates a magnet. Electricity and magnetism are intrinsically connected.

This relationship between a star's internal structure—and in particular, the depth of its outer convection zone—and the strength of its magnetic field also explains the pattern of flaring of X-ray emission that we observe in other stars. Perhaps counterintuitively, the stars with the strongest magnetic flares and the most intense bursts of X-ray emission are the lowest mass stars. But this makes perfect sense when we recall that the lowest mass stars are also the ones with the deepest convective zones, and therefore have the strongest dynamos with which to generate the strongest magnetic fields.

As a consequence of this, massive stars tend to be rapidly spinning, and to remain so throughout their lives. Without a strong magnetic field to drive a wind to remove angular momentum, they cannot slow down. On the other hand, less massive stars do slow down with time, just like the Sun. This effect has led to the development of a promising technique for inferring the ages of stars from their spins. This technique, known as gyrochronology, is important and useful because the ages of stars during their long middle years are, in general, quite difficult to determine accurately if using only the stars' appearances. This is similar to guessing the age of a person. When a human is an infant or a toddler, you can be just about 100% sure that you are looking at a very young person. Or when a human is elderly, you can generally be very confident in that assessment. But when a person is an adult in middle age, it can be difficult to know with certainty whether you're looking at someone in their 40s or 50s.

Stars in middle age don't change much in terms of their appearances—their temperatures or luminosities—because as main sequence stars they are very stable. However, their spins do change, and they change steadily, spinning ever more slowly as their winds remove their angular momentum bit by bit. Like a clock winding down, a star's spin conveys the amount of time it's been ticking, and this gives astronomers one of the only means we have to pin an accurate age on a star during its middle years.

Besides causing the stars to spin more slowly over time, what are some of the other effects of these magnetic fields? When a star's magnetic field is very strong, it can cause the star's surface to become cooler than it would normally be, and this, in turn, can cause the star to have a larger diameter than it otherwise would have. Think about a sunspot on the Sun. A sunspot is a locally very cool region on the Sun caused by the suppression of upwelling heat by the strength of the magnetic field at that spot.

On a star with an extremely strong magnetic field, the same effect can cause the entire surface to be cool like a sunspot. But with a decreased temperature, the star must readjust to have a larger surface in order to be able to radiate the energy that has been generated deep in the star's core. This phenomenon is an important one for astronomers to take into account when attempting to infer the dimensions of stars or of any bodies orbiting those stars, such as planets.

When stars are very young, still going through the birth process within the clouds of gas and dust from which they are formed, the effects of the fledgling stars' magnetic fields can be even more important. For a long time, one of the fundamental mysteries in understanding the stellar birth process is what is referred to as the angular momentum conundrum. It was realized that as a baby star accretes mass from its cocoon of gas and dust, it will also accrete angular momentum from that large rotating mass. This accretion of angular momentum should cause the baby star to spin so fast— at the maximum speed, also known as breakup speed—that they should fling material off their surfaces. But here's the conundrum; most young stars are, in fact, observed to be spinning much more slowly than that. They do spin faster than old stars like the Sun, but nonetheless, they spin much more slowly than the accretion of angular momentum would predict. So somehow, a baby star manages to quickly remove most of the angular momentum that it inherits from its parent cloud of gas and dust.

A couple of mechanisms for this rapid angular-momentum removal have been proposed. Warning, this is a hotly debated area of research, and it is not yet clear which if any of these proposed mechanisms are correct. One proposed mechanism is referred to as the disk-locking mechanism. The idea here is that the young star's strong magnetic field allows it to grab on to the protoplanetary disk of gas and dust encircling it and to slow itself down. Think of this like sitting in a pool; you're spinning around on a float, and you stick your hands in the water on either side of you. The resistance of the water on your hands will slow down your spin. This is an attractive mechanism, so long as the star has a protoplanetary disk of sufficient mass that the star can stick its hands into. But one problem with this picture is that the protoplanetary disks of young stars generally go away after just a few million years into the star's life, whereas the star needs to continue to get rid of excess angular momentum for much longer, perhaps 50 million years or so until it becomes a main-sequence star.

So another mechanism proposed by my own research group is that young stars may shoot off a large enough number of extreme coronal mass ejections, just like those observed on the Sun, but much more powerful and more frequent, so as to fling enough mass with a long enough lever that the star slows down its spin. This mechanism has the advantage that it can operate for as long as necessary to ensure that the star slows down by the time that it becomes a main-sequence star. But a challenge with this mechanism is that it's not clear whether young stars actually shoot off such powerful ejections often enough to make it work. When we observe young stars, such as in the Orion Nebula stellar nursery, in X-ray light, we do indeed observe very powerful X-ray flares indicative of powerful coronal mass ejections. But the very most extreme of these flares happen relatively infrequently.

So the jury is still out on exactly how young stars resolve the angular momentum conundrum in their birth, how they manage to avoid spinning so fast that they tear themselves apart. Perhaps it is through a combination of the mechanisms we've discussed; the disk-locking idea early on, while the protoplanetary disk is still available for the star's field to grab hold of, and the extreme coronal mass ejection idea after that. But the role of magnetism in the birth of stars remains one of the frontiers of research.

Magnetic fields matter at the ends of stars' lives, too. Indeed, some of the most extreme phenomena we have observed from stars are from the highly magnetized corpses of dead stars. The best example of this is what we refer to as pulsars. Pulsars are neutron stars. But they are neutron stars that we happen to view from Earth at just the right angle so that the intense light that they emit from their extremely strong magnetic poles aims right at us as the neutron star spins around. This occurs because the magnetic poles are offset somewhat relative to the spin axis of the neutron star, and so the magnetic poles swing around as the neutron star spins. It's like the beacon of a light house. The light beam is always on, but it spins around, and so as the beam swings past us, we perceive a flash of light that repeats with a steady cadence corresponding to the rotation of the beam.

One of the first pulsars to be discovered was the neutron star at the center of the Crab Nebula supernova remnant. That pulsar spins about 30 times per second, and in fact, that pulse can even be seen in visible light. Some individuals with extremely sensitive vision have reportedly sensed the rapid flickering of the Crab Nebula pulsar when observing it visually through a telescope.

But most pulsars radiate their pulse beams primarily in radio light. The first pulsar ever discovered came about 30 years after the invention of radio telescopes and was such a surprise that for a while nicknamed LGM-1, which was short for Little Green Men number 1. It took astronomers a while to prove that such a perfectly repeating pulsing light signal did not necessarily represent a technologically built signal from another civilization.

Since then, many pulsars have been discovered, and they all spin very fast. The slowest among them spin about once per second, and the fastest, known as the millisecond pulsars, spin about 1000 times per second. Let's take a moment to listen to what such pulsars might sound like. The sounds you're about to hear are tones that were synthesized on a computer using the steady cadence of pulses from different pulsars. Each time the pulsar was observed by the radio telescope to pulse, the computer recorded a click sound. Let's hear a few examples together.

First, we have pulsar B0329. [Star sound.] That's a neutron star spinning around about once per second, producing a light pulse that we see once per second, and so what we just heard was a corresponding click about once per second. Next we have the Crab Pulsar in the Crab Nebula, spinning about 30 times per second. [Star sound.] Finally, we have pulsar B1937. [Star sound.] You didn't hear any clicks that time, did you? Actually, that tone is the sound resulting from clicks occurring about 600 times per second. If you're musically inclined, that's close to the pitch of C above Middle C. Can you think of anything else that makes that kind of pitch? Think of the spinning blades in your kitchen blender on puree. That's how fast this neutron star spins, an object weighing more than the Sun, about the size of a city on Earth, spinning like the blades of a kitchen blender.

The regularity of a pulsar's pulse is so perfect that it can be used as one of the most exact timekeeping devices known. As a result of this exquisite timekeeping ability, pulsars have been used to prove Einstein's prediction of gravitational radiation, in which a pair of pulsars orbit one another so fast that they radiate away gravitational energy and are spiraling in to one another. This led to the Nobel Prize in physics in 1993. The pulsar regularity also led to the discovery of the first exoplanet. That's right; the first planets discovered outside of our own solar system were discovered not around another living star, but around a pulsar. Astronomers observed periodic modulations of the pulsar's pulses and determined that this was induced by the gravitational wobble of the neutron star induced by the planets orbiting it!

In this lecture, we've explored the magnetic natures of the stars. The stars, because they spin like dynamos, generate magnetism. They are magnets, literally. And their magnetic nature produces a range of interesting and useful phenomena, the spin-down of stars, like the Sun, allowing us to infer their ages; the survival of baby stars from prematurely spinning themselves apart; and the amazing pulses of pulsars, not little green men, but nature's perfect clocks in the sky.

Solar Storms—The Perils of Life with a Star
Lecture 22

I n this lecture, you will learn that the Sun governs the space weather of its environs and, in turn, exerts a direct influence over weather on Earth, the health of our astronauts in space, and the durability of the satellite platforms upon which our technologically built world increasingly depends. However, although the Sun's weathering of our planet might seem harsh now, it is nothing compared to the intense weathering that the Sun exerted on the proto-Earth, a weathering that might have been an important part of how the building blocks for life as we know it came to be.

Aurorae

- One of the most basic—and beautiful—manifestations of space weather on Earth is the aurorae. The word "aurora" derives from the Latin word for dawn. Aurorae are the eerie curtain-like sheets and long streamers of shimmering light that appear near the north and south poles. These wonderful, if strange, apparitions have been noticed by people going all the way back to the most ancient civilizations, particularly those who resided at far northern latitudes.

- The aurora borealis occurs above the north pole, and its counterpart, the aurora australis, occurs above the south pole. But with no early civilizations living near Antarctica, only the northern lights (as the aurora borealis is also known) were recorded in ancient times. It was noticed early on that the northern lights did not occur constantly but, rather, sporadically. And when they occur, they tend to occur strongly, like a gust or storm.

- Today, our basic picture for how the aurorae are produced is as follows. Energetic flares erupt on the Sun's surface, driven by the twisting and kinking of the Sun's magnetic field. Those flares, representing the sudden snap of the Sun's kinked magnetic field and signaled by a burst of X-ray light, fling a strong gust of energetic, charged particles away from the Sun and into interplanetary space.

Most often, these so-called coronal mass ejections and the solar energetic particles that they carry are launched by the Sun in a direction other than toward Earth.

- But in some cases, by chance, the solar energetic particles are directed at Earth. After a short travel time from the Sun to Earth, those particles impact the Earth's magnetic field—which is good news for us because these particles, which would otherwise directly impinge on the Earth's surface and cause biological damage, are instead deflected.

- A charged particle in the presence of a magnetic field cannot move across a magnetic field but can effortlessly move along a magnetic field line, like a bead on a string. So, the incoming solar particles ram into the Earth's magnetic field and then free-fall slide along the magnetic field toward the two points from which the Earth's magnetic field emanates—the north and south poles.

© Joe Rainbow/iStock/Thinkstock.

The aurora borealis is found in the Northern Hemisphere of Earth.

- This is where we see the aurorae, as those energetic particles, sliding along the Earth's magnetic field, at last impinge on the Earth's atmosphere, heating it, and causing the atoms in the air to fluoresce. The same basic process occurs on the other planets in our solar system, too.

- Earth's magnetic field provides some protection from otherwise damaging energetic radiation. However, for astronauts in space, such gusts of space weather pose a very real danger. For example, the class of very energetic solar flares that occur approximately once per year would cause a level of radiation poisoning to an astronaut on the Moon sufficient to cause death. Even airlines monitor the occurrence of such solar storms in order to modify flight plans to protect their crews if needed.

Impacts of Solar Storms
- One of the most extensively studied impacts of solar storms on biological life is how they affect the navigational abilities of some animals. For example, homing pigeons have been shown to have their navigational abilities disrupted, as have dolphins and potentially whales as well.

- It has been suggested that this may be caused by minerals in the animals' heads or beaks being triggered by geomagnetic currents generated by the solar storm. However, more recent research suggests that the sensory effect is caused by some other mechanism that has yet to be identified.

- Another major impact is on our communications systems, which increasingly rely on satellites in Earth's orbit to relay and transmit long-distance and cellular signals. Direct damage to the electronics aboard these satellites is an obvious concern.

- Another concern is the effect of the changing height of the Earth's atmosphere. When a strong shower of solar energetic particles strikes the Earth's atmosphere, those energetic particles deposit their energy into the atmosphere, which has the effect of puffing it out a bit. This puffing of the atmosphere increases the amount of drag on the atmosphere that low-altitude communications satellites experience, which can cause them to prematurely spiral in to the atmosphere and begin to burn up.

- The effect of space weather on satellites that we increasingly rely on for navigation—such as GPS satellites—has understandably become a major concern. During particularly severe solar storms, positions from GPS satellites have been known to be off by as much as a few miles.

- If electrical disruptions of telecommunications satellites seem bad, consider what such disturbances might do to an electrical power grid on the ground. Indeed, among the biggest ongoing concerns highlighted in recent research reports is the potential for significant negative impact on major metropolitan electrical distribution systems.

- The Earth's magnetic field moves in response to the impact of energetic particles from a solar storm, and that motion of the Earth's magnetic field generates electricity in any electric circuits that might be present nearby.

- So, those large electric circuits on the ground—our electric grid—have strong electric currents induced in them by the Earth's magnetic field moving in response to the solar storm. These induced currents can be quite strong and, most importantly, can spike unexpectedly.

- Similar concerns apply to the major pipelines that move petroleum across vast stretches. These pipelines, while of course not intended as electrical circuits, can have electrical currents induced in them in the same way that circuits in the electrical grid can. And because these pipelines were never intended to carry electricity, there generally has not been the same degree of safety mechanisms built in.

- Because of all of these concerns, there are now major research efforts underway to improve our ability to predict bad space weather. For example, the National Science Foundation's Center for Integrated Space Weather Modeling led by Boston University has been developing sophisticated computer models that are now being implemented by the National Weather Service.

Chondrules in Meteorites

- Stars do more weathering of their environs when they are young. From X-ray studies of young stars, we now know that the Sun, when it was young, produced 10,000 times harsher space weather than it does now.

- In addition to the extreme weathering driven by powerful gusts from the Sun, there is good evidence that the young Earth was subjected to powerful electrical disturbances directly from the Sun's magnetic field. A number of the meteorites that have been found include clear evidence of having been subjected to lightning that penetrated deep into their interiors even before they encountered any effects of falling through the Earth's atmosphere.

- The evidence for this is the presence of so-called chondrules within the meteorite—small nodules of rock embedded within the meteorite that indicate a portion of the meteorite was flash heated, melting it. Then, it resolidified within the meteorite as distinct chondrule inclusions.

- How would that happen? A young star's magnetic field can dig into its surrounding protoplanetary disk and channel material from the disk onto the star. That same magnetic connection between the star and its disk can channel highly energetic particles from the young star into the disk and onto any small rocky bodies that may be beginning to coalesce into planets.

- As these energetic particles from the young star stream onto these rocks—future meteorites—there can be electrical discharges around and on these rocks. Such electrical discharges are a form of lightning, so this would explain the features that we see in some meteorites that have fallen to Earth with signatures of having been exposed to a kind of lightning in the very distant past, when the solar system and the Earth were first forming.

- Clearly, the proto-Earth was in an extremely hostile environment, at least as far as life as we know it now would be concerned. But that doesn't mean that it was necessarily a bad thing then. In fact, it may have been crucially important to the very emergence of life on Earth.

- The class of meteorites known as carbonaceous chondrites are among the most pristine examples known of the material that comprised the environs of the Sun's protoplanetary disk, within which the Earth was formed. Embedded within these meteorites are various types of organic material.

- For example, more than 70 different amino acids, the building blocks of proteins, have been identified. Laboratory experiments dating back to the 1950s have shown that zapping basic organic compounds with high-energy particles or with lightning can synthesize such amino acids.

- In fact, more recent lab experiments have shown that it is possible for lightning or high-energy X-rays from space weather to stimulate the creation of adenine, which is one of the four types of bases that make up DNA and the genetic code to all of life.

- Ideas have been proposed for harnessing the Sun's weather as a means of propulsion for future space flight. The idea here is to build spaceships that have enormous sails made of Mylar or some other strong but lightweight material. These sails would be pushed by the gusts of wind of particles streaming from the Sun, and the gustier the better as far as getting more propulsion is concerned.

Suggested Reading

Moldwin, *An Introduction to Space Weather*.

Schopf, *Life's Origin*.

Questions to Consider

1. What are the most important ways in which the Sun affects life on Earth, both now and when the Earth was very young?

2. How might space weather potentially help counteract the effects of human-caused climate change? How might it exacerbate the problem?

Solar Storms—The Perils of Life with a Star
Lecture 22—Transcript

The Sun and stars impact their surroundings in ways beyond just the light and heat that they produce. This weathering of their environments can be both harmful and conducive to life. Indeed, as we'll see, the extreme weathering of the environment in which life first evolved on Earth was extremely harsh, not the type of planetary environment that would be conducive to life today. Yet there is evidence that such extreme space weather around the young planet Earth may, in fact, have been necessary for the first life forms to evolve and for the proliferation of life more generally to take hold.

In this lecture, we'll learn about the phenomenon of space weather, specifically as we experience it in our solar system today, and we'll look at the types of sophisticated computer models that space scientists are developing to better improve our ability to predict when dangerous solar storms might occur. We'll also use what we've learned about the space weathering of very young suns to infer what the space weather of our solar system must have been like early on. And we'll use information from the Earth's geologic record to consider what space weather around the Earth might be like in the future.

To begin, let's look at one of the most basic and beautiful manifestations of space weather on Earth, the aurorae. The word aurora, with its plural aurorae, derives from the Latin word for dawn. In Roman mythology, Aurora was also the name given to the goddess of the dawn. But to us, Aurorae are the eerie curtain-like sheets and long streamers of shimmering light that appear near the north and south poles. These wonderful, if strange, apparitions have been noticed by people going all the way back to the most ancient civilizations, particularly those who resided at far northern latitudes.

The aurora borealis occurs above the north pole, and it has a counterpart, the aurora australis, which occurs above the south pole. But with no early civilizations living near Antarctica, only the northern lights, as the aurora borealis is also known, were recorded in ancient times. It was noticed early on that the northern lights did not occur constantly but rather occurred sporadically. And when they occur, they tend to occur strongly, like a gust or storm. You could say that aurorae are just like rainstorms; when it rains it pours.

Today, our basic picture for how the aurorae are produced is as follows. Energetic flares erupt on the Sun's surface, driven by the twisting and kinking of the Sun's magnetic field that we've discussed in previous lectures. Those flares, representing the sudden snap of the Sun's kinked magnetic field and signaled by a burst of X-ray light, fling a strong gust of energetic, charged particles away from the Sun and into inter-planetary space. Most often, these so-called coronal mass ejections and the solar energetic particles that they carry are launched by the Sun in a direction other than toward Earth.

But in some cases, by chance, the solar energetic particles are directed at Earth. After a short travel time from the Sun to Earth, approximately one day, those particles impact the Earth's magnetic field. Note that this travel time of about a day is considerably longer than the mere eight minutes it takes for light from the Sun to reach Earth—the Earth being eight light-minutes away from the Sun. That's because while these energetic particles from the Sun can be flung toward the Earth at relatively high speeds of hundreds to a thousand kilometers per second, that is still a very small fraction of the speed of light. In any case, it's good news for us that these energetic particles impact the Earth's magnetic field when they arrive, because these particles, which would otherwise directly impinge on the Earth's surface and cause biological damage, are instead deflected.

To understand the reason for this deflection, think back to our discussion about magnetism in the Sun and other stars. We said that a charged particle in the presence of a magnetic field cannot move across a magnetic field but can effortlessly move along a magnetic field line, like a bead on a string. So the incoming solar particles ram into the Earth's magnetic field and then free-fall slide along the magnetic field toward the two points from which the Earth's magnetic field emanates, the north and south poles.

And so this is where we see the aurorae, as those energetic particles sliding along the Earth's magnetic field at last impinge upon the Earth's atmosphere heating it and causing the atoms in the air to fluoresce. The same basic process occurs on the other planets in our solar system also. For example, the aurorae on the planet Jupiter have been seen by the Hubble space telescope. The rings of light that you see in this amazing image are the direct manifestation of the northern and southern aurorae on Jupiter, its northern and southern lights, if you will. The rings of light directly reveal where Jupiter's magnetic field enters the Jovian surface. So you see that the planet's magnetic field lines do not come together at a single point at the poles but rather are still somewhat spread out into a circular ring where they come together at the planet's surface.

Notice also that Jupiter's aurorae are very nearly coincident with its north and south poles. You can, in fact detect a small offset appearing as a mild tilt of those auroral rings of light, but it's a small tilt. That's because Jupiter's magnetic field is very nearly aligned with its rotational axis. This is in contrast to the Earth's aurorae, which are, in fact, displaced from the Earth's north and south poles by approximately 20 degrees in latitude, resulting from the fact that the Earth's magnetic field has its magnetic poles offset by a fairly large amount, about 20 degrees, from the rotational poles of the Earth. It's that offset of the Earth's magnetic poles that in fact allow us to see the northern lights even from the northern United States, the Earth's north magnetic pole being somewhere over Canada, instead of directly over the north pole. In any event, the aurorae of Jupiter are a direct confirmation that the effects of the Sun's weather are not restricted to just our own planet.

Coming back to Earth, our planet's magnetic field provides some protection from otherwise damaging energetic radiation. But for astronauts in space, such gusts of space weather pose a very real danger. For example, the class of very energetic solar flares that occur approximately once per year, would cause a level of radiation poisoning to an astronaut on the moon sufficient to cause death. Even airlines monitor the occurrence of such solar storms in order to modify flight plans to protect their crews if needed.

So, let's look at some of the other ways in which geomagnetic storms, triggered by energetic solar events, can impact life and civilization here on Earth. To put this discussion in perspective, let's first look at some of the most memorable events that have been recorded historically. One of the most powerful space weather events ever recorded was on September 1st, 1859, named the Carrington Event, after the pioneering solar physicist,

Richard Carrington, who spent his life studying the Sun. This event caused a disruption to worldwide telegraph communications. It is reported that some telegraph operators received electrical shocks from their equipment; and some short-circuited equipment started fires. Aurorae associated with this storm were seen by people as far south as Mexico. Such an extreme event is a once in 500-year occurrence on Earth. In 1989, a storm occurred with sufficient power to temporarily knock out a Canadian power grid, leaving 6 million people without power for 9 hours. Indeed, as we'll see, disruptions to power grids are one of the most important types of modern dangers from such powerful storms.

A particularly famous storm occurred on July 14, 2000, the so-called Bastille Day storm. A powerful flare on the Sun launched a coronal mass ejection right at Earth, which impacted the Earth's atmosphere a day later. Fortunately there were no major power grid disruptions reported. But the Voyager satellites, then already at the outer reaches of the solar system, recorded that gust from even that great distance, showing that these powerful storms can be solar-system wide phenomena.

Finally, between October 29 and November 4 of 2003, the so-called Halloween storm occurred, instigated by the most intense X-ray flare recorded from the Sun since the era of space-borne X-ray solar detectors in Earth orbit. The GOES satellite network—GOES stands for Geostationary Operational Environmental Satellite—has been recording solar X-ray flares since the 1970s. The purpose of these satellites is to provide nearly continuous monitoring of weather on Earth and its various causes, including monitoring the Sun at X-ray wavelengths, especially for the purpose of detecting magnetic flares and understanding how often these events occur and ultimately what their relationship is to storms impacting the Earth.

The GOES satellites use a system of classification based on the intensity of the X-rays emitted by the flare. The weakest flares are A class; then B, which are 10 times stronger; then C, which are 10 times stronger still. After that are the M-class flares, 10 times stronger than C class; finally are the X class, 10 times more powerful than M class, or 10,000 times stronger than A class. An X10 is ten times stronger still. The Bastille Day flare I mentioned a moment ago was an X6. The Halloween event in 2003, the GOES satellite recorded the X-ray flare as a class X38!

One of the most extensively studied impacts of solar storms on biological life on Earth is the navigational abilities of some animals. For example, homing pigeons have been shown to have their navigational abilities disrupted, as have dolphins and potentially whales as well. It has been suggested that this may be caused by minerals in the animals' heads or beaks being triggered by geomagnetic currents generated by the solar storm. But more recent research suggests that the sensory effect is caused by some other as-yet unidentified mechanism.

Another major impact is on our communications systems, which increasingly rely on satellites in Earth orbit to relay and transmit long-distance and cellular signals. Direct damage to the electronics aboard these satellites is an obvious concern. Another, less obvious, concern is the effect of the changing height of the Earth's atmosphere. You see, when a strong shower of solar energetic particles strikes the Earth's atmosphere, those energetic particles deposit their energy into the atmosphere, which has the effect of puffing it out a bit. This puffing of the atmosphere increases the amount of drag on the atmosphere that low-altitude communications satellites experience, which can cause them to prematurely spiral in to the atmosphere and begin to burn up.

In fact, damage to satellites due to the increased height of the Earth's atmosphere is a very general issue for all types of orbiting hardware. The orbiting scientific platform, Skylab, went into premature de-orbit in 1979 as a result of solar storms puffing up the atmosphere. During a particularly strong solar storm in 1989, four of the U.S. Navy's navigational satellites had to be taken out of service for a week until new orbit-boosting commands could be uploaded to the satellites. And perhaps ironically, a satellite called the Solar Maximum Mission, intended to study activity on the Sun, was caused to fall out of orbit that same year.

The effect of space weather on satellites that we increasingly rely upon for navigation, such as GPS satellites, has understandably become a major concern. During particularly severe solar storms, positions from GPS satellites have been known to be off by as much as a few miles. One of the ways in which navigation systems are specifically affected is through what is known as scintillation. This is the same effect of the Earth's atmosphere that causes stars to appear to twinkle. Basically, as starlight reaches the Earth's atmosphere on its way to the ground, it passes through pockets of air of different temperatures and density. Each of those pockets of air deflects the light slightly, similar to the way a spoon in a glass of water appears bent at the point where it enters the water. Those multiple little deflections of the light along its path causes the star to appear to shift ever so slightly this way and that as seen from the ground, and this gives rise to the shimmer, or twinkle, of the starlight.

The same thing happens with the light signals from GPS satellites. Scintillation at the wavelengths that GPS operates at increases during strong solar storms. This makes the GPS signals twinkle, which in turn produces less accurate information. So it has become increasingly important to develop alert mechanisms to let pilots and other navigators know about impending solar storms. And obviously, scientists and engineers are working hard to find ways to improve the robustness of GPS and other navigation systems.

For these reasons, research into radiation hardening of microchips has become a big and important research area for telecommunications, navigation, and national security. There are several issues specific to the harsh space weather environment of Earth's orbit that have to be dealt with. One issue is that, as computer microchips become smaller and smaller, computers onboard satellites become susceptible to impacts by individual highly energetic solar particles that can briefly short circuit these tiny circuits and even cause the satellite to misinterpret a command from its onboard programming. Another issue is that different parts of a satellite can become electrically charged by different amounts—one part more electrically charged, another less so. And if this differential charge builds up sufficiently, internal lightning can occur as the built-up electricity discharges. As you can imagine, lightning within a satellite is not a good thing.

If lightning within a telecommunications satellite sounds bad, consider what such an electrical disruption might do to an electrical power grid on the ground. Indeed, among the biggest ongoing concerns highlighted in recent research reports is the potential for significant negative impact on major metropolitan electrical distribution systems. Some studies have estimated that a solar storm as severe as the 1859 event that we discussed earlier could cause several trillion dollars' worth of damage to millions of electrical transformers and knock out power for over 100 million people.

How does that happen? It's essentially a chain reaction involving magnets and electric currents. Recall that the Sun's own magnetism is generated through a dynamo process in which electrical currents within the Sun, together with the Sun's rotation, create the Sun's magnetic field. Those magnetic fields get twisted and kinked, occasionally erupting, producing the coronal mass ejections that can impinge on the Earth as a solar storm. Those solar energetic particles impact the Earth's magnetic field, jostling it about.

Now, down on the ground, we have large electrical circuits covering the globe. What do I mean? I mean, our entire network of wires that comprise our electrical distribution system. These are, essentially, enormous loops of wire that run from large electricity generators to our homes and businesses and back again in big circuits. Now, recall in our discussion of magnetism on the Sun that the way the Sun's magnetism is generated is through a dynamo, in which electric currents in the Sun, moving as a result of the Sun's rotation, create the Sun's magnetic field. And that's a consequence of Maxwell's laws of electromagnetism, which link electricity and magnetism into a unified force in nature, such that changes in one cause changes in the other. Motion of the electrical currents in the Sun causes the creation of its magnetic field.

Well, the same thing happens in a magnetic storm on Earth, only the other way around. The Earth's magnetic field moves in response to the impact of energetic particles from a solar storm. And that motion of the Earth's magnetic field generates electricity in any electric currents that might be present nearby. So those large electric circuits on the ground—our electric grid—have strong electric currents induced in them by the Earth's magnetic field moving in response to the solar storm. These induced currents can be quite strong and, most importantly, can spike unexpectedly. The storm-induced currents are also of a direct-current, or a DC nature, like the current from a battery, whereas the electrical grid is based on alternating current, or

AC. A very strong DC current in an AC circuit system can fry generators and transformers.

Similar concerns apply to the major pipelines that move petroleum across vast stretches. These pipelines, while of course not intended as electrical circuits, can have electrical currents induced in them in the same way that circuits in the electrical grid can. And because these pipelines were never intended to carry electricity, there generally have not been the same degree of safety mechanisms built in. Here the concerns are not as severe as with the electrical grid. The primary concerns have to do with accuracy of flow meters and of increased corrosion to the pipeline. But the point is that major space weather can negatively impact the built world around us in significant and multi-faceted ways.

Because of all these concerns, there are now major research efforts underway to improve our ability to predict bad space weather. For example, the National Science Foundation's Center for Integrated Space Weather Modeling (CISM) led by Boston University has been developing sophisticated computer models that are now being implemented by the National Weather Service. In these computer models, the Sun's magnetic field is linked to the Earth's magnetic field, representing an interplanetary conduit, whereby solar storms travel to and interact with the Earth. By integrating these space weather transport models with observations of eruptions on the Sun, the severity of the storm at Earth, and even its time of arrival and localities of strongest impact, can be predicted. Over the coming decade you can expect to hear more about the emergence of space weather meteorology becoming an ever-more important component of weather forecasting.

If you think space weather sounds bad now, it's nothing compared to what it was like when the solar system was young. Stars do more weathering of their environs when they are young. From X-ray studies of young stars, such as in the Orion Nebula stellar nursery, we now know that the sun, when it was young, produced 10,000 times harsher space weather than it does now.

Remember the classification system we discussed earlier in this lecture, by which the power of X-ray flares on the Sun are classified; A-type flares are the weakest, and the strongest recorded was an X38? Well, Sun-like stars in the Orion Nebula stellar nursery have X-ray flares that, if observed from the Earth-Sun distance, would be classified as X10,000! So when the Earth was very young, it was bombarded with solar storms and subjected to a magnitude of space weather that is difficult to fathom today.

In addition to the extreme weathering driven by powerful gusts from the Sun, there is good evidence that the young Earth was subjected to powerful electrical disturbances directly from the Sun's magnetic field. A number of the meteorites that have been found include clear evidence of having been subjected to lightning that penetrated deep into their interiors even before they encountered any effects of falling through the Earth's atmosphere. The evidence for this is the presence of so-called chondrules within the meteorite, small nodules of rock embedded within the meteorite that indicate a portion of the meteorite was flash heated, melting it, and then it re-solidified within the meteorite as the distinct chondrule inclusions that we see.

How would that happen? Well, remember in a previous lecture when we talked about newly formed stars and the ways in which they feed from their protoplanetary disks of gas and dust, we talked about how the young star's magnetic field can dig in to its surrounding protoplanetary disk and channel material from the disk onto the star. Well that same magnetic connection between the star and its disk can channel highly energetic particles from the young star into the disk and onto any small rocky bodies that may be beginning to coalesce into planets. As these energetic particles from the young star stream onto these rocks—future meteorites—there can be electrical discharges around and on these rocks. Such electrical discharges are a form of lightning, and so this would explain the features that we see in some meteorites that have fallen to Earth with signatures of having been exposed to a kind of lightning in the very distant past, when the solar system and the Earth itself was first forming.

So, clearly the proto-Earth was in an extremely hostile environment, at least as far as life as we know it now would be concerned. But that doesn't mean that it was necessarily a bad thing then. In fact, it may have been crucially important to the very emergence of life on Earth. The class of meteorites known as carbonaceous chondrites are among the most pristine examples known of the material that comprised the environs of the Sun's protoplanetary disk within which the Earth was formed. Embedded within these meteorites are various types of organic material. For example, more than 70 different amino acids, the building blocks of proteins, have been identified. Laboratory experiments dating back to the 1950s have shown that zapping basic organic compounds with high energy particles or with lightning can synthesize such amino acids.

In fact, more recent lab experiments have shown that it is possible for lightning or high-energy X-rays from space weather to stimulate the creation of adenine, which you may recognize as one of the four types of bases that make up DNA and the genetic code to all of life.

Space weather has also influenced life on Earth in other ways related to global, geological, and meteorological changes over time. For example, measurements of the number of sunspots on the Sun over time have found two important trends. First, the number of sunspots undergoes a regular cycle every 11 years, during which time the number and intensity of solar storms also varies. In other words, the 11-year sunspot cycle corresponds to an 11-year storm cycle.

Second, long-term sunspot records show that there was an unusual, extended period of time from about 1645 to about 1715 called the Maunder Minimum when the Sun was virtually devoid of sunspots for 50 years or so. Weather records on Earth show that this same period saw lower-than-average temperatures around the world, sometimes referred to as the Little Ice Age. This goes to show that solar weather and weather on Earth have an intimate relationship. As we enter an era during which human-caused climate change is occurring at an accelerating pace, it will be increasingly important for us to understand the ways in which the Sun might change again, and how changes in space weather may bode well or ill for weather and life on Earth.

On a more hopeful, if perhaps fantastical, note, there have been ideas proposed for harnessing the Sun's weather as a means of propulsion for future space flight. The idea here is to build spaceships that have enormous sails made of Mylar or some other strong but lightweight material. These sails would be pushed by the gusts of wind of particles streaming from the Sun, and the gustier the better, as far as getting more propulsion is concerned.

How fantastic to imagine that, if our planetary home should ever become less comfortably habitable due to climatological changes or other reasons, that we might seek to venture to another planetary home by way of a clipper-ship-like vehicle, pushed along by the wind—the solar wind!

In this lecture, we've seen an entirely new perspective on the connection between us and the stars. We live with a star, the Sun. Our star governs the space weather of its environs, and in turn exerts a direct influence over weather on Earth, the health of our astronauts in space, and the durability of the satellite platforms upon which our technologically built world increasingly depends. Yet as we've seen, though the Sun's weathering of our planet may at times seem harsh now, this is nothing compared to the intense weathering that the Sun exerted on the proto-Earth, a weathering that may, in fact, have been an important part of how the building blocks for life as we know it came to be.

The Stellar Recipe of Life

Lecture 23

What is the recipe of life? The most important ingredients—the flour, salt, baking powder, and yeast—are carbon, nitrogen, oxygen, and hydrogen. From our pantry, we also need a handful of other ingredients in smaller amounts, but these are nonetheless crucial to the success of the recipe: calcium, sodium, sulfur, chlorine, phosphorus, potassium, and iron. Finally, we need the spice rack—smidgens of trace elements that are the spice of life. These elements are copper, zinc, selenium, and cobalt.

The Main Ingredients

- Imagine that we are cosmic chefs and that the recipe we're preparing is the recipe of life. What ingredients does the recipe call for? In our pantry, we have the major elements, and in our spice rack, we have the trace elements. Our kitchen is the galaxy, with tools such as blenders in the form of supernovae that stir up the batter. Our oven is a star. And gravity is the cake form, holding the ingredients together while they bake.

- The ingredients that we would mainly require are hydrogen, carbon, nitrogen, oxygen, calcium, sodium, sulfur, chlorine, phosphorus, potassium, and iron. These elements make up almost everything around us and are the bulk elements of our bodies. Sure enough, aside from hydrogen provided by the creation of the universe itself, these are the elements that stars "cook" during their lives.

- All living things on Earth have just 4 elements in common: carbon, nitrogen, oxygen, and the hydrogen that bonds to them. Ignoring helium, those 4 elements are the 4 most abundant elements in the universe. Hydrogen is about 75% of all matter, a direct inheritance from the big bang.

- Carbon is made by all stars, including massive stars as one stage of fusion, and by lower-mass stars as the final stage of fusion before their deaths. So, no wonder it's so abundant. It's the one element, helium aside, that every star in the universe will manufacture. Carbon is the flour in our pantry, so it makes sense that just about anything we might bake will not only involve it, but also use it as a base ingredient.

- Nitrogen and oxygen are the first elements after carbon to be fused by massive stars, and they are also the elements that, after carbon, provide stars with the most oomph in the form of additional fusion energy. So, it's not surprising that these are the next most abundant elements in the universe. These are the sugar and salt in our pantry, and it's no wonder that anything we might bake will include these ingredients also in good measure.

Other Crucial Ingredients

- Next in our pantry are the elements calcium, sodium, sulfur, chlorine, phosphorus, potassium, and iron. Calcium is an essential component in human bones and teeth. So, if you're going to cook up a recipe of life, calcium is the natural choice as a leavening agent. Every atom of calcium was produced in massive stars through the fusion of sulfur and 2 helium atoms.

- We immediately associate sodium with salt, so let's think of it as the salt in our cosmic pantry. This is an ingredient that we're definitely going to need as a foundation for the flavor of our recipe for life, but we don't want to overdo it. Every atom of sodium in the universe was made in high-mass stars through the fusion of 2 carbon atoms. In our bodies, sodium is an essential nutrient that regulates blood volume, blood pressure, fluid equilibrium, and pH balance.

- Phosphorus is, after calcium, the next most abundant element in the human body. Most of it is just locked up in our bones and teeth, where it helps form the protective enamel layer, but a tiny bit of it is so essential that life as we know it simply would not exist or function without it. This is the baking soda in our cosmic pantry.

which make up our lifeblood as well as the most massive and substantial products of human ingenuity.

The Spice Rack

- If we've just gone through the pantry of our cosmic kitchen, representing the most important and foundational ingredients of life, we'll also want to make sure that we have the right spices in our spice rack. These are the trace materials that, used in pinches, make up a tiny but nonetheless crucial part of the overall recipe.

- Indeed, life depends on smidgens of certain rare elements: copper, zinc, selenium, cobalt. These elements show up in our hair and blood. These are the elements that stars forge in their fiery deaths.

- The first element in our spice rack is copper. In our bodies, just a trace amount of the stuff—about 2 parts per million—is crucial for proper iron uptake. And because iron is so important in our blood, copper deficiency can in turn lead to anemia-like symptoms, bone abnormalities, osteoporosis, and impaired metabolism.

- Another essential trace element in all animals is zinc, which is a component of hundreds of different enzymes. There are just a few grams of zinc distributed throughout the adult human body. Most of it is found in the brain, muscles, bones, kidney, and liver, with the highest concentrations in the prostate. There may be no other element produced in the fiery supernova deaths of stars that is so intimately connected with the conception of new human life.

- In addition to copper and zinc, the element cobalt is essential to all animals. It is a key constituent of cobalamin, also known as vitamin B12, which is essential to the health of our nervous system. It is made by bacteria in the guts of ruminant animals, which convert cobalt-based salts into B12.

- The final element in our metaphorical spice rack is selenium, which makes up less than 0.01% of our body mass but plays an important role in the functioning of the thyroid gland and in every cell that uses thyroid hormones.

Stocking Our Cosmic Kitchen

- Where do we get the ingredients for our cosmic pantry and spice rack? Hydrogen is essentially free, having been provided by the big bang. However, every bit of carbon requires a star to make it. Stars like our Sun make it as red giants shortly before their deaths. And nitrogen and oxygen require massive stars to fuse them from carbon and helium atoms. These are the main elements that stars make, so our pantry has them in relative abundance.

- Every single atom of calcium, sodium, sulfur, chlorine, phosphorus, potassium, and iron comes to us from the later stages of fusion as massive stars flail to keep themselves alive in their dying breaths. The stars don't get much bang for the buck with these stages of fusion, so they produce them in relatively smaller abundances than carbon, nitrogen, and oxygen.

- As for the spices, these rare elements can only be obtained from very special places—the fiery supernovae that are the violent deaths of massive stars. These elements are flash forged in just an instant, and very little of them is produced. Indeed, these are rare spices, but without their seasoning, the recipe of life will fail.

- The stars, both in their lives and in their deaths, make every ingredient in the recipe of life. Through their deaths, they blend these ingredients together throughout the galaxy. Through gravity, they gather up these blended ingredients. And in the course of their struggle to live, they bake that batter to just the right temperature.

- We are made of the stars: We breathe them, we consume them, they course through our veins, and we tingle with them. We live on a world made of them, and we move through a material world built of them. In our contemplations of the stars, in our comprehension of them—indeed, in our very sentience—we are the stars contemplating and comprehending themselves.

Suggested Reading

Darling, *Life Everywhere*.

Gray, *The Elements*.

Questions to Consider

1. If you had to choose, which one element would you regard as the single most important element in the universe, from the standpoint of the "recipe of life"?

2. What other elements might form the basis for life? Can you think of scenarios in which these less abundant elements might be preferred as a basis for life?

The Stellar Recipe of Life
Lecture 23—Transcript

In this course, we've discussed the lives of stars in terms of a life cycle, from birth to death and to new stellar birth. One of our consistent themes has been that, through their lives, stars perform a service of alchemy, whereby the material of the universe, initially simply hydrogen and helium, is steadily enriched with the elemental products of the previous generations of stars. What was nothing more than hydrogen and helium, increasingly becomes other, heavier elements—carbon and oxygen, silicon and iron, indeed, all of the elements of the periodic table. From this chemical perspective, we gain an appreciation for the life cycles of stars as living entities with a kind of cosmic purpose, a higher calling to create the very substance of life and of the material world around us.

In this lecture, we'll be looking at the periodic table of elements with fresh eyes. We'll look specifically at the elements that stars produce as those elements relate to us. Some of these are obvious, such as carbon as the basis for all life and oxygen as the stuff we breathe. Others will be less familiar, such as the vital importance of potassium neurologically and the surprising importance of various trace elements. My hope is that through this tour of the elemental products of the stellar life cycle, you'll forever rethink the world around you at a granular level, specifically in reference to what stars make. When you see the world this way, then your—our—intimate connection to one another as inheritors of the chemical legacy of the stars becomes clear.

So let's start by looking at the periodic table, and let's do a simple division of the table into four broad groups of elements based on the stars. The first group of elements would be the first five: Hydrogen, helium, lithium, beryllium, and boron. These are the elements that came to us from the big bang. In fact, most of what the big bang delivered were just the first two of those five elements—hydrogen and helium—in a proportion of roughly 75% and 25%, with just a smidgen of the other three elements.

Consequently, hydrogen and helium are the most abundant elements in the universe, and in all stars appear in approximately this same proportion. Stars don't make lithium, beryllium, or boron, because making them from helium would actually represent a loss of energy for stars. Recall from a previous lecture that we can use the so-called binding energy curve to see why this is. Fusing heavier elements from lighter ones generally produces energy in the process, but only if the product of the fusion is higher up on the binding energy curve than the elements that went in to the fusion. And lithium, beryllium, and boron are lower on the binding energy curve than helium.

The second major group of elements is actually just one, carbon. This is the element made by relatively low-mass stars, stars like the Sun, when they near the ends of their lives to become red-giant stars. This is as far through the periodic table as a low-mass star like the Sun will get before ending its life as a planetary nebula and leaving behind a white dwarf corpse.

The third major group of elements includes the elements from nitrogen through iron. These are the elements that, in addition to carbon, high-mass stars produce through their lives as they proceed through successive rounds of fusion, creating successively heavier elements from the lighter ones, as they more and more desperately try to extract additional energy to hold themselves up against gravity. Remember that the heaviest element a massive star may produce before it finally dies in a supernova explosion is iron, because that is the heaviest element that may be fused from lighter ones and produce some amount of energy for the star.

The final major group of elements is everything heavier than iron, from cobalt—element 27—all the way to livermorium—element 116—and perhaps some even heavier elements only now being discovered in laboratories. In fact, the discovery of element 117 has been claimed, and since it was discovered by my colleagues at Vanderbilt University and at Oak Ridge National Laboratory, both in Tennessee, it could potentially be called tennessium. These are the elements that cannot be forged by the stars during their lives. However, in the fiery deaths of massive stars as supernovae, releasing all at once more energy than all of the stars in the galaxy shining together, an abundance of free energy permits these energy-costly elements to be forged in a flash.

By the way, even though we've included the noble gases in these groupings, in fact, we'll mostly ignore those elements here. That's because the noble gases, also known as inert gases, cannot form compounds. Those elements always exist as isolated elements even when in the presence of other elements or compounds. For example, in Earth's air, the element argon makes up about 1%. But it doesn't do anything; it doesn't combine with anything. So in terms of an imaginary chef cooking up the recipe of life, elements like argon might be thought of as bulking agents that carry no functional or nutritional value and that contribute no flavor.

So with this basic grouping of the elements, let's now imagine ourselves as a cosmic chef. The recipe we're preparing is the recipe of life itself. What ingredients does the recipe call for? In our pantry we have the major elements, and in our spice rack we have the trace elements. Our kitchen is the galaxy, with tools such as blenders in the form of supernovae that stir up the batter. Our oven is a star. And gravity is the cake form, holding the ingredients together while they bake.

How do we make the best use of the ingredients we've got? We want the bulk of our recipe to involve staples that we have lots of, the equivalents of things like flour and sugar and salt. But we also want to give this recipe some zest and nuance and sublime flavor. So we'll want to make sure to include dollops of this and to sprinkle in pinches of that. This is going to be a concoction unlike any other, and it has to be just right.

So what are the ingredients that, as a chef, we would mainly require? What's in our pantry? Those would be hydrogen, carbon, nitrogen, oxygen, calcium, sodium, sulfur, chlorine, phosphorus, potassium, and iron. These elements make up almost everything around us and are the bulk elements of our bodies. And sure enough, aside from hydrogen provided by the creation of the universe itself, these are the elements in those two major groups of elements that stars cook during their lives. Let's get to know these elements.

All living things on Earth have just four elements in common. They are, carbon, nitrogen, and oxygen, and the hydrogen that bonds to them. And guess what, ignoring helium, being one of those inert bulking agents, those four elements are the four most abundant elements in the universe. Hydrogen is about 75% of all matter, a direct inheritance from the big bang. Carbon is made by all stars, including massive stars as one stage of fusion, and by lower mass stars as the final stage of fusion before their deaths. So no wonder it's so abundant. It's the one element, helium aside, that every star in the universe will manufacture. Carbon is the flour in our pantry. And so it makes sense that just about anything we might bake will not only involve it, but use it as a base ingredient.

Nitrogen and oxygen are the first elements after carbon to be fused by massive stars, and they are also the elements that, after carbon, provide stars with the most oomph in the form of additional fusion energy. So it's not surprising that these are the next most abundant elements in the universe. These are the sugar and salt in our pantry, and it's no wonder that anything we might bake will include these ingredients, also in good measure.

Carbon is the backbone of all life. From the spiral lattice of DNA to the basic structure of proteins, carbon is the base, the foundation. Every cell of every living thing has it. It is in a very real, physical, literal sense what you and I are made of. When plants perform the magic of photosynthesis to grow the biomass of their bodies, what they are doing is taking in carbon dioxide from the air, and using the energy of sunlight to incorporate the carbon from that carbon dioxide into their bodies and to return the oxygen back to the air. When animals consume those plants, they are incorporating the carbon that the plants took in from the air. Then, when carnivores consume those animals, they are, in turn, taking in that same carbon and incorporating it into their bodies. Chemically speaking, we are not that different from a lump of coal.

What is an enormous sprawling oak made of? It's made of carbon, carbon that the tree has sucked in from the air and has bound to itself with the energy of sunlight. Burn that tree, and the fire produced is that same sunlight energy pouring back out. Those carbon atoms, every last one of them, came from a long-dead star, sprinkling its ashes into the cosmos. Those dead stellar ashes now make up at the atomic level the fiber of our beings, stellar ashes held together with the energy of sunlight.

But if carbon is the foundation of life, oxygen is what propels life. Oxygen has an uncanny ability to react with just about any organic compound, meaning any compound containing carbon. And these reactions are the essential functions of life, including respiration, which as you probably know involves the taking in of oxygen and the exhalation of carbon dioxide—oxygen bound to carbon.

But what about nitrogen? It is made in stars through the fusion of carbon and hydrogen. By the way, do you remember how that works? To remind you, fusion involves sticking together atoms of light elements to create new, heavier elements. And those heavier elements weigh slightly less than the sum of the parts that went into making them, a little bit of the mass having been converted into lots of pure energy, through the power of Einstein's $E = mc^2$.

Returning to nitrogen, you may feel less connected to this essential element, but in fact, it's everywhere. Perhaps the most common and important form of nitrogen is in ammonia, that's one nitrogen bound to three hydrogens, which is a main ingredient in fertilizer. Prior to the large-scale production of ammonia-based fertilizers in the 1940s, farms routinely rotated out crops, like corn, for crops like alfalfa or beans that could return nitrogen to the soil, as needed by those important cash crops. But with the invention of ammonia-based fertilizers, the ability of farming, and in particular, the continual large-scale production of corn and grains, has revolutionized agriculture. So while our bodies don't make use of nitrogen directly, it is central to many of the processes by which our most basic carbon-based foods grow.

So we've looked at hydrogen, carbon, oxygen, and nitrogen. Next in our pantry are the elements calcium, sodium, sulfur, chlorine, phosphorus, potassium, and iron. We'll start with calcium, because it is actually the next most abundant element on the Earth's surface. And you probably already associate that element with being an essential component in human bones and teeth. Calcium is the stuff that helps us stand up, to push with our joints against Earth's gravity pulling us down, so that we might move through the world and regard the cosmos upright. So if you're going to cook up a recipe of life, calcium is the natural choice as a leavening agent. Every atom of calcium was produced in massive stars through the fusion of sulfur and two heliums. By the way, calcium also plays a very important function in signaling within our cells, helping to regulate the action of nerves and muscles. That signaling is so important, in fact, that the body will sooner dissolve the bones than permit calcium levels in the blood to drop too low.

So, indeed, that white, chalky substance scratching out formulae such as $E = mc^2$ on a chalkboard is a highly essential ingredient in our recipe of life. By the way, calcium, together with carbon and oxygen, in the form of calcium carbonate, is also the basis for cement. So you could say that while calcium helps to form our backbones, it also forms the backbone, if you will, of our built-up material world.

Let's turn to sodium. We immediately associate sodium with salt, so let's just go ahead and think of it as the salt in our cosmic pantry. This is an ingredient that we're definitely going to need as a foundation for the flavor of our recipe for life, but we don't want to overdo it. Every atom of sodium in the universe was made in high-mass stars through the fusion of two carbon atoms. In our bodies, sodium is an essential nutrient that regulates blood volume, blood pressure, fluid equilibrium, and pH balance.

The human body must have at least 500 milligrams of sodium per day to function, with a recommended amount of two to three grams per day. It's no wonder the stuff tastes so good to us. Sodium is also a critically important component of neuron function in our brains. Hyponatremia is a condition in which sodium levels in the blood drop too low, and one of the main symptoms is confusion and disorientation. Without sodium, we cannot think straight, let alone ponder where we come from. We need stellar ash in our brains to contemplate the stars.

Phosphorus is, after calcium, the next most abundant element in the human body. Most of it is just locked up in our bones and teeth, where it helps form the protective enamel layer. But a tiny bit of it, about one-tenth of 1% of our body weight, is so essential, that life as we know it simply would not exist or function without it. This is the baking soda in our cosmic pantry.

Phosphorus is required by all known forms of life. It plays a major role in biological molecules, like DNA and RNA, literally forming a part of their structural backbone. Importantly, phosphorus is also the key component of a molecule called adenine triphosphate, or ATP, in which a group of three phosphorus atoms provide that molecule's key features and functions. You probably recognize ATP from what you've learned about how biological processes use and store food for energy. Nearly every cellular process that uses energy obtains it in the form of ATP. Phosphorus is also one of the key components of all cellular membranes, which separate and protect the cell from its surroundings. Every atom of phosphorus that exists was made in massive stars through the fusion of silicon and helium.

Potassium is the next most abundant element in the body. Massive stars make potassium through the fusion of two oxygen atoms; every atom of potassium in the universe was made this way. In our bodies, potassium is crucial for proper nerve transmission. Without enough potassium, fingers freeze up, and eventually cardiac arrest follows, because the heart's beating function is regulated by this important element.

And what about sulfur? It is made in massive stars through the fusion of silicon and helium. Because of its nasty smell, it has through history been associated with hell-fire burning. But, smelly though it may be, we cannot live without it. Sulfur is an essential component of all living cells. It is present in all proteins and enzymes that contain certain essential amino acids. As nasty as sulfur may smell, it does not compare to chlorine in pure nastiness, at least in its pure gaseous form. Pure chlorine gas is about as horrific a poisonous substance as you could imagine. A whiff of pure chlorine gas is like a blast of a blowtorch to the sinuses and lungs, and breathing more than just a small amount can cause instant death. But together with sodium, chlorine forms regular table salt—tasty! And in our bodies, chlorine serves an important role in nerve signaling and digestion.

Finally in this group of elements, the main ingredients in our cosmic pantry, we have iron. Iron is the heaviest element that massive stars produce before they explode as supernovae. So you might say that stars barely are able to produce iron. But it's a good thing they do. Iron is abundant in biology. Iron-based proteins are found in all living organisms, from the evolutionarily most primitive, right up to us humans. The color of our blood is due to hemoglobin, an iron-containing protein. And iron together with sulfur is found pervasively throughout our bodies, especially in enzymes responsible for many cellular functions, perhaps most importantly, the transport of oxygen in blood to cells throughout our bodies. In fact, the major theories of evolution invoke these iron sulfides as a fundamental driver of the evolution of life on Earth.

Beyond its role in our bodies, coursing through our very veins, iron is, of course, truly foundational to our built modern world. From rebar in concrete, to massive I-beams in skyscrapers, iron represents all that is substantial in our material world. Every atom of iron in the universe was smelted in the crucibles that are the cores of massive stars in the final, dying gasps of their lives. Iron atoms, sprinkled through the galaxy by supernova explosions spreading these dead stars' material up through galactic geysers and across the galaxy, these stellar ashes make up our life-blood and make up the most massive and substantial products of human ingenuity.

If we've just gone through the pantry of our cosmic kitchen, representing the most important and foundational ingredients of life, we'll also want to make sure we've got the right spices in our spice rack. These are the trace materials that, used in pinches, make up a tiny but nonetheless crucial part of the overall recipe. Indeed, life depends on smidgens of certain rare elements: copper, zinc, selenium, cobalt. These elements show up in our hair and blood. These are the elements that stars forge in their fiery deaths. Let's get to know these elements, these spices. We'll sample just a little bit of each one with the tips of our tongues, and sprinkle just a dash of each into our cosmic recipe.

The first element in our spice rack is copper. In our bodies, just a trace amount of the stuff—about two parts per million—is crucial for proper iron uptake. And because iron is so important in our blood, copper deficiency can, in turn, lead to anemia-like symptoms, bone abnormalities, osteoporosis, and impaired metabolism.

Another essential trace element in all animals is zinc. When your daily multi-vitamin refers to being complete from A to Z, that Z is zinc. Zinc is a component of hundreds of different enzymes. There are just a few grams of zinc distributed throughout the adult human body. Most of it is found in the brain, muscles, bones, kidney, and liver, with the highest concentrations in the prostate. Semen, the base of human sperm, is particularly rich in zinc, as zinc plays a key role in the proper function of the prostate gland and in the development of our reproductive organs. There may be no other element produced in the fiery supernova deaths of stars that is so intimately connected with the conception of new human life.

In addition to copper and zinc the element cobalt is essential to all animals. It is a key constituent of cobalamin, which also known as vitamin B12, which you'll probably recognize as one of the major vitamins you may take daily. Vitamin B12 is essential to the health of our nervous system, and it is made by bacteria in the guts of ruminant animals, which convert cobalt-based salts into B12. So it's crucially important to have a certain amount of cobalt in the soil where these animals graze, so that they can make the B12 that is in turn so essential to us.

The final element in our metaphorical spice rack is selenium. Selenium makes up less than 0.01% of our body mass, but it plays an important role in the functioning of the thyroid gland and in every cell that uses thyroid hormone.

So what is the recipe of life? What are the ingredients that we'll need to mix into our cake mix? The most important ingredients, let's call them the flour, salt, baking powder, and yeast, are carbon, nitrogen, oxygen, and hydrogen. From our pantry, we also need a handful of other ingredients in smaller amounts, but nonetheless crucial, to the success of the recipe: calcium, sodium, sulfur, chlorine, phosphorus, potassium, and iron. Finally, we need the spice rack too, smidgens of trace elements that are the spice of life. We need copper, zinc, selenium, and cobalt.

So that's what the recipe for life requires. But how do we stock our cosmic kitchen? Where do we get the ingredients for the pantry and for the spice rack? Hydrogen is essentially free, having been provided by the big bang itself. But carbon, every bit of that carbon requires a star to make it. Stars like our Sun make it, as red giants shortly before their deaths. And nitrogen and oxygen, those require massive stars to fuse them from carbon and helium atoms. These are the main elements that stars make, and so our pantry has them in relative abundance.

Calcium, sodium, sulfur, chlorine, phosphorus, potassium, and iron, every single atom of these comes to us from the later stages of fusion as massive stars flail to keep themselves alive in their dying breaths. The stars don't get much bang for the buck with these stages of fusion, and so they produce them in relatively smaller abundances than carbon, nitrogen, and oxygen. And as for the spices, well, just like vanilla beans from Madagascar, these rare elements can only be gotten from very special places, the fiery supernovae that are the violent deaths of massive stars. These elements are flash forged in just an instant, and very little of them is produced. So these are rare spices indeed. But without their seasoning, the recipe of life will fail.

In earlier lectures, we've often noted that the stellar life cycle is driven by each star's unconscious drive to preserve itself against gravity. But we've also remarked that it's not too much of a stretch to find a kind of purpose in this life cycle. The stars, both in their lives and in their deaths, make every ingredient in the recipe of life. Through their deaths, they blend these ingredients together throughout the galaxy. Through gravity, they gather up these blended ingredients. And in the course of their struggle to live, they bake that batter to just the right temperature. We are made of the stars; we breathe them, we consume them, they course through our veins, and we tingle with them. We live on a world made of them, and we move through a material world built of them. In our contemplations of the stars, in our comprehension of them, indeed, in our very sentience, we are the stars contemplating and comprehending themselves.

At the beginning of this course, I invited you to recall the birth of a child or grandchild, or a niece or nephew. And I shared with you the moments of my own sons' births, cradling them in my hands, and recalling the transcendence of those moments as I beheld their new, living, breathing bodies, made of the stars, and humbled by the enormity of the cosmic investment in them. My boys are older now, and on a recent family trip to the beach I vividly recall another one of those transcendent experiences. As my boys ran through the sand and splashed in the waves on that sun-filled day, I could not help but regard them as for the first time, these children of mine, children of the stars, birthed in a gush of saline, now bathing in a salty sea, sodium from the stars; erected in bones of calcium, now pressing through the sand and gathering up chalky sea shells, ashes of the stars; illuminated in sunlight, warmed and energized by it, the oxygen of the air filling their lungs, oxygen exhaled by the stars; climbing the dunes, their toes gripping the very earth of which they are made, all of it made of stars; held down by gravity, but defying it as they leapt; not a care in the world, these boys rich in their stellar heritage, these motes of dust, these inheritors of the stars.

A Tale of Two Stars
Lecture 24

This lecture will bring together the essential aspects of the life cycle of stars, personified through a tale of two stars: one like our Sun and one 10 times more massive. The relative masses of these two stars is important because the DNA of the stars, the one thing that determines their life course—from how long they will live to the manner in which they will die—is their mass, their birth weight. Mass is more than destiny for a star; it is fate. The two stars presented in this lecture have very different masses and, therefore, very different fates.

The Stellar Nursery Stage

- Recall that the mass of a star determines everything about it, including its physical properties at every stage of its life, the manner in which it dies, the type of corpse it leaves behind, and, importantly, the types of chemical elements that it can synthesize and return to the cosmos.

- Massive stars live lives that are fast and furious, burning out quickly, leaving behind neutron stars or black holes, but forging in their lives and deaths all of the elements of the periodic table, including the heavy elements that are required for life.

- In contrast to the massive stars, less massive stars—stars like our Sun—live more modestly. They live very long lives, and after they die, their white dwarf corpses are like diamonds in the sky, made of the same carbon that comprises all of life, including our own bodies. Indeed, in their deaths, stars like our Sun gently sprinkle their carbon ashes into the cosmos through majestically beautiful planetary nebulae.

- The two stars in our tale, like all stars, were born in stellar nurseries, enormous clouds of gas and dust that usually weigh about 10,000 times the mass of our solar system. Within stellar nurseries, hundreds to thousands of individual stars take shape and light up under the influence of gravity, which condenses and collapses the material in the cloud into individual stars.

- Let's imagine that the two stars in our tale are born in the same nursery—they're not siblings, but think of it as if they were born in the same hospital. Of these two stars—the one weighing 1 solar mass or the one weighing 10 solar masses—the more massive one would likely be born first.

- Through their powerful radiation and winds, the most massive stars in the hearts of nurseries exert an enormous influence on the surrounding nursery, weathering and eroding and stripping the surrounding gas and dust and creating the dramatic pillars of creation. In the process, they further compress the material, nudging gravity along and triggering the birth of less massive stars like our Sun within that compressed material.

- What about the planets that will be formed from that protoplanetary disk encircling the Sun-like star? Because the young star is gulping from that disk and will gobble it up in just a few million years, the planet-formation process has a very small window of opportunity to play out. So, just as the smaller star in our tale is getting through its own growing pains, it very quickly has to figure out the challenge of building its own family—and all of this in the dynamic, tumultuous environment of a stellar family, the star cluster, of which the star is but one member.

- Indeed, the Sun-like star in our story is unlikely to be an only child. Most stars like the Sun are not born single; rather, they are typically born with at least one close sibling. So, let's imagine that the Sun-like star in our tale does have a sibling and that this sibling is similar to the Sun-like star but is a bit less massive. How will our star and its sibling influence one another?

- If there are just two siblings, then the two stars will generally leave one another alone, orbiting at a comfortable distance, each capable of building its own family of planets from its own protoplanetary disk of material. However, if there is a third sibling, a powerful sibling rivalry will play out.

- For the purposes of our tale, let's imagine that our Sun-like star has the lower-mass sibling but not a third. Our Sun-like star and its sibling are part of a larger stellar family, the star cluster of hundreds to thousands of stars all birthed in the same stellar nursery. Even if our two Sun-like siblings treat one another gently, that doesn't mean that the rest of the extended family is going to be so hands-off. However, let's imagine that our Sun-like star manages to avoid the worst consequences of cosmic roughhousing.

- While our Sun-like protagonist has been through all kinds of trouble, our more massive stellar protagonist has been steadily doing its duty, entering the responsibility of adulthood early. Even as the massive star has started the process of nuclear fusion, synthesizing heavier elements in its core, it is taking on a large responsibility for caring for the young stars throughout the stellar nursery and establishing the foundations for the extended stellar family.

- As the family of stars leaves the nursery after about 50 million years, it is a full-fledged cluster, leaving its breeding grounds to join the rest of the galaxy. And just as the extended stellar family moves on, the massive star in our tale is preparing to end its life in dramatic fashion. The massive star quickly goes through the various stages of nuclear fusion.

- With each round of fusion, the massive star in our tale is resuscitating itself for briefer and briefer periods of time. Finally, once our massive star protagonist has fused all the way to iron in the core, its beating heart comes suddenly and tragically to a halt, for good, a little more than about 100 million years into its life.

- Perhaps morbidly, the massive star in our story dies in a violent supernova explosion in full view of the rest of its extended family of stars in its cluster. Its core collapses down to an unimaginably dense neutron star, and as the rest of its hulking body implodes onto that neutron star at its center, that material bounces off and explodes into the surrounding space with an energetic brilliance unmatched anywhere in the galaxy—indeed, in the whole universe. At that moment, in the fireball of the supernova, all of the rest of the elements of the periodic table are flash forged.

- These elements—indeed, all the elements forged over the life of this massive star—are now scattered throughout the stellar cluster. The blast wave of the supernova pushes a supernova remnant, a hot expanding bubble of gas and ash, beyond the extended family of stars in this one cluster and out into the surrounding galactic medium. It even carves a chimney up out of the galaxy and sends its ashes like a geyser into the halo of the galaxy.

- That material finally rains back down throughout the galaxy, sprinkling the dead star's ashes everywhere. In its superlative death, our massive star protagonist provides a sort of manna from heaven, in the form of gas and ash enriched in all of the elements of the periodic table, from which new generations of stars and planets may form, enhanced in their ability to create and sustain life.

- In the outskirts of the cluster where the massive star exploded, the shockwave of its expanding supernova remnant may compress and heat the gas and dust in the surrounding space, directly triggering and fertilizing a new stellar breeding ground.

- In the wake of this incredible event, the Sun-like star in our tale is perhaps a light-year away from the massive star that exploded. Still residing in the stellar family that is a few light-years across overall, it goes on about the business of growing up and starting its own solar system. Little by little, our Sun-like protagonist grows up, gradually growing apart from the rest of its extended family and eventually establishing its own place with its young solar system in the larger galaxy.

The Adult Stage

- Our Sun-like star enters the long stage of middle life. As a young adult, the star jumps straight into the responsibility of hard work, fusing hydrogen in its core into helium. Those early fruits of its labors may not be worth much in and of themselves, but the true value of those labors is in the foundation that these products will provide in the later stages of the star's adulthood to create true wealth—in the form of carbon.

- As a young adult, the star is full of vim and vigor; it spins rapidly. Its magnetic field, generated in the dynamo action of its rapid spin, lurches and occasionally erupts in fits of rage that the star will reign in as it continues to mature and slows down a little bit.

- These bursts of magnetism, which shower the orbiting planets with intense X-rays, might be essential to spur on early biogenesis. They may also have a profound effect on nascent life forms on one or more of those planets by promoting the types of rapid mutations that lead to the proliferation of life in all its varieties.

- As the adult star ages and as its system of planets enters its own stable state, the star gradually winds down, its space weather becoming gentler—though with occasional stern reminders. Meanwhile, its vital signs, as measured through its sunquakes and the neutrinos that it puts out, clearly indicate that fusion continues strong and steady in its core.

- Then, suddenly, when the star is about 10 billion years old, the star experiences the equivalent of a severe cardiac arrest as it completes its task of fusing the hydrogen in its core completely to helium. And at that moment, gravity pounces, collapsing the core toward oblivion.

- But this star is not done yet. The core heats up hotter and hotter under the crush of gravity, until it reaches about 100 million degrees. At that temperature, the core can start the process of fusing helium into carbon. So, the star resuscitates as a red giant. Overall, the star is cooler now, yet it shines more brightly than before, having swelled in size 100-fold.

- This is an elderly parent, still glowing with life, knowing that it is banking up the wealth for the next generation that it had prepared for throughout its life—storing up entire worlds worth of carbon, the stuff of life. The process of fusing helium to carbon lasts a relatively short time—perhaps no more than 1 billion years for a star with the mass of the Sun.

- Finally, after a long life of about 10 billion years, our red giant protagonist, ringed by its stellar sibling and its planetary children, lying still and short of breath, reveals the inheritance that it has in store. This Sun-like star cannot initiate another round of nuclear fusion. It simply doesn't have the mass that would allow it to

© Sergii Tsololo/iStock/Thinkstock.

Only a very small percentage of the many, many trillions of stars that exist in the universe is visible to the naked eye.

generate a sufficient temperature. The star's mass dictating its fate, it is doomed now to die.

- As it breathes its last, it returns itself whence it came, a quiet but beautiful and graceful planetary nebula sprinkling its carbon ashes like rich seeds of potential for future generations. And for a time, the grave marker of the star is fresh and gleaming—a white dwarf, a tiny but white-hot diamond of a corpse. The star will now only fade into the ages, forever cooling, dimming, and eventually becoming a black dwarf, inert and still and dark.

- The star's sibling has been aging and tending its own, too. It was a bit less massive than our Sun-like star, which means that it will live longer. Even so, it too is destined to die. And when its time comes, it too will swell into a red giant as it enters its second-wind stage of fusing helium to carbon. But the sibling's experience as a red giant will be quite different.

- As this red giant swells and comes into proximity with its now long-dead sibling, the red giant transfers some of its mass onto the white dwarf, slowly building up the white dwarf's heft to the breaking point. The white dwarf collapses and produces a white dwarf supernova.

- Both of our story's protagonists are notable for their steadfast commitment throughout their lives—one brief, one long—to hold strong against gravity long enough to produce a legacy for the subsequent generation and beyond. Indeed, our story doesn't end so much as mark the new beginning of those next generations of stars.

Suggested Reading

Kippenhahn, *100 Billion Suns*.

Lang, *The Life and Death of Stars*.

1. Which types of stars—stars less massive than the Sun or stars more massive than the Sun—are the most likely likely to harbor long-term habitable planets?

2. With which type of star—a star like our Sun or a star much more massive—do you identify with most, in terms of the experience of the circle of life?

A Tale of Two Stars
Lecture 24—Transcript

"It was the best of times, it was the worst of times." Charles Dickens may not have had the life cycles of stars in mind when he penned those words, but he might as well have. As we've discussed throughout this course, a star's existence resembles human existence in a number of striking and profound ways. From the miracle of their births, surely the best of times, to the violence of their deaths, perhaps the worst of times, the stars live out lives at once majestic and beautiful, yet full of drama. And as we saw especially in our last lecture, there is an intimacy that we share with the stars, being made, literally, of them, of the nuclear products of their life's work. My two sons, to whom I've introduced you, they, like all of us, are inheritors of this vital nuclear legacy of the stars.

Today we reach the end of our course, and what I'd like to do in this final lecture is to bring together the many pieces of the stellar life cycle into a unified picture. To accomplish that, I'm going to take my cue from Dickens, but instead of telling a tale of two cities, I'm going to offer you a tale of two stars. In other words, I'm going to synthesize what we've learned by comparing and contrasting the life cycles of two stars; one, a star like the Sun, and the other, a star 10 times more massive.

Now, the relative masses of these two stars is important. Because remember, the DNA of the stars, the one thing that determines their life course from how long they will live to the manner in which they will die, is their mass, their birth weight. Mass is more than destiny for a star; it is fate. And so let's bring our exploration of the stars full circle, by comparing the life cycles of two stars with very different masses, and therefore with very different fates. Now, I'm going to tell this story with a fair amount of anthropomorphizing. Throughout the lecture I will consistently use metaphors drawn from human life to describe the experience of these two stars. But I really do think that, even as scientists, we can better remember the various pieces of the stellar life cycle story through the analogy of the human experience. So permit me some creative license as we personify these two stellar actors, and we'll use the powerful themes of the human life cycle to draw parallels between our own experiences and those of the stars.

Let's begin with a quick overview. As we've said, the mass of a star determines everything about it, including its physical properties at every stage of its life, the manner in which it dies, the type of corpse it leaves behind, and importantly, the types of chemical elements that it can synthesize and return to the cosmos. So think about high-mass stars for a moment, stars much larger than our Sun. If I asked you to give me a thumbnail sketch of the life cycles of massive stars, what would you say? Well, here's the sketch I'd give. Massive stars live lives that are fast and furious, burning out quickly, leaving behind neutron stars or black holes, but forging in their lives and deaths all of the elements of the periodic table, including the heavy elements that are required for life.

And what about less massive stars, like our Sun? How, in general, do their life cycles differ? Well, in contrast to the massive stars, stars like the Sun live more modestly. They live very long lives, and after they die, their white dwarf corpses are like diamonds in the sky, made of the same carbon that comprises all of life, including our own bodies. Indeed, in their deaths, stars like our Sun gently sprinkle their carbon ashes into the cosmos through majestically beautiful planetary nebulae. So with that quick sketch in mind, let's now go through the different stages of the stars' lives, piecing them together from beginning to end, and with an eye to comparisons and contrasts in this tale of two stars.

The two stars in our tale, like all stars, had to be born someplace. And where are stars born? That's right, stellar nurseries. Stellar nurseries are enormous clouds of gas and dust, usually weighing in at about 10,000 times the mass of our solar system, within which hundreds to thousands of individual stars take shape and light up under the influence of gravity, which condenses and collapses the material in the cloud into individual stars. Maybe you remember some of the stellar nurseries we've talked about, such as the Orion Nebula and the remarkable Horsehead Nebula. Let's imagine that the two stars in our tale are born in the same nursery. Not siblings, mind you—let's not complicate the story too much. Think of it more like born in the same hospital. Which of these two stars, the one weighing 1 solar mass or the one weighing 10 solar masses, do you think is most likely to be born first?

If you said the more massive star is the one more likely be born first, you'd be right. In our exploration of stellar nurseries, one of the common aspects that we saw in those nurseries was the most massive stars in the heart of the nursery, and we saw that through their powerful radiation and winds, those massive stars exert an enormous influence on the surrounding nursery. Think of Eta Carinae, the intensely luminous star within the Carina stellar nursery. Massive stars like Eta Carinae sculpt their nurseries, weathering and eroding and stripping the surrounding gas and dust, creating the dramatic pillars of creation that we saw in many of the nurseries. And in the process, they further compress the material, nudging gravity along, and triggering the birth of less massive stars like our Sun within that compressed material.

And even before these baby stars within their cocoons are revealed to us in visible light, we can peer inside the birth clouds with infrared light, like an obstetrician's ultrasound revealing the baby within the womb. Or to use another image, just as we can detect a kicking baby by placing a hand on the mother's belly, so can astronomers detect a kicking baby star in the form of Herbig-Haro jets that the star ejects into the surrounding the nebula. What causes the baby star to kick so?

Remember that these young stars are encircled by disks of gas and dust made of the same material from which the stars themselves are made. So let's imagine the Sun-like star in our story as a baby star surrounded by a protoplanetary disk. That disk will become the star's solar system of planets. But first, the baby star gulps material from that disk, feeds from it, its magnetic field functioning as a kind of bottle that guides the food—the gas and dust—from the disk into the star and helping it grow. But that same magnetic field also functions as a spray nozzle, and the more our baby star feeds, the more it spits up, producing those Herbig-Haro jets, which in turn further sculpt and erode the stellar nursery.

What about the planets that will be formed from that protoplanetary disk encircling the Sun-like star? Well, because the young star is gulping from that disk, and will gobble it up in just a few million years, the planet formation process has a very small window of opportunity to play out. So just as the smaller star in our tale is getting through its own growing pains, it very quickly has to figure out the challenge of building its own family, and all of this in the dynamic, tumultuous environment of a stellar family—the star cluster—of which the star is but one member.

Indeed, the Sun-like star in our story is unlikely to be an only child. Let's recall what we previously discussed about the commonality of stellar siblings. First of all, most stars like the Sun are not born single. Rather, they are typically born with at least one close sibling. So let's imagine that the Sun-like star in our tale does indeed have a sibling, and let's imagine that this sibling is similar to the Sun-like star but a bit less massive. How will our star and its sibling influence one another?

If you're thinking that it depends on whether there's a third sibling, you'd be right. If there are just two siblings, then the two stars will generally leave one another alone, orbiting at a comfortable distance, each capable of building its own family of planets from its own protoplanetary disk of material. But if there is a third sibling, a powerful sibling rivalry will play out. Do you remember the essential aspects of the three-sibling dynamic in stars? The most important aspect is that three stars will proceed through a complex gravitational dance, what we referred to as a *pas de trois*, in which two of the stars are brought tightly close together at the expense of the third, which is cast out into a distant, wide orbit, if not ejected altogether. For the purposes of our tale, let's imagine that our Sun-like star has that lower mass sibling but not a third. This is not a family drama.

Okay, maybe it is a family drama after all. Remember that our Sun-like star and its sibling are, in fact, part of a larger stellar family, the star cluster of hundreds to thousands of stars all birthed in the same stellar nursery. Even if our two Sun-like siblings treat one another gently, that doesn't mean that the rest of the extended family is going to be so hands off. Think of it like cousins roughhousing at a large family gathering. The dynamics within the larger star cluster can send stars reeling and careering through the cluster, jostling the other stars, potentially disrupting their protoplanetary disks and thereby rendering them incapable of making planets. But let's give our tale a happier twist and imagine that our Sun-like star manages to avoid the worst consequences of this cosmic rough-housing.

Before we move on in our story from the stellar nursery stage, let's not forget about the other main protagonist in our story, the massive star. Remember, that star was the firstborn of the entire nursery; it helped to sculpt the nursery as a whole and even to trigger the birth of the Sun-like protagonist in our tale. But while our Sun-like protagonist has been through all kinds of trouble—sibling rivalry, roughhousing—our more massive stellar protagonist has been steadily doing its duty, entering the responsibility of adulthood early.

Even as the massive star has started the process of nuclear fusion—synthesizing heavier elements in its core—it is taking on a large responsibility for caring for the young stars throughout the stellar nursery and establishing the foundations for the extended stellar family, the cluster of hundreds to thousands of stars now becoming fully revealed as that massive star clears out the detritus from the surrounding nursery.

As the family of stars leaves the nursery after about 50 million years, it is a full-fledged cluster, leaving its breeding grounds to join the rest of the galaxy. And just as the extended stellar family moves on, the massive star in our tale is preparing to end its life in dramatic fashion. The massive star quickly goes through the various stages of nuclear fusion. A main-sequence star of type B with a temperature of about 20,000 degrees Celsius and a luminosity of about 5000 times the Sun's luminosity, it initially fuses hydrogen to helium in its core. Then, as the hydrogen in its core is depleted after about 100 million years, converted entirely to helium, that helium becomes the fuel for a next round of fusion, producing carbon. After that, subsequent rounds of fusion produce increasingly heavier elements through oxygen and silicon, all the way up to iron.

The source of the star's power—the light and heat it generates to sustain itself against the crush of gravity—is $E = mc^2$. Each round of fusion creates heavier elements from the lighter ones, and in the process, trades a small amount of elemental mass for pure energy. How does that work? Recall that Einstein's $E = mc^2$ tells us that mass is really just another form of energy. And so mass can be used, traded, to make energy in the form of light and heat, the total energy, of course, always being conserved. In the process of making heavier atoms from lighter ones, our massive star is always trading a little bit of mass for the pure energy that keeps gravity at bay. But remember also that each successive round of fusion to heavier and heavier elements extracts less and less energy. So with each subsequent round of fusion, the massive star in our tale is resuscitating itself for briefer and briefer periods of time.

Finally, once our massive star protagonist has fused all the way to iron in the core, its beating heart comes suddenly and tragically to a halt, for good, a little more than 100 million years or so into its life. Why does that happen? Recall that iron is the heaviest element that can be fused from lighter elements and still generate a net amount of energy in the process. If the star were to attempt to fuse its iron into a heavier element, it would need to inject energy into the fusion, which would only sap it of the energy it desperately needs in its battle with gravity. So with no more energy to produce, the star at last gives up the ghost, sacrifices itself to gravity, that merciless, relentless grim reaper.

Perhaps morbidly, the massive star in our story dies in a violent supernova explosion in full view of the rest of its extended family of stars in its cluster. Its core collapses down to an unimaginably dense neutron star, and as the rest of its hulking body implodes onto that neutron star at its center, that material bounces off and explodes into the surrounding space with an energetic brilliance unmatched anywhere in the galaxy, indeed in the whole universe. At that moment, in the fireball of the supernova, all of the rest of the elements of the periodic table are flash-forged.

These elements, indeed, all the elements forged over the life of this massive star, are now scattered throughout the stellar cluster. Indeed, the blast wave of the supernova pushes a supernova remnant—a hot expanding bubble of gas and ash—beyond the extended family of stars in this one cluster and out into the surrounding galactic medium. It even carves a chimney up out of the galaxy and sends its ashes like a geyser into the halo of the galaxy. That material finally rains back down throughout the galaxy, sprinkling the dead star's ashes everywhere. In its superlative death, our massive star protagonist provides a sort of manna from heaven in the form of gas and ash enriched in all of the elements of the periodic table from which new generations of stars and planets may form enhanced in their ability to create and sustain life.

Indeed, in the outskirts of the cluster where the massive star exploded, the shockwave of its expanding supernova remnant may compress and heat the gas and dust in the surrounding space, directly triggering and fertilizing a new stellar breeding ground. In the wake of this incredible event, the Sun-like star in our tale is perhaps a light-year away from the massive star that exploded. Still residing in the stellar family that is a few light-years across overall, it goes on about the business of growing up and starting its own solar system. Little by little, our Sun-like protagonist grows up, gradually growing apart from the rest of its extended family, remember, that's what stars in open clusters do, and eventually establishing its own place with its young solar system in the larger galaxy.

And so we come to a point in our story where it's now all about the Sun-like star. Remember that we set up our tale so that the Sun-like star is orbited by a sibling star. Perhaps like a brother or sister with whom there is a warm bond of family but with whom contact is only made when something urgent comes up. We will continue the tale of our Sun-like star with only passing reference to that sibling until later, when that sibling will make its presence dramatically known once again. For now, our Sun-like star enters the long stage of middle life. It is a responsible, mature adult, setting about the business of doing its life work, being productive, saving its earnings in order to bequeath a rich inheritance to its future children and grandchildren. The star, having been nurtured and protected in an extended family, hardened through perhaps a bit of sibling rivalry, is now ready to do its part.

As a young adult, the star jumps straight into the responsibility of hard work, fusing hydrogen in its core into helium. Those early fruits of its labors may not be worth much in and of themselves. But the true value of those labors is in the foundation that these products will provide in the later stages of the star's adulthood to create true wealth, in the form of carbon. As a young adult, the star is full of vim and vigor. A main sequence star, squarely in the middle of the pack as far as its physical attributes are concerned—with a temperature of 6000 degrees Celsius, a luminosity of one solar luminosity, and a radius of one solar radius—the star fits right in to its generational cohort of young families out there in the galactic suburbs, focused on family and work, its solar system of planets and the daylong business of fusion.

- Phosphorus is required by all known forms of life. It plays a major role in biological molecules like DNA and RNA, forming a part of their structural backbone. Importantly, phosphorus is also the key component of adenine triphosphate, or ATP, in which a group of 3 phosphorus atoms provide the molecule's key features and functions. Nearly every cellular process that uses energy obtains it in the form of ATP. Every atom of phosphorus that exists was made in massive stars through the fusion of silicon and helium.

- Potassium is the next most abundant element in the body. Massive stars make potassium through the fusion of 2 oxygen atoms; every atom of potassium in the universe was made this way. In our bodies, potassium is crucial for proper nerve transmission. Without enough potassium, fingers freeze up and, eventually, cardiac arrest follows because the heart's beating function is regulated by this important element.

- Sulfur is made in massive stars through the fusion of silicon and helium. Because of its nasty smell, it has throughout history been associated with hell-fire burning. However, we cannot live without it. Sulfur is an essential component of all living cells. It is present in all proteins and enzymes that contain certain essential amino acids.

- Pure chlorine gas is about as horrific a poisonous substance as you could imagine. A whiff of pure chlorine gas is like a blast of a blowtorch to the sinuses and lungs, and breathing more than just a small amount can cause instant death. But together with sodium, chlorine forms regular table salt. And in our bodies, chlorine serves an important role in nerve signaling and digestion.

- Iron is the heaviest element that massive stars produce before they explode as supernovae. Every atom of iron in the universe was smelted in the crucibles that are the cores of massive stars in the final, dying gasps of their lives. Iron atoms—sprinkled through the galaxy by supernova explosions spreading these dead stars' material up through galactic geysers and across the galaxy—are stellar ashes

The star spins rapidly, maybe too much Starbucks coffee? And that exuberance some days makes the star seem a bit stressed. Its magnetic field, generated in the dynamo action of its rapid spin, lurches and occasionally erupts in fits of rage that the star will reign in as it continues to mature and slows down a little bit. Interestingly, through these bursts of magnetism, which shower the orbiting planets with intense X-rays, may be essential to spur on early biogenesis. They may also have a profound effect on nascent life forms on one or more of those planets by promoting the types of rapid mutations that lead to the proliferation of life in all its varieties.

As the adult star ages and as its system of planets enters its own stable state, the star begins to show its age a bit. The star gradually winds down, its space weather becoming gentler, though with occasional stern reminders. Meanwhile, it continues to receive a clean bill of health, as its vital signs, as measured through its sunquakes and the neutrinos that it puts out, clearly indicate that fusion continues strong and steady in its core.

And then, suddenly, now about 10 billion years old, everything changes. The star experiences the equivalent of a severe cardiac arrest as it completes its task of fusing the hydrogen in its core completely to helium. And at that moment gravity pounces, collapsing the core toward oblivion. But this star is not done yet, not if it can help it. The core heats up hotter and hotter under the crush of gravity until it reaches about 100 million degrees. At that temperature, the core can start the process of fusing helium into carbon. And so the star resuscitates as a red giant. Overall, the star is cooler now, yet it shines more brightly than before, having swelled in size 100-fold. This is an elderly parent, still glowing with life, having been there and done that, now comfortably Zen in the knowledge that it is banking up the wealth for the next generation that it had prepared for throughout its life, storing up entire worlds worth of carbon, the stuff of life.

The process of fusing helium to carbon lasts a relatively short time, perhaps no more than one billion years for a star with the mass of the Sun. And so finally, after a good, long life of some 10 billion years, our red giant protagonist, ringed by its stellar sibling and its planetary children, lying still and short of breath, reveals the inheritance that it has in store. This Sun-like star cannot initiate another round of nuclear fusion. It simply doesn't have the mass that would allow it to generate a sufficient temperature. The star's mass dictating its fate, it is doomed now to die. And as it breathes its last, it returns itself whence it came, a quiet but beautiful and graceful planetary nebula sprinkling its carbon ashes like rich seeds of potential for future generations. And for a time, the grave marker of the star is fresh and gleaming. A white dwarf, a tiny but white hot diamond of a corpse, a reminder to the rest of the galaxy that says, "I was here."

If the star had ever entertained fantasies of more heroic deeds, or at least of acts that might garner more fanfare, unfortunately, the star will now only fade into the ages, forever cooling, dimming, eventually becoming a black dwarf, inert and still and dark. But wait; there's the star's sibling! It has been aging and tending its own, too. But remember we said that it was a bit less massive than our Sun-like star, which means that it will live longer. Even so, it too is destined to die. And when its time comes, it too will swell into a red giant as it enters its second-wind stage of fusing helium to carbon. But the sibling's experience as a red giant will be quite different. What will happen as this red giant swells and comes into proximity with its now long-dead sibling?

The red giant transfers some of its mass onto the white dwarf, slowly building up the white dwarf's heft to the breaking point. Recall that white dwarfs are held up by degeneracy pressure that can only withstand a certain total amount of gravitational weight. That limit, known as the Chandrasekhar limit, is 1.4 times the mass of the Sun. So when the red giant's spilling of mass causes the white dwarf to violate that limit, what happens? The white dwarf instantly collapses and produces a white dwarf supernova, achieving considerable fame after all, if posthumously.

And so as we draw our tale of two stars to a close, we see that while the two stars were fated to a certain extent by their DNA—their mass—the experiences of the stars with their surroundings, and especially with their siblings, has allowed one of the stars in our story to defy its fate. Where the Sun-like star in our telling would have otherwise ended up in obscurity, through its sibling interaction it has ended up known and felt throughout the galaxy. But both of our story's protagonists are notable for their steadfast commitment throughout their lives—one brief, one long—to hold strong against gravity long enough to produce a legacy, indeed a wealth, for the subsequent generation and beyond. And indeed, our story doesn't end so much as mark the new beginning of those next generations of stars. The old gospel song asks the question, "Will the circle be unbroken?" When it comes to the stellar life cycle, the answer seems to be an emphatic, yes!

Today, we've brought together the essential aspects of the life cycle of stars, personified through a tale of two stars, one like our Sun, one 10 times more massive. But coming back to our quote from Dickens at the beginning of this lecture, for which type of star would you say the stellar life cycle represents the best of times? For which is it the worst of times? Maybe, as in the tale of two cities, it is for all stars, as it is for us, a bit of both.

Well, it's been a real pleasure exploring the stars with you these past 24 lectures. Together we've traced the amazing cycle of the life and death of stars. It's been a dramatic tale, hasn't it? And I hope you'll agree that in the physical understanding of how the stars work—the microscopic physical processes that explain how and why these behemoths do what they do—there is an extraordinary beauty.

At the same time, I hope you've come to see a star's existence as resembling human existence in profound ways. There are the growing pains of youth. There are siblings with their rivalries, and extended families with complex dynamics. There is the Sisyphean existential struggle to perform a function that, while perhaps monotonous in the day in and day out, can be seen through the perspective of purpose. There is the building of a legacy that will be inherited by future offspring never to be known but whose own lives are enriched by the products of that labor. And of course, there is the simple fact of the life cycle itself, birth, life, and inevitably death—ashes to ashes and dust to dust. And yet, from the ashes, from the dust, new life arises.

Bibliography

Archinal, Brent A. *Star Clusters*. Willmann-Bell, 2003. A compendium of various star clusters, including both open clusters and globular clusters, and detailed physical information about each, as well as useful guides to observing them with backyard telescopes.

Arnett, David. *Supernovae and Nucleosynthesis (Princeton Series in Astrophysics)*. Princeton University Press, 1996. A highly technical volume describing the physics of nuclear fusion in supernova explosions by which the elements heavier than iron come to be made. This should be of interest to readers with strong mathematics backgrounds.

Astrobiology Magazine. "The Stuff Stars Are Made Of." http://www.astrobio.net/exclusive/1617/the-stuff-stars-are-made-of. A beautiful collection of images of Bok globules in various stellar nurseries and an accessible discussion of these in the broader context of stellar nurseries and stellar birth.

Astronomy Picture of the Day. "Stellar Nurseries." http://apod.nasa.gov/apod/stellar_nurseries.html. Collection of breathtaking astronomical images of different stellar nurseries, with detailed explanations for the appearances and structure of each, along with interpretations of each in terms of the stellar birth process.

Bartusiak, Marcia. *The Day We Found the Universe*. Vintage Books, 2010. A superb read on the role of Edwin Hubble in the use of Cepheid variable stars to at last reveal the true size of the universe.

Big Bear Solar Observatory. "Solar Movies." http://www.bbso.njit.edu/movies.html. A collection of scientific observations of the Sun's surface from the Big Bear Solar Observatory, including movies of granulation on the Sun, transits of the Sun by the inner planets, and a number of other stunning and informative visuals.

Bodanis, David. E = mc²: *A Biography of the World's Most Famous Equation*. Berkley Trade, 2001. A fun and informative account of how Einstein's most famous equation, $E = mc^2$, came to be; a discussion of its broad significance; and, overall, an enjoyable and memorable read.

Boss, Alan. *International Astronomical Union Working Group on Extrasolar Planets Definition of a 'Planet.'* http://www.dtm.ciw.edu/boss/definition. html. A brief but interesting peek into how astronomers attempt to formalize definitions, this is the current working definition for how to distinguish a planet from a brown dwarf.

Brandner, W., and H. Klahr. *Planet Formation: Theory, Observations, and Experiments (Cambridge Astrobiology)*. Reissue edition. Cambridge University Press, 2011. A relatively technical volume, but should be accessible to readers with a grasp of college-level physics and mathematics. A detailed account of our understanding of planet formation a full decade after the discovery of the first exoplanets.

Burgasser, Adam. "Brown Dwarfs: Failed Stars, Super Jupiters." *Physics Today* 61, no. 6 (2008): 70–71. http://link.aip.org/link/phtoad/v61/i6/p70/ s1. A modestly technical, popular account of what brown dwarfs are, what we've learned about them, and what we still don't know about them.

Choi, Charles Q. "Alien Planets Circling Pulsing Stars May Leave Electric Trails." *SPACE.com* October 1, 2012. http:// www.space.com/17848-alien-planets-pulsars-electric-trails.html. Brief article but with a number of hyperlinks to additional information on the few known planets orbiting pulsars.

Darling, David. *Life Everywhere: The New Science Of Astrobiology*. Basic Books, 2001. An introduction to the science of astrobiology, this volume links what we know about elements created in the stars to the processes of life as we know it, to possible alternative forms of life and the different elements upon which they might be based.

DeVorkin, David H. *Henry Norris Russell*. Princeton University Press, 2000. A fascinating biographical account of Henry Norris Russell, one of the codiscoverers of the Hertzsprung–Russell diagram, who lived in two worlds as a Presbyterian and as an astronomer and helped transform the entire discipline of astronomy into the modern discipline of astrophysics.

European Space Agency. "Born in Beauty: Proplyds in the Orion Nebula." http://www.spacetelescope.org/news/heic0917/. A stunning collection of images from the Hubble Space Telescope of different infant stars still surrounded by their proplyd cocoons, disks of gas and dust from which they feed and from which solar systems are made.

Giant Magellan Telescope Observatory. "Giant Magellan Telescope (GMT)." http://www.gmto.org/index.html. The website of the Giant Magellan Telescope (GMT), a next-generation concept for an extremely large ground-based telescope with an effective primary mirror diameter of 80 feet, includes technical details for the concept and the types of discoveries that such a large telescope will enable.

Gray, Theodore. *The Elements: A Visual Exploration of Every Known Atom in the Universe.* Reprint edition. Black Dog & Leventhal Publishers, 2012. A beautiful, visual guide to every element of the periodic table, including examples from everyday life as well as little-known facts about each.

HubbleSite. "Star Clusters." http://hubblesite.org/gallery/album/star/star_cluster/. An online gallery of stunning photographs of star clusters obtained with the Hubble Space Telescope, including all types and spanning all ages of clusters, with detailed information about each.

HubbleSource. "Orion Nebula Fly-Through." http://hubblesource.stsci.edu/sources/video/clips/details/orion.php. Computer simulation of the Orion Nebula nursery, with physically accurate representation of the different types of stars, their relative locations within the nebula, and interactions between them.

IceCube Neutrino Observatory. "IceCube Neutrino Observatory." http://icecube.wisc.edu/. The website of the IceCube neutrino observatory, one of the longest running experiments to directly observe and quantify neutrinos from the Sun.

Jastrow, Robert. *Red Giants and White Dwarfs*. 3rd edition. Readers Library, 1990. A lay reader's account of stars at the ends of their lives, written authoritatively but accessibly by the former director of the Mount Wilson Observatory.

Kallrath, J., and E. F. Milone. *Eclipsing Binary Stars: Modeling and Analysis (Astronomy and Astrophysics Library)*. Softcover reprint of hardcover 2nd edition (2009 edition). Springer, 2012. A detailed volume for the truly technically oriented that provides detailed information on the types of models that astronomers use to determine the physical properties of stars from measurements of eclipsing binary systems.

Kippenhahn, Rudolf. *100 Billion Suns: The Birth, Life, and Death of the Stars*. Princeton University Press, 1993. A more technical but authoritative supplement (unaffiliated) to the material of this course.

Kwok, Sun. *Cosmic Butterflies: The Colorful Mysteries of Planetary Nebulae*. Cambridge University Press, 2001. A beautiful pictorial summary of over 100 planetary nebulae, with stunning images and detailed explanations.

Lang, Kenneth R. *The Life and Death of Stars*. Cambridge University Press, 2013. A book with the same title as this course, but unaffiliated, this provides a good supplemental overview of some of the topics discussed in this course.

Large Synoptic Survey Telescope. "Large Synoptic Survey Telescope (LSST)." http://www.lsst.org/lsst/public. This is the website of the Large Synoptic Survey Telescope (LSST), a telescope planned for construction around 2015 that will create "movies" of the sky, including the ability to watch how millions of stars across the sky vary with time, allowing discovery of binary star systems, exoplanets, supernovae, and many other phenomena pertaining to how stars evolve throughout their lives.

Larson, Richard B., and Volker Bromm. "The First Stars in the Universe." *Scientific American* Jan. 19, 2009. http://www.scientificamerican.com/article.cfm?id=the-first-stars-in-the-un. A popular account of the first stars, what we know about them, and current frontiers of research in this area.

Las Cumbres Observatory Global Telescope. "Hertzsprung–Russell Diagram Simulator." http://lcogt.net/files/flash/hr-diagram/main.html. An interactive Hertzsprung–Russell diagram showing the stages of stars' lives as depicted through their positions in the H–R Diagram.

Lowell Observatory. "Navy Precision Optical Interferometer." http://www. lowell.edu/npoi/index.php. The website of the Navy Precision Optical Interferometer includes technical details on this cutting-edge facility for the highest angular resolution imaging currently possible in visible light, including making direct images of the surfaces of other stars.

Mann, Alfred K. *Shadow of a Star: The Neutrino Story of Supernova 1987A.* W. H. Freeman & Co, 1997. A first-person account of the observation of the supernova 1987A with a neutrino observatory. This was the first supernova in human history to have neutrinos directly detected from it.

Marschall, Laurence. *The Supernova Story.* Princeton University Press, 1994. A very nicely written, engaging account of how our understanding of supernovae has developed, from the earliest observed by humans to the most recent.

McNamara, Geoff. *Clocks in the Sky: The Story of Pulsars (Springer Praxis Books/Popular Astronomy).* Praxis, 2008. An excellent, popular book on the story of the discovery of pulsars and of some of the many exciting insights into physics and astronomy that these objects have enabled.

Mobberley, Martin. *Total Solar Eclipses and How to Observe Them (Astronomers' Observing Guides).* Springer, 2007. A "field guide" to observing total eclipses of the Sun, including information on when future eclipses will occur, the best places to observe, and lots of interesting information about solar eclipses past, present, and future.

Moldwin, Mark. *An Introduction to Space Weather.* Cambridge University Press, 2008. A technical treatment of all aspects of space weather, intended for readers with a moderately technical mathematics background.

NASA Goddard Space Flight Center. "Electromagnetic Spectrum." http://imagine.gsfc.nasa.gov/docs/science/know_11/emspectrum.html. An overview of the electromagnetic spectrum, examples of different types of light, and examples of different types of objects that emit different types of light.

NASA Jet Propulsion Laboratory. "Near Earth Object Program: Orbit Diagrams." http://neo.jpl.nasa.gov/orbits/. See actual orbits of all known bodies in the solar system, including comets and asteroids. Not directly about stellar orbits, but these many real (and fun) orbital examples convey much of the diversity of orbits in nature, including in stars.

Nicolson, Iain. *Dark Side of the Universe: Dark Matter, Dark Energy, and the Fate of the Cosmos*. Johns Hopkins University Press, 2007. Going beyond the role of dark matter in the lives and deaths of stars, this comprehensive treatment of the subject of dark matter in the universe covers all aspects of how this mysterious substance came to be discovered and the current limits of our understanding of it.

O'Dell, C. Robert. *The Orion Nebula: Where Stars Are Born*. Belknap Press of Harvard University Press, 2003. This popular book on the Orion Nebula by Professor Robert O'Dell is a wonderful overview of what we know about this magnificent stellar nursery.

Ptable. "Periodic Table." http://www.ptable.com/. Online periodic table with interactive information about each element, including each element's visible light spectrum.

Schopf, J. William. *Life's Origin: The Beginnings of Biological Evolution*. University of California Press, 2002. An engaging read, this volume treats the evolution of life on Earth through the perspective of the stars, the emergence of organic compounds as captured in meteorites, and the geologic record.

Shapiro, Stuart L. *Black Holes, White Dwarfs and Neutron Stars: The Physics of Compact Objects*. Wiley-VCH, 1983. An excellent technical treatise on the mathematics and physics of white dwarfs, neutron stars, and black holes.

Warning: This is a highly technical volume intended for individuals with at least college-level physics.

Simon, Seymour. *The Sun.* 1ˢᵗ Mulberry Ed edition. HarperCollins, 1989. A dramatic and pictorially stunning account of the Sun, this reference book is a must-have for any enthusiast of the Sun and offers a clear, visual account of our nearest star.

Sky & Telescope Magazine. "Video: Evolution of the Solar System." http://www.skyandtelescope.com/skytel/beyondthepage/123632899.html. An animation of the Nice model for the formation of our solar system and a lay account of how the model came to be developed.

Smith, N., K. G. Stassun, and J. Bally. "Opening the Treasure Chest: A Newborn Star Cluster Emerges from Its Dust Pillar in Carina." *Astronomical Journal* 129, no. 2 (2005): 888–899. http://adsabs.harvard.edu/cgi-bin/nph-bib_query?bibcode=2005AJ....129..888S&db_key=AST&high=3d3c3a078102266. This scholarly article by Stassun and his colleagues on the stars being birthed within the pillars of the Carina Nebula is modestly technical and contains some lovely visuals.

Steel, Duncan, and Paul Davies. *Eclipse: The Celestial Phenomenon That Changed the Course of History.* Joseph Henry Press, 2001. A comprehensive treatment of eclipses throughout history and how the ability of astronomers to predict these events became a major focus of astronomical study for much of human history.

Tyson, Neil DeGrasse. *Death by Black Hole: And Other Cosmic Quandaries.* Reprint edition. W. W. Norton & Company, 2007. An enjoyable collection of essays from arguably the most popular astrophysicist of our time, including (among other things) a discussion of black holes in a fun but informative and illuminating way.

University of Nebraska at Lincoln. "Doppler Shift Demonstrator." http://astro.unl.edu/classaction/animations/light/dopplershift.html. Animated demonstration of the Doppler effect as it pertains to light.

———. "Eclipsing Binaries Simulator." http://astro.unl.edu/naap/ebs/animations/ebs.html. Interactive simulator for exploring how eclipsing binary star systems behave in terms of their orbits and brightness changes and how these permit measurement of the stars' properties.

Wikipedia. "N-Body Simulations." http://en.wikipedia.org/wiki/N-body_simulation. An accessible discussion of the general problem of N-body simulations, of the sort required to understand the complex dynamics in gravitational systems involving more than two bodies.

———. "Stellar Classification." http://en.wikipedia.org/wiki/Stellar_classification. Examples of spectra of stars of different temperatures, including the history of the stellar classification system.

Notes

Notes

Notes

Notes